Additive and Subtractive Manufacturing Processes

This reference text discusses the fundamentals, classification, principles and applications of additive and subtractive manufacturing processes in a single volume.

The text discusses 3D printing techniques with the help of practical case studies, covers rapid tooling using microwave sintering and ultrasonic assisted sintering process, and covers different hybrid manufacturing techniques like cryo-MQL and textured cutting inserts. It covers important topics including green manufacturing, ultrasonic assisted machining, electro-thermal based non-conventional machining processes, metal based additive manufacturing, laser based additive manufacturing, indirect rapid tooling and polymer based additive manufacturing.

The book

- Discusses additive and subtractive manufacturing processes in detail.
- Covers hybrid manufacturing processes.
- Provides life cycle analysis of conventional machining.
- Discusses biomedical and industrial applications of additive manufacturing.

The topics on the sustainability aspects of conventional machining in reducing the carbon footprint of machining by adopting different hybrid and non-conventional machining processes will be useful for senior undergraduates, graduate students and academic researchers in areas including industrial and manufacturing engineering, mechanical engineering and production engineering.

Additive and Subtractive Manufacturing Processes

Principles and Applications

Edited by
Varun Sharma
Pulak Mohan Pandey

CRC Press
Taylor & Francis Group
Boca Raton London New York

CRC Press is an imprint of the
Taylor & Francis Group, an **informa** business

First edition published 2023
by CRC Press
6000 Broken Sound Parkway NW, Suite 300, Boca Raton, FL 33487-2742

and by CRC Press
4 Park Square, Milton Park, Abingdon, Oxon, OX14 4RN

CRC Press is an imprint of Taylor & Francis Group, LLC

ISBN: 978-1-032-05451-3 (hbk)
ISBN: 978-1-032-35555-9 (pbk)
ISBN: 978-1-003-32739-4 (ebk)

DOI: 10.1201/9781003327394

Typeset in Sabon
by SPi Technologies India Pvt Ltd (Straive)

Contents

Preface

Manufacturing has become an integral aspect of any industrial scenario. It can be broadly classified in terms of subtractive, formative and additive methodologies. Subtractive manufacturing has been the most accepted strategy for producing finished products. It has a very conventional approach that deals with material removal. With the rise in industrialization and liberalization, the whole world has turned into a marketplace. This competition has driven the quest of industries towards rapid manufacturing with limited inventories and lead-time. This has paved the way for the rise of additive manufacturing in rapid industrialization with limited resources and minimal wastage. The motivation for this book emerged from the needs of today's global industry. It explores the topics related to both prominent manufacturing strategies, namely subtractive and additive manufacturing. The concept of additive manufacturing is still unknown to a large section of researchers and academics. This lack of acceptance may be due to the high costs of installment and insufficient exposure. This book aims to bridge this gap of knowledge by providing more insight into the various concepts related to additive manufacturing. The presence of subtractive manufacturing is an added advantage, enabling a comparative study of both strategies for fabricating parts.

The book starts with the basic concepts of manufacturing, such as history, classifications and principles. Chapters 1–3 focus on various subtractive manufacturing methods, while Chapter 4 deals with the sustainability assessment of machining and related works. A few non-conventional material removal techniques are discussed in Chapters 5 and 6. The concepts of additive manufacturing are presented in Chapters 7–11 based on material, source of energy, and supply chain management. Finally, the concluding Chapters 12 and 13 deal with futuristic approaches by combining subtractive and additive manufacturing strategies, forming a revolutionary hybrid manufacturing. The future of Industry 4.0 is also discussed in relation to these important concepts.

In this book we have attempted to present the selected topics in an understandable and systematic manner. The authors believe the book will help future researchers not just as a piece of information but also to motivate them to embrace additive manufacturing and hybrid manufacturing and the existing subtractive manufacturing. Any form of constructive feedback regarding this publication will be acknowledged with a deep sense of gratitude.

Varun Sharma, Pulak Mohan Pandey

Notes on the Editors

Dr. Varun Sharma completed his B.Tech. degree at Guru Nanak Dev Engineering College in 2011, where he also obtained a master's degree in production engineering specialization in 2013. He joined IIT Roorkee as a faculty member in 2018 and presently serves as an assistant professor in the Department of Mechanical and Industrial Engineering at the Indian Institute of Technology, Roorkee, Uttarakhand, India. He has eight years of experience in research and teaching. He has published 35 research papers in peer reviewed journals and 11 at national and international conferences. His research interest includes conventional machining processes, non-conventional machining processes, machining and process optimization, ultrasonic assisted machining, additive manufacturing/3D printing, and mechanical and biomedical applications.

Dr. Pulak Mohan Pandey completed his B.Tech. degree at H.B.T.I. Kanpur in 1993, where he also obtained a master's degree in 1995 in manufacturing science specialization. He served H.B.T.I. Kanpur as a faculty member for approximately eight years; he completed a Ph.D. in the area of additive manufacturing/3D printing at IIT Kanpur in 2003. He joined IIT Delhi as a faculty member in 2004 and presently serves as professor. At IIT Delhi, Pandey diversified his research areas in the field of micro- and nano-finishing, and micro-deposition, and also continued working in the area of 3D printing. He has supervised 41 Ph.D.s and 36 M.Tech. theses over the last ten years and also filed 22 Indian patent applications. He has 201 international journal papers and 48 international/national refereed conference papers to his credit. He received the Highly Commended Paper Award made by the *Rapid Prototyping Journal* for his paper "Fabrication of Three Dimensional Open Porous Regular Structure of PA 2200 for Enhanced Strength of Scaffold Using Selective Laser Sintering", published in 2017. Many of his B.Tech. and M.Tech. supervised projects have received awards from IIT Delhi. He is a recipient of an Outstanding Young Faculty Fellowship (IIT Delhi) sponsored by the Kusuma Trust, Gibraltar and a recipient of the J. M. Mahajan Outstanding Teacher Award from IIT Delhi. His students won the Gandhian Young Technological Innovation Award in 2013, 2015, 2017, 2018 and 2020.

Contributors

Khalid Abdelghany
Additive Manufacturing
  Department, Central
  Metallurgical Research and
  Development
Institute (CMRDI)
Cairo, Egypt

Rajesh Babbar
Dr. B. R. Ambedkar National
  Institute of Technology
Jalandhar, Punjab, India

Hardik Beravala
Birla Vishvakarma
  Mahavidhyalaya,
  Vallabhvidyanagar
Gujarat, India

Vikrant Bhakar
Birla Institute of Technology, Pilani
Pilani, Rajasthan, India

Paramjit Singh Bilga
Guru Nanak Dev Engineering
  College
Ludhiana, Punjab, India

Neha Choudhary
Department of Mechanical and
  Industrial Engineering,
  IIT Roorkee
Roorkee, Haridwar, India

Vivek Dhimole
Mechanical Engineering Discipline,
  PDPM IIITDM Jabalpur
Jabalpur, Madhya Pradesh,
  India

Uday Shanker Dixit
Indian Institute of Technology
  Guwahati
Godhuli Gopal Path
Guwahati, India

Haytham Elgazzar
Additive Manufacturing Department,
  Central Metallurgical Research
  and Development
Institute (CMRDI)
Cairo, Egypt

Nitish P. Gokhale
Birla Institute of Technology, Pilani
Pilani, Rajasthan, India

Vishal Gupta
Thapar Institute of Engineering &
  Technology (Deemed to be
  University)
Patiala, Punjab, India

Prashant K. Jain
Mechanical Engineering Discipline,
  PDPM IIITDM Jabalpur
Jabalpur, Madhya Pradesh, India

A. N. Jinoop
Laser Technology Division
Raja Ramanna Centre for
    Advanced Technology
Indore, Madhya Pradesh, India

Prateek Kala
Birla Institute of Technology,
    Pilani
Pilani, Rajasthan, India

Kamal Kishore
National Institute of Technology
    Hamirpur
Hamirpur, Himachal Pradesh,
    India

Narendra Kumar
Department of Industrial and
    Production Engineering
Dr. B. R. Ambedkar National
    Institute of Technology Jalandhar
Jalandhar, Punjab, India

Raman Kumar
Guru Nanak Dev Engineering
    College
Ludhiana, Punjab, India

Shivendru Mathur
Department of Mechanical and
    Industrial Engineering, IIT
    Roorkee
Roorkee, Haridwar, India

Aviral Misra
Dr. B. R. Ambedkar National
    Institute of Technology
Jalandhar, Punjab, India

S. K. Nayak
Laser Technology Division
Raja Ramanna Centre for
    Advanced Technology
Indore, Madhya Pradesh, India

Pulak Mohan Pandey
Indian Institute of Technology Delhi
Delhi, India

C. P. Paul
Homi Bhabha National Institute
Mumbai, India
Laser Technology Division
Raja Ramanna Centre for
    Advanced Technology
Indore, India

Kedarnath Rane
Faculty of Engineering
University of Strathcylde
Glasgow, United Kingdom

Kuldip Singh Sangwan
Birla Institute of Technology
Pilani, Rajasthan, India

Dinesh Setti
Indian Institute of Technology
    Palakkad
Palakkad, Kerala, India

Maitrik Shah
Department of Mechanical and
    Industrial Engineering
IIT Roorkee
Roorkee, India

Kshitij Sharma
Department of Mechanical and
    Industrial Engineering
IT Roorkee
Roorkee, India

Pawan Sharma
Department of Mechanical
    Engineering
Sardar Vallabhbhai National
    Institute of Technology
Surat, India

Varun Sharma
Department of Mechanical and
    Industrial Engineering
IIT Roorkee
Roorkee, India

S. Shiva
Department of Mechanical
    Engineering
Indian Institute of Technology Jammu
Jammu, India

Nitesh Sihag
Birla Institute of Technology
Pilani, India

Buta Singh
Institute of Energy Engineering and
    Chemical Machinery
University of Miskolc
Hungary, UK

Gurminder Singh
School of Mechanical and Materials
    Engineering
University College Dublin
Dublin, Ireland

Narinder Singh
DICIV
University of Salerno
Fisciano SA, Italy

Nishant K. Singh
Hindustan College of Science and
    Technology
Mathura, India

Ravinder P. Singh
Maharishi Markandeshwar Deemed
    to be University
Mullana, Ambala, Haryana,
    India

Sehijpal Singh
Guru Nanak Dev Engineering
    College
Ludhiana, Punjab, India

Sunpreet Singh
Mechanical Engineering
    Department, National University
    of Singapore
Singapore

Manoj Kumar Sinha
National Institute of Technology
    Hamirpur
Hamirpur, Himachal Pradesh,
    India

Girish C. Verma
Indian Institute of Technology
    Indore
Indore, Madhya Pradesh, India

Chapter 1

Evolution of Manufacturing
Growing on a Circular Track

Uday Shanker Dixit
Indian Institute of Technology Guwahati, India

CONTENTS

1.1 INTRODUCTION

Manufacturing can be defined as the activity of converting raw material into a more valuable form by human beings. The word "manufacture" originated from the Latin words "*manu*" and "*factum*", which together mean "made by hand". Today most products are manufactured with the help of tools and machines. Hence, "made by hand" can be interpreted as "made by human beings". Manufacturing has been an integral part of society. Modern humans are called *Homo sapiens*, and the species evolved around 0.3 to 0.4 million years before the present. The term *Homo sapiens* means "wise man", which is reflected by the tremendous progress made by modern day humans in science and technology. *Homo habilis*, one of the earliest humans, appeared around 2.0 to 2.5 million years ago. Because of their use of stone tools, they have been named "handy man" or *Homo habilis*. However, stone tools were used in the Eolithic period also, about 10 million years ago (McNeil, 1990). Thus, it is clear that the history of manufacturing is very old.

DOI: 10.1201/9781003327394-1

It is very difficult to say when manufacturing emerged as an economic activity. Handicrafts such as carpentry, smithying, pottery and weaving have been practiced for ages. In 17th-century western Europe, the domestic or putting-out system of production was in vogue (Lazerson, 1995). In the domestic system, merchants used to "put out" material into the houses of their employees. Employees used to work from home and convert raw material to useful products. The marketing of the product was done by the merchants. Sometimes the workers labored in their homes or a location in nearby sheds. This system was very convenient for the employees as they did not have to leave their homelands. The First Industrial Revolution (1760–1840) started the factory system of production. It mainly began with the setting up of various textile mills. One of the earliest factories was John Lombe's water-powered silk mill at Derby, which was operational by 1721. However, the development of the steam engine by James Watt gave impetus to the development of large factories (Ghosh, 2021). In this period, a number of textile and iron mills were opened.

The factory system has a lot of pros and cons. Mass production reduced the cost of goods. It led to the urbanization of society. A number of cities were born and developed. In most cases, the living standard of the workers improved. The division of labor enhanced skill in one or two areas. However, employees lost their freedom and had to work for several hours a day. Sometimes, the factory owners employed children. Often the living conditions in the factory as well as in the residences were unhealthy. It was common to have frequent accidents in some factories.

Although the factory system almost eliminated the domestic system, traces of it remained. The domestic system is highly decentralized in nature, while the factory system is centralized. The centralized system of a big factory causes a lot of problems due to inertia and inefficiency. An intermediate system is the "ancillary unit" system (Hamaguchi, 1985). Here, a large factory outsources the work to a number of smaller factories called ancillary units. In this system, at least the owner of the ancillary unit gets some flexibility, although the employees do not get any substantial change in their working style.

With the advent of additive manufacturing, popularly called 3D printing, there is the possibility of going back to the domestic system. Additive manufacturing offers the possibility of distributed manufacturing. The product design can be carried out at one central place, which can be transmitted to various locations, where the physical avatar of the virtual product can appear. It is also possible to carry out minor modifications in the design within a very short time. Thus, tailor-made products can be manufactured economically depending on the need of customers. This could ease the constraint of mass production, where several copies of the product need to be manufactured to reach break-even point. As of now, additive manufacturing technology is sufficiently developed and customized products can be manufactured easily, but perhaps not economically. The challenge is to fabricate the products through additive manufacturing in an economical manner. The issue of the environment is also pertinent. The process, raw material and auxiliary material used during the process should not cause pollution. Addressing the nation on its 68th Independence Day on 15 August 2014, the Indian Prime Minister urged industry to manufacture goods with "zero defect, zero effect". Zero defect pertains to quality assurance and zero effect connotes that there should not be any adverse impact on the environment. Against this backdrop, researchers involved in the development and refinement of additive manufacturing technology have to face the challenge of developing economical and environmentally friendly technologies. In addition to the economic and environmental aspects, the technologist should also consider the social aspects. All these aspects constitute sustainability.

In 1987, the Brundtland Commission (a commission set up by the United Nations in 1983 with the leadership of Norwegian Prime Minister Gro Harlem Brundtland) released its final report entitled "Our Common Future" (World Commission on Environment and Development, 1987, www.ask-force.org/web/Sustainability/Brundtland-Our-Common-Future-1987-2008. pdf). This provides a simple definition of sustainable development as "development that meets the needs of the present without compromising the ability of future generations to meet their own needs". Over the last four decades, governments and social organizations have placed strong emphasis on sustainable manufacturing. In the early phase of the First Industrial Revolution, the scarcity of natural resources was not felt. The entire focus was on developing efficient and efficacious technologies. The introduction of disruptive technologies drastically changed the lifestyle of human beings and many unhealthy practices became the order of the day. Now, there is general awareness of the need to adopt a lifestyle in compliance with nature.

As already mentioned, manufacturing is an age-old activity that evolved with the human race. It emerged as an instinctive activity. After some time, it developed as an art. Until the introduction of machines around the dawn of the First Industrial Revolution, the quality of manufactured goods largely depended on the skill of the artisan. The introduction of machines brought tremendous benefit to humanity at large but also caused some adverse effect on the life of the artisans. This highlights the fact that the basic processes of manufacturing have been the same since time immemorial. Developments in technology and science play an assistive role in basic processes. A machine was supposed to do the same task as a skilled artisan would have done; it would of course have an edge over an artisan in terms of speed, uniformity in quality and endurance. Nevertheless, machines were operated by human beings and fatigue, mood and skill-variation of the operators had their own effect.

At the beginning of the 20th century, J. A. Fleming (1849–1945) obtained a patent for a thermionic valve in the United Kingdom. This consisted of two electrodes—a heated cathode and an anode—and was called a diode. A diode was used as a rectifier. Lee de Frost (1873–1961) developed another thermionic valve, called a triode, in the USA. The triode found application in radios and musical instruments. The first electronic computer called ENIAC (Electronic Numerical Integrator and Calculator) was developed in 1946 at the University of Pennsylvania. It had 18,000 thermionic valves and consumed 150 kW of electrical power. It weighed 25,000 kg and filled a room, occupying a 15 m × 9 m floor area with its 40 panels of 2.4 m height. It is obvious that such types of computer could not be of use in controlling machines for manufacturing. The breakthrough came with the invention of the transistor in 1947–1948 by John Bardeen (1908–1991), Walter Brattain (1902–1987) and William B. Shockley (1910–1989) at Bell Laboratories. Compared to the triode, transistors had a very long life, a small size and were lightweight. This motivated their use in controlling machine tools. The first numerical control (NC) machine tool was developed at Massachusetts Institute of Technology (MIT), USA, in 1952; it used vacuum tubes instead of the transistors that we use today. The first computer numerical control (CNC) was developed at MIT in 1957. CNC machines started gaining popularity in the 1970s, with the development of electronics and computer science. Initially, some machine tool industries started retrofitting existing machines with programmable logic controllers. Gradually, CNC machines began dominating the market. Developments in the area of computer science and information technology motivated manufacturing professionals and researchers to use data for modelling, forecasting and planning. Data had long been used in manufacturing, but the advent of computers accelerated its use for enhancing the performance of manufacturing systems.

It is evident that developments in electronics and computer science have changed the face of manufacturing. Similarly, developments in material science have influenced manufacturing sectors. In the recent past, several new materials were developed and found application in engineering. Some examples include smart materials and high-entropy alloys. High-entropy alloys have a number of (usually more than four) elements in large proportions. One challenge for the manufacturing sector is to find sustainable techniques for the manufacturing and processing of these materials. The other challenge is to utilize them effectively for enhancing the performance of manufacturing.

Automation in the manufacturing sector is not new, but started germinating 300 years ago. Presently, apart from technological issues, societal and ethical issues play a role in deciding the level of automation. What is the future of manufacturing? What level of automation will be commonly used? In what direction should research be focused? These are some questions that naturally arise in the minds of manufacturing professionals. It is prudent to understand the growth of the manufacturing sector with the help of some historical input. One can learn a lot from history for the better planning of the future.

The present chapter discusses the evolution of the manufacturing sector and its impact. Since the emergence of humans on earth, the manufacturing sector has been growing, but it is not moving on a linear path, in which one moves ahead and leaves the past behind. Many times, one has to look back at traditional practices and readopt old methods. This chapter discusses different aspects of manufacturing, highlighting this point. Imagine moving on a circular track, with continuously increasing speed; one reaches the original point with increased velocity. In the same way, the enhancement of technology makes it possible to adopt the old good practices in their efficient avatars. In the following sections, a historical journey of manufacturing, focusing on various aspects, is presented. This is not an exhaustive coverage of the history of manufacturing, which would need a thick tome; rather this chapter is just an attempt at a critical analysis of the past, present and future of manufacturing.

1.2 TRANSFORMATION OF THE MANUFACTURING SYSTEM: DOMESTIC–FACTORY–DOMESTIC

Before the First Industrial Revolution, the domestic system was prevalent in England. In this system, the business owners used to supply the raw material to workers at their homes and collect the finished goods. It was different from a handicraft system, where the craftsman himself used to procure raw material and market the product. Thus, in the domestic system, craftsmen became free from the procurement of raw materials and marketing; he could better concentrate on production only. The payment to workers was based on the quantity of production. The domestic system had the following advantages:

- The worker had enough flexibility. He could decide the number of hours he wished to work; the hours of work were also flexible.
- The working and living conditions, in general, were hygienic. He did not have to relocate to a city or town.
- There was better gender equality in jobs. The housewife could easily share the work.
- The employment of children was secured because they could easily adopt the profession of their parents.

However, the domestic system could not remain competitive with the development of machines and automation. A number of large factories were set up, which utilized big and expensive machines. The rate of production was drastically enhanced, which resulted in cheaper goods. It became necessary to keep the workers in the same shed, which further reduced the transportation costs. In the domestic system, the businessman used to transport the material from one stage of production to another; usually, different workers were engaged for different stages of the work. Thus, the factory system was more efficient and increased the overall production of the nation. Its impact on the workforce was mixed. Now, workers lost flexibility, and sometimes they were physically abused by factory managers (Pollard, 1963). In particular, a large number of children were employed and often subjected to corporal punishment. The emoluments and facilities for workers varied from factory to factory. At many places, the workers had to live in unhygienic conditions and there was not enough social security. On the positive side, the factory system led to urbanization and an uplift to society. Gradually, the workers realized the importance of literacy and an awareness of their rights.

The factory system started with the textile sector. It is often said that the basic necessities of humans are bread, clothes and a house. The bread is obtained from the farm sector and houses have to be constructed at different locations. It was not possible to produce houses in a factory, so it was quite natural to start with the production of clothes in factories. In 1771, Richard Arkwright (1732–1792) set up Cromford Mill, which was a water-powered cotton spinning mill; he had patented a water frame for spinning the cotton in 1769. Before him John Lombe (1693–1722) had set up a water-powered silk mill at Derby in 1721. James Hargreaves (1720–1778) invented the multi-spindle spinning frame in 1764–1765. However, yarn produced in the mill of Richard Arkwright was far better in strength.

Apart from the textile sector, several machine industries were also set up. In fact, a large number of steam engines could be produced due to developments in the machine tool industry. Matthew Boulton (1728–1809) was a famous industrialist in Birmingham who collaborated with James Watt (1736–1819), a Scottish inventor, and produced many steam engines. Boulton's father was already a manufacturer of small parts. In the 18th century, Birmingham was a center of the ironworks industry, which had expanded after a technique for smelting iron had been developed that used coke instead of charcoal. Charcoal is obtained from the incomplete combustion of wood; but too much use of charcoal leads to deforestation. On the other hand, coke is made from mineral coal and has a high calorific value, although it also causes more pollution.

In the period 1775–1800, Boulton sold 500 steam engines. The entire manufacturing process was not carried out at his factory: Boulton used to obtain components from a number of ancillary units and most of the assembly was done in his factory. The bored cylinders of steam engines were supplied by English industrialist John Wilkinson, who invented a precision boring machine for cast iron. In Wilkinson's boring machine, the shaft that held the tool was supported at both the ends, instead of being a cantilever shaft, the practice which was prevalent at that time. The English manufacturer Henry Maudslay (1771–1831) is called the founding father of the machine tool industry. Around 1800, he invented a lathe that could manufacture standard screw threads, which later helped in mass production.

The Industrial Revolution in America came slightly later compared to that in England, by two decades. Samuel Slater (1768–1835) set up a textile mill in America in 1790. David Wilkinson (1771–1852) is called the father of the American machine tool industry. He designed a screw-cutting lathe in 1794. He built a steam engine to propel a boat and produced a large general purpose lathe in 1806. He may also have invented the first centerless grinder.

Eli Whitney (1765–1825) was the American inventor of the cotton gin (Woodbury, 1960), though he made some losses due to patent issues. To make up for the loss, he started making muskets for the newly formed United States Army. He developed a milling machine in 1820 and propagated the concept of interchangeable parts.

In 1812, American inventors Paul Moody (1779–1831) and F. C. Lowell (1775–1817) produced cloth by spinning raw cotton in a mill. American inventor Frederick W. Howe (1822–1891) of the Robbins & Lawrence company designed a milling machine in 1848 (Smith, 1973). In his milling machine, a cutter could move vertically, whereas in early machines, such vertical motion of cutter and work table was not possible. Stephan Fitch built the first turret lathe in America in 1845 (Guergov, 2018). And in 1854, a turret with replaceable tools was developed. Subsequently, cam operated automatic turret lathes were developed. Prominent machine tool inventors in Europe were James Fox (1780–1830) in England, British inventor Henry Maudslay (1771–1831), Scottish engineer James Nasmyth (1808–1890), who developed a steam hammer, and English engineer Sir Joseph Whitworth (1803–1887), who is famous for inventing a quick return mechanism for shapers.

While a lot of developments pioneered by artists, craftsman, technocrats and inventors created tremendous growth during the First Industrial Revolution, it also created a negative impact on traditional craftsmen. Manual production could not compete with machine-assisted production. There was also some unrest among the working class because of exploitation by the mill owners. For the welfare of the working class, a series of acts were passed by the British Parliament, but the most effective act would not come until 1833. This provided for a provision of factory inspectors for enforcing the law and was mainly for improving the conditions of children working in factories.

The main developments during the First Industrial Revolution were (i) the popularization of the factory system, (ii) the establishment of the textile industry, (iii) the use of steam power and (iv) the invention of machine tools. The pace of innovation slowed down after 1825. Industries faced stiff competition and there was a need for some revolutionary inventions, so that industries could come out of recession. The Second Industrial Revolution started in 1870, which continued up to 1914, before the First Word War started. This Revolution is known for the following features:

- The Bessemer Process for steel production was invented that reduced the cost of steel production by 50%. In this process impurities from molten pig iron are removed by oxidation of the air. As the oxidation is an exothermic reaction, it also raises the temperature of the steel and helps in keeping it in a molten state. The process was invented by Henry Bessemer in 1856; several plants based on this technology were set up in the period 1870–1890.

- Many nonferrous metals gained popularity. For example, the electrolytic refining process for the extraction of aluminum was developed. Aluminum was considered more precious and rare than gold until a cost-effective process was developed in 1889. Celluloid was invented in 1855. Polyvinyl chloride (PVC) was synthesized in 1872 by German chemist Eugen Baumann. Bakelite was developed in 1907.
- Many applications of electricity and magnetism were developed. The first electric lighting company was formed in 1852. In 1866, C. W. Siemens developed self-excited generators. Incandescent light bulbs were developed independently by T. A. Edison (1847–1931) in America and by J. W. Swan (1828–1914) in England around 1879; before that lighting was done by electric arc. Electric trains were used for public transport after 1880. Nikola Tesla developed the alternating current motor in 1888.
- The production of oil became cheaper. Automobiles running with oil became popular. In many industries, internal combustion engines replaced steam engines.
- Several techniques were developed to enhance mass production. Henry Ford (1863–1947) is remembered for developing the concept of the assembly line technique in the automobile industry. In his Ford Motor Company, he produced Ford Model T cars, which could be afforded by the middle class. During the Second Industrial Revolution, production and industrial engineering developed as an important discipline.

The period from the First World War to the end of the Second World War was not good for the economic growth of nations. However, many new technologies were developed for meeting the needs of the Second World War. During this War, a lot of fighter planes were used, which forced the aerospace industry to manufacture aircraft at a cheaper rate. During this period, mathematicians developed operations research for optimizing scarce resources during war time. It can be said that mechanical engineering played a major role during the First Industrial Revolution, and that electrical and industrial engineering played a major role during the Second Industrial Revolution. The Third Industrial Revolution (1970–2000) was triggered by developments in electronics, computer science and information technology. Although CNC machines were developed in the 1950s, they found wide use after 1970. During that period, robotics also gained popularity in factories. Japanese engineer Tetsuro Mori of the Yaskawa Electric Corporation coined the term "mechatronics", for which Yaskawa obtained the trademark rights, though they were abandoned in 1982. Email was developed by Ray Tomlinson in 1971. Internet came into existence in 1973. Tim Berners-Lee, a British scientist, invented the World Wide Web (WWW) in 1989. In the 1990s, web based manufacturing started gaining popularity.

The term "Industrie 4.0" was introduced in Germany in 2011. In 2013, the German government launched its Industry 4.0 plan. Similarly, other nations launched their plans for upgrading industry—"Made in China 2015" by China, "Horizon 2020" by the European Union and "Make in India" by India. Many call this era the Fourth Industrial Revolution, although it is debatable if it has already started or is yet to come. This Revolution is triggered by massive developments in electronics, computer science and information technology. It is the age of the digital transformation of industry. Now, not only computers, but machines, devices and products will be interconnected; this is called the Internet of Things (IoT). This means that physical objects will be able to collect information through sensors, store that information and also transmit it to the desired location; they will also be able to receive commands remotely. Thus, it will be possible to have manufacturing plants at several locations but controlled from only one place. Often the term "cyber-physical system" is used interchangeably with IoT. There will be drastic change in manufacturing processes. Instead of using traditional processes of machining, casting, forming and joining in a mechanical factory, only one process, that is additive manufacturing, will produce all types of components. This may also help in the seamless integration of design and manufacturing. As additive manufacturing processes are a computer controlled process, all the manufacturing information of the product can be in digital form. Digital manufacturing may become the norm of the day. It may be possible to control production when sitting kilometers away. Thus, "work from home" may become feasible not only for the software industry but for the hardware industry also. This may bring the domestic system back, but this time the system will use advanced technologies.

In Table 1.1, some landmark events and their impact on manufacturing systems are listed. In particular, it is highlighted how developments affected the system of production. A careful reading of the table shows how the main source of power to factories was water, steam, oil and electricity, in that order. In the early factories, there was dominance of mechanical engineering. In the second half of the 19th century, electrical engineering also became important thanks to the development of generators and motors. In the second half of the 20th century, first electronics and then computer science helped in the automation of factories. Automation helped in bringing neatness and comfort to industrial jobs; instead of sweating in a harsh working environment, operators could control the machining sitting comfortably in air-conditioned rooms. In the 21st century, computers and information technology plays a dominant role. With the help of manufacturing technologies like additive manufacturing and IoT, companies are heading towards distributed manufacturing. It will not be surprising if the factory system becomes almost extinct in favor of a domestic system.

Table 1.1 Landmark Events in the History of Manufacturing

Year	Event	Impact on Manufacturing System
1733	John Kay developed a flying shuttle for weaving.	Productivity improved in textile industry.
1765	James Hargreaves invented a multi-spindle spinning mill named the Spinning Jenny.	This was an early trigger for industrialization. The economy improved with some harm to the handicraft sector.
1769	Richard Arkwright developed a water-powered spinning mill.	The factory system started replacing the domestic system of production.
1774	John Wilkinson invented a boring machine capable of boring precise cylinders.	This paved the way for the economical manufacture of steam engines.
1776	First commercial Boulton and Watt steam engine was built.	Era of steam power began.
1794	Eli Whitney got a patent for a cotton gin in the USA.	This helped in the development of the cotton industry in America.
1801	Eli Whitney demonstrated the interchangeability of the components of muskets.	This provided impetus to the concept of mass production in the mechanical manufacturing industry.
1830	George Stephenson constructed a railway line between Liverpool and Manchester.	Construction of railways and steam locomotives encouraged centralized production.
1893	Rudolph Diesel invented the diesel engine.	Internal combustion engines emerged as an alternative to the steam engine.
1866	C.W. Siemens developed self-excited generators.	Generators were used for arc-lighting, an impetus to work at night.
1876	Alexander Graham Bell invented the telephone.	Communication helped the growth of industry.
1879	Thomas Edison invented the electric bulb.	Working hours in industry increased.
1888	Nikola Tesla developed the AC motor.	Compared to engines, a more compact drive became available.
1913	Henry Ford introduced the assembly line production technique.	Mass production became important.
1971	First microprocessor was developed.	Era of automation started. Many blue-collar jobs became white-collar.
1989	Tim Berners-Lee, a British scientist, invented the World Wide Web (WWW).	A big impetus was given to global manufacturing.
2005	Adrian Bowyer started the RepRap project for developing a self-replicating 3D printer. The Alfred P. Sloan Foundation launched the Synthetic Biology Initiative.	An era of Industry 4.0 and "work from home" was in the making. A revolutionary change in the manufacturing system is expected.

1.3 CUSTOMIZATION TO MASS PRODUCTION TO MASS CUSTOMIZATION AND AGAIN TO CUSTOMIZATION

In the early days, manufacturing was mainly a manual, and often menial, activity. However, it was possible to produce the goods as per the requirement of customers. When manufacturing industry received the assistance of machines, it became possible to carry out mass production. Mass production could reduce costs drastically, because the fixed component of the manufacturing cost was distributed across a number of units of production. It is also possible to employ a better technology, when the production volume is large. Before mass production, many factories used to follow job or batch production. Job production is the manufacturing of a customized product, one unit at a time. Even now this procedure is followed in some industries, for example, in ship manufacturing. In fact, during the product design phase, usually one or a few units of a prototype are manufactured. For helping the faster development of a prototype, rapid prototyping technologies became popular after 1987. Batch production stands in between job and mass production. It is useful for mass customization, in which a customer is able to choose the object of his or her liking amongst the available varieties. For example, the footwear industry may produce four to five designs of shoes and, in each design, there may be four to five standard sizes.

Henry Ford was the pioneer at propagating the concept of mass production in the automobile industry. In 1908, he launched the "Ford Model T" car from his Ford Motor Company. The car was affordable to middle class people in the USA. In 1913, he introduced the assembly line production technique, which resulted in an enormous increase in production volume. It is said that 50% of cars in America were Ford Model Ts by 1918.

This system of mass production could develop because of the ability of machine tools to produce interchangeable parts. In this regard, Eli Whitney is considered a pioneer (Woodbury, 1960). He completed his study of mechanical engineering at Yale. The US government signed a contract with Whitney to produce 10,000 muskets in 1798. He invented jigs for producing interchangeable parts. Although the target was to produce 10,000 muskets in two years, it was only achieved after ten years, in around 1807. The use of jigs and fixtures helped in the mass production. Simeon North (1765–1852) supplied pistols to the US government following a mass production system. He had developed a milling machine in America in 1818 for making interchangeable parts. Also the inventor and clock-maker Eli Terry Sr. (1772–1852) in Connecticut had also developed a milling machine in 1795. Sir Samuel Bentham (1757–1831) was a British engineer, who worked for the navy in Russia and England. He developed a mass production factory called Block Mill. By 1805, it was producing 130,000 pulley blocks per year. The machinery in the factory was designed by a French civil engineer Mark Brunel (1769–1849); Brunel is also famous for designing a locomotive named "The Lord of Isles" for the Great Western Railway. The fabrication of the factory was carried out by Henry Maudslay (1771–1831), who is called the father of the machine tool industry by some historians. Thus, it is clear that the seeds of mass production were sown long before the time of Henry Ford. Frederic Winslow Taylor (1856–1915), Henry Robinson Towne (1844–1924), Henry Laurence Gantt (1861–1919), Frank B. Gilbreth (1868–1924) and Lillian Gilbreth (1878–1972) made significant contributions to scientific management, which greatly helped in mass production.

Mass production may not always be desirable. Sometimes it is better to produce in batches. In many chemical industries, chemicals are produced in batches. Clay bricks are produced in batches, because in traditional kilns, only a limited number of bricks can be fired at a time. However, in many modern factories, bricks are fired in a tunnel kiln, through which the bricks can move on a conveyor. Castings are usually produced in batch mode. Pig iron is produced in a blast furnace and cast in molds. In the 1950s, continuous casting was developed where the molten metal flows from a tundish to a mold and rolling mill in a continuous manner.

In the late 1960s, the flexible manufacturing system (FMS) was conceived. In FMS, several machine tools are numerically controlled by a central system; a job shifts from one machine to another with the help of robots and automated guided vehicles. Many flexible manufacturing systems contain an automatic storage and retrieval system. By changing the sequence of operations, a variety of components can be produced without making changes in the physical layout of the machines. Figure 1.1 shows a schematic diagram of a representative FMS. The figure shows two CNC lathes, one milling machine and a 3D printer, but several different types of machines can be incorporated depending on the range of products.

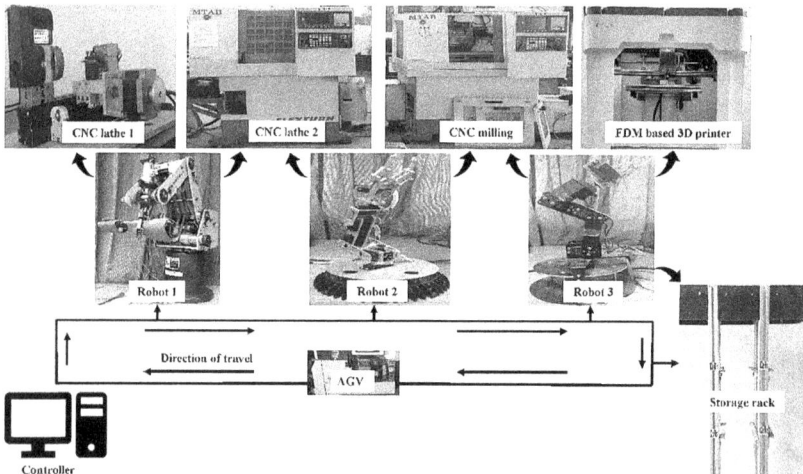

Figure 1.1 A schematic representative flexible manufacturing system.

In the 1990s, the term "agile manufacturing" was coined (Thilak et al. 2015), which indicates the ability of an organization to adopt to changes depending on customer requirements. The concept of lean and agile manufacturing is very important for batch production. While agile manufacturing imparts the ability to respond to changing requirements from batch to batch, lean is essential for adopting to the changing production volume. Taiichi Ohno (1912–1990) developed the Toyota Production System (TPS), which became popular as a lean manufacturing system in the USA. Ohno developed the kanban system for the just-in-time (JIT) system of production. Kanban is a pull type of system, in which a message for production is sent in the upstream side of the supply chain, whenever there is consumption of the product and another unit needs to be produced. A message is transferred by means of a card called a kanban. However, in this digital age, the kanban can be replaced with a virtual card.

Ohno identified seven types of *muda* (waste), that is defects in production, overproduction of goods, inventory, unnecessary processing, unnecessary movement of people, unnecessary transport of goods and waiting by employees. Womack and Jones (2013) added an eighth *muda*, that is design of goods and service that do not meet the user's needs. Apart from eliminating *muda*, in lean manufacturing, *mura* (inconsistencies) and *muri* (overburdens) should also be eliminated. The combination of lean and agile is called "leagile". Leagile manufacturing is very suitable for mass-customization (Putnik and Putnik, 2012).

Developments in additive manufacturing technologies are orienting factories towards customized production. It is possible to develop a product of the customer's liking at economical rates. Customers can be involved in the design of products right from the beginning. In 1997, AeroMat produced the first 3D metal printer, which used a high power laser to fuse titanium alloy powder. The main issue is how 3D printing can be made more economical and faster. While assessing the production of components through metal powders, the technology for producing the raw material, that is the powder, should also be assessed. In 2005, the RepRap open source concept was developed by which a 3D printer can create its own replicate. The word "RepRap" is an abbreviation for a replicating rapid prototype (Jones et al., 2011). In 2008, Darwin became the first commercially available self-replicating printer. Starting from 2009, several patents related to additive manufacturing were expiring, one-by-one, due to the time limit of the licenses. Thus, several technologies became free, which drastically reduced the cost of 3D printers. Since 2015, bio-printing has also been gaining popularity. There is hope for the printing of vital human organs by 3D printing. In a nutshell, additive manufacturing brings back the handicraft era, where a craftsman used to make specialized items for a special customer. In the present era, the varieties of products can be manufactured economically without worrying about the scale of production.

1.4 IMPORTANCE OF SUSTAINABILITY IN THE MANUFACTURING SECTOR

Manufacturing invariably needs raw materials, which require energy to process them. Energy can be obtained from natural sources or it has to be produced, which again needs some raw materials. We are dependent on nature for raw materials. Nature does not have an infinite source of these raw materials. Mahatma Gandhi said, "Earth provides enough to satisfy every man's need, but not for every man's greed." It is important to use natural resources judiciously, so that enough are left for future generations. There is a famous Aesop tale of a hen that used to lay a golden egg every day. Not satisfied with one golden egg a day, the owners cut open the stomach of the hen to get all the golden eggs at once. The hen was killed, but not a single egg was stored in the stomach. The moral of the story is that we should consume natural resources only at the rate at which they will be replenished. While most of the great civilizations followed this policy, development took place at too fast a pace from the 16th century in Europe. There was too much deforestation due to the excessive consumption of wood, which was used as a raw material as well as fuel (Ghosh, 2021). The need was felt to search for alternatives. One alternative was coal. Hence, mining activities started. However, coal is a depleting resource and mining activities are not benign to the environment. Once the internal combustion engine was developed, petroleum oil was used as fuel. Petroleum oil is also a depleting resource and its burning produces harmful products. Nowadays, engineers are trying to exploit non-depleting resources such as solar energy.

In manufacturing, problems concern not only the consumption of raw material and energy. Most manufacturing processes cause pollution, which affects human beings as well flora and fauna, in a very adverse manner. Burning wood, for example, produces a lot of irritating smoke. The search for neater and cleaner sources of energy has been going on for ages. One good source of energy is wind. Machines operating on wind power have been in use since the first millennium AD. Windmills were used in Persia, Afghanistan, India and China in the 9th century AD. They were used for milling and the pumping of water. They became popular in 12th-century England and France. Up to the middle of the 18th century, they were commonly used for sawing, grinding, milling and running fans in mines. The development of steam engines diminished the use and importance of windmills. In the last quarter of the 19th century, wind power increased in prominence again. James Blyth built the first wind turbine in Scotland in 1887. And a wind turbine constructed by J. B. Wilbur of MIT and Palmer C. Putnam in Vermont was operational in 1941. Today India has the fourth largest installed capacity of wind power after China, the USA and Germany. In future, its use is likely to go up.

Another clean source of energy was the watermill, which was in use in the Greek civilization around the 3rd century BC. According to Greek historians, the technology of watermills was transferred from the Roman Empire to India in the 4th century AD. Watermills were used for grinding grains, sawing marble and carrying out other industrial tasks. Several industries were set up near rivers to exploit the use of water. This caused the development of cities near rivers, but resulted in their pollution. With the development of the steam engine, the importance of watermills diminished. However, in the 19th century water turbines were developed and soon their use for generating electricity was quite common, which continues today (Viollet, 2017).

Most steam engines used steam generated from coal-fired boilers. Coal was obtained from mines. As the consumption of coal increased, the need arose to extract it from much deeper mines. The burning of the coal in boilers caused pollution. In the 19th century, the production of mineral oil became cheaper, which encouraged engineers to use the internal combustion engine instead of the steam engine. During this period, gas turbines were also developed and electrical motors were invented.

It is interesting to note that the German engineer Rudolf Diesel (1858–1953) developed an internal combustion engine that ran on peanut oil; it was demonstrated at the World's Fair in Paris in 1900. Diesel's first prototype, built in 1893, used petrol as fuel. Today the compression ignition engine runs with diesel fuel. However, attempts are being made to run it on biodiesel. It is also interesting to note that the first successful electric car was developed by William Morrison in 1890. However, electric vehicles would not become popular until 1997, when Toyota released a hybrid vehicle. Today many nations are planning to replace existing vehicles by electrical vehicles. Thus, we can clearly see that we are going to have to take a fresh look at several older technologies. Why did engines running on biofuel, which were in fact developed in the early 20th century, not become popular then? Why did electrical vehicles not gain popularity in the 20th century? One answer is that the technologies were not mature; however, this is only partially true. Technologies mature with extensive use and simultaneous research to improve them. A more convincing answer is that those technologies were not sustainable. Three pillars of sustainability are the economy, environment and society (Purvis et al., 2019). These pillars are equally important and all have to be equally strong. For most of the 20th century petroleum oil was available at a relatively cheaper rate. Hence, it was economical to use internal combustion engines running on petroleum based fuels. As the number of vehicles was not so high, the adverse impact of automobiles on the environment was not apparent. With the rising prices of oil and the realization about global warming, nations have started paying attention towards developing sustainable technologies. Various nonconventional energy sources such as solar and wind energy are becoming popular. Realizing the importance of environmental protection, several countries are providing subsidies to the nonconventional energy sector, thus making this sector economically sustainable.

Apart from the economy and environment, society plays a role in deciding the direction of technological development. In the past, several materials were used in manufacturing because of their easy availability, ease in manufacturing and their useful properties. However, some of them were found to be detrimental for the health of human beings. One such material was lead. This has a density of 11,340 kg/m^3, heavier than steel. Its melting point is 327 °C and its tensile strength is 18 MPa. It has been used on its own for ages as well as in combination with other metals. Lead pipes were commonly used for the supply of drinking water. However, lead is toxic and affects the brain and nervous system (Verstraeten et al., 2008). In 1986, the use of lead water pipes was banned in the USA. Thomas Midgley, Jr. (1889–1944), an American chemist, developed the tetraethyl lead (TEL) additive for petrol to reduce knocking in internal combustion engines. From the 1970s, several countries started phasing it out. Today the use of leaded petrol is prohibited in most countries; India banned it in 2000. In metal joining, solder used to be an alloy of lead and tin. Realizing the adverse impact of lead fumes, lead-free solders were developed. One such solder is SAC, which is an alloy of tin (Sn), silver (Ag) and copper (Cu). Similar to lead, asbestos is also detrimental to health; its fibers are carcinogenic. Thus, the production of asbestos-cement sheets has stopped. Aluminum cooking vessels are also being rejected by society because of health concerns, although it is easier to produce aluminum vessels by metal forming.

Today additive manufacturing is gaining popularity, but sustainability aspects need to be looked into. In fused deposition modeling, various polymers are melted; but they may have an adverse effect on health. Some chemicals used for post-processing are also harmful. Moreover, many materials used in additive manufacturing are not biodegradable. For cyber-physical systems also, there is a need to investigate the health aspect. Many of these systems will use 4G and 5G technologies. Many researchers have pointed out the health hazards associated with exposure to radio-frequency radiation (Miller et al., 2019). Overall, it is a challenging task to keep pace with sustainable development.

1.5 ROLE OF DATA AND ANALYTICS IN MANUFACTURING

The first significant use of data in manufacturing seems to have been by Fredrick Winslow Taylor (1856–1915), an American mechanical engineer. It is said that he carried out 50,000 experiments over a period of more than two decades to find out the optimum process parameters in metal cutting. He consumed more than 350 metric tons of steel and iron. He had joined the Midvale Steel Company as a shop clerk and, within six years, became the chief engineer. In 1889, he joined Bethlehem Steel but continued his project. In 1906, he presented his findings at conferences and his paper entitled "On the Art of Cutting Metals" was published by the American

Society of Mechanical Engineers (AMSE) in 1907. In the process of conducting the experiments, Taylor developed a high speed steel in association with Maunsel White of Bethlehem Steel which was an alloy with the major constituents of tungsten, molybdenum and chromium.

Taylor is also known as the Father of Scientific Management. He studied and analyzed the different methods for carrying out tasks. He made extensive use of time and motion studies and introduced the piece rate system. He quantified the performance of workers. His work laid the foundation of industrial engineering. Furthermore, Frank B. Gilbreth and his wife Lillian Gilbreth also worked more on "time study". They carried out time and motion studies with the help of motion pictures and identified the most productive and efficient method of executing a task. The quote by Frank Gilbreth "they come cheaper by the dozen" became a popular phrase for mass production.

The famous quality guru W. Edwards Deming (1900–1993), an American electrical engineer, advocated statistical methods for the analysis and control of quality in Japanese industries. Walter Andrew Shewhart (1891–1967), an American physicist, is sometimes called the father of statistical quality control. He provided the concept of identifying variations due to assignable and chance causes. Suppose a machine is producing some component in large quantities. Over a period of time, the dimensions of the component may start to differ due to tool wear. In this case, tool wear is the assignable cause. Dimensions may differ due to several other random causes, such as occasional human error in setting the tool, a random high spot in the raw material that might trigger vibration during processing, an occasional rise in the temperature due to random obstruction in the flow of the coolant and so on. Based on the analysis of data, it is easy to identify assignable causes but not the chance causes. Shewhart provided a scientific method to find out the assignable causes through so-called control charts. These, also called Shewhart charts, depict the change in process characteristics over time. Genichi Taguchi (1924–2012) also propagated the use of statistics or data science for the control of quality.

Many consider F. W. Taylor to be the pioneer in the modeling of metal cutting. In his experiments, he considered 12 variables and studied their influence on tool life. However, his modeling was based only on data. M. Eugene Merchant (1913–2006) and his group provided the first physics-based model of metal cutting. Astakhov (2006) pointed out that the attempt to model metal cutting was made by I. Time in 1870 and H. Tresca in 1873. Zorev (1966) pointed out that the single-shear plane model proposed by Merchant was actually developed by Zvorykin (1896). Ernst and Merchant had published their paper on the single-shear plane model in 1941 (Ernst and Merchant, 1941). Concurrently Vaino Piispanen had described the cutting process as like the movement of a deck of cards, where one card slides over the other (Piispanen, 1937). However, his paper was published in Finnish; hence, the majority of the research community was unaware of it.

Lee and Shaffer (1951) provided a slip-line solution to metal cutting, and several researchers applied the concept of fracture mechanics to the process. Considering the limitation of physics-based models, several researchers employed mechanistic models. In these models, the complex cutting edge of a cutting tool is discretized into a number of elements. The force components on each discretized element are computed based on simple empirical relations and integrated to provide the overall force. Most researchers took the cutting force as proportional to the chip area. The proportionality constant is empirically determined. In a similar way, the thrust component is also computed. Thus, this approach is semi-empirical in nature.

Starting from the 1970s, the finite element method (FEM) has been applied to model machining processes. FEM is a physics-based numerical technique and does not rely on data except for material and environmental properties. However, until now the FEM modeling of machining processes has not reached maturity. First of all, it requires a huge amount of computational time to model the processes. Hence, most of the time, instead of simulating the whole process, a simplified model of the problem is solved. Second, all the required performance indicators are not predicted properly. There are several articles in the literature that report a reasonable prediction of the main cutting force, albeit by fine-tuning the material parameters, but the thrust force is either not properly predicted or not even reported. There is hardly any paper that predicts surface roughness in machining with the help of a FEM model. It is apparent that the physics of the process is still not understood; however, practicing engineers do need a solution, which has motivated researchers to use data-based modeling in machining.

Starting from a pioneering article by Rangwala and Dornfeld (1989), a large number of papers have been published on the applications of neural networks for the prediction of machining performance. Other machine learning techniques have also been applied to model machining processes, and the research is continuing. Machine learning techniques require a lot of data. Thanks to development in the field of computer science and information technology, there is no scarcity of data these days, though to use the data in a judicious manner for making proper inference remains a challenging task. The term "big data" was used in a journal paper in 1999 to refer to gigabyte-size data (Bryson et al., 1999). It is difficult to define big data, but easy to feel it. In a simple way, big data refers to the processing of huge amounts of data, which cannot be handled in a simple way. According to Gartner Inc., "**Big data** is high-volume, velocity, and variety information assets that demand cost-effective, innovative forms of information processing for enhanced insight and decision making. This definition uses subjective terms, but they indicate that big data requires unconventional processing techniques. Big data can be structured as well as non-structured. Processing difficulty is not only dependent on the volume of the data but also on the variety. It may be difficult to process data comprising a variety information even if in small volumes in comparison to the processing of large amounts of structured data. Moreover, managing dynamic data can be an even tougher task. Thus, it is not a single factor that characterizes big data, rather it is the combination of the volume, velocity and variety of the data. With the advent of the internet in the 1990s and Web 2.0 (also called the Social Web), data applications in manufacturing have increased. Cloud computing has further assisted the application of data science in manufacturing. "Cloud computing", a termed coined in 1996–1997, refers to the on-demand availability of computer system resources, which includes free as well as paid services. Overall, it can be seen that earlier applications of empirical techniques in manufacturing have appeared in the form of data-based modeling and data analytics.

1.6 INFLUENCE OF EVOLUTION IN MATERIAL SCIENCE

In prehistory, human beings used stone and wood for making various artifacts. The first metal to be practically utilized was copper; some artifacts dating back to 9000 to 7000 BC have been found in Western Iran. The Chalcolithic or Copper Age is the transitional period between the Neolithic and Bronze Ages. The addition of tin to copper makes bronze, which is harder than copper. The addition of tin to copper also lowers its melting point; hence, manufacturing with bronze becomes easier. The Bronze Age spans from 3300 to 1200 BC, followed by the Iron Age (1200 to 550 BC). The addition of zinc to copper produced brass, which might have been produced from about 500 BC. The original home for manufacturing brass was Asia. Metallic zinc was produced at Zawar in Rajasthan, which is presently in Udaipur district. Today, a public sector company, Hindustan Zinc limited, is situated there, which extracts zinc and lead. There are historical records that show that artisans of Zawar used to produce zinc by a distillation process. In China, during the Ming Dynasty (1368–1644 AD), zinc coins were manufactured. The technology for the extraction of zinc came to Europe much later.

In 1742, the French physicist and chemist P. J. Malouin (1701–1778) observed that zinc coating on iron utensils improved their resistance to corrosion. Later, the French civil engineer and chemist Stanislas Sorel (1803–1871) developed a galvanizing process. In the early 1900s, the inventor Sherard O. Cowper-Coles (1866–1936) invented the sherardizing process for applying a zinc coating. In this process, the iron object is heated along with the zinc powder in a closed container and heated below the melting point of the zinc. An alloy of zinc and iron is formed with a zinc coating on top.

From the Iron Age up until the First Industrial Revolution, wood and iron were the most widely employed engineering materials. Wooden lathes were in use around the 3d century BC in the Greek and Roman civilizations. They were used for the turning of wood and bone. An Etruscan wooden bowl found in the Tomb of the Warrior at Cornetto dates back to 700 BC. The Etruscan civilization flourished in Central Italy from 800 to 300 BC and was conquered by the Roman civilization. With the progress of industrialization in 16th and 17th-century Europe, the serious problem of deforestation arose due to the excessive use of wood. Wood was a good engineering material as well as a fuel. Due to the scarcity of wood, people started using coal as fuel and hence a lot of mining work began. In fact, the mining of coal provided the impetus to develop the steam engine for pumping out water from deep mines. On the material front, people started replacing wood with iron. Although steel was produced in several civilizations as early as the 6th century BC, Wootz steel produced in South India and Sri Lanka was exported all over the word; however, it took a long time to discover an economical process for the mass production of steel. The breakthrough came when English inventor Henry Bessemer invented the Bessemer process, in which impurities are removed from the molten air by oxidation in the presence of flowing air. From then on steel became the main engineering material and its processing by machining and metal forming became an important manufacturing activity. American chemist Charles Martin Hall (1863–1914) invented an electrolysis technique for the production of aluminum in 1886; French scientist Paul Héroult (1863–1914) had independently invented this process too. And both did so at the young age of 22–23. It is an interesting coincidence that both of them lived from 1863 to 1914. With the invention of the Hall–Héroult process, aluminum became affordable; before its price was almost the same as that of silver, and sometimes more. Being a precious metal in those days, aluminum jewelry was prestigious, but the Hall–Héroult process reduced its status!

Alexander Parkes (1813–1890) presented "Parkesine", the first man-made plastic in the world, in 1862. Leo Baekeland (1863–1944) created the first synthetic plastic in 1907, called Bakelite, a thermosetting plastic that becomes hard after heating, which found application as an insulator, as kitchenware and in toys. With the invention of plastics, several manufacturing techniques were developed that were specifically suited for producing plastic components. One such process was injection molding, developed in the 1870s for manufacturing celluloid components. James Hendry (1921–2014) developed the first screw injection molding machine in 1948, and polymers were developed in the 1920s. In 1939, American Company DuPont developed nylon that was used by the U.S. military for making parachutes and ropes. Nylon is a thermoplastic that is softened with heat. Thermoplastics can be melted also, whereas thermosets cannot. Although both types of plastics found several applications, the dominance of steel in engineering components remained, because plastics were not able to provide sufficient strength. In most mechanical engineering products, their role was limited to providing covering. For example, plastic panels were used in automobile and machine tools. The main quality of plastics was their light weight, while their disadvantage was their lack of strength. However, plastics did have other interesting properties to justify their application as an engineering material: they provided good thermal and electrical insulation. They were also well resistant to corrosion and wear. In 1938, polytetrafluoroethylene (PTFE) was invented, which was sold on the market under the trade name Teflon. This is a very slippery plastic and can be used as a lubricant. In fact, early CNC machines used PTFE coated guideways to provide linear motion for the slides.

In 1951, acrylonitrile butadiene styrene (ABS) was produced. ABS is a thermoplastic that now finds application in the housings of some machines. Since 1953, General Motors started using glass reinforced plastic for making car bodies. In the mid-1980s, the company started making the panels of the Saturn Corporation (a subsidiary of General Motors) by injection molding using a blend of polycarbonate and acrylonitrile butadiene styrene (PC/ABS). Nowadays, several composite materials are used in engineering that have some metallic or nonmetallic reinforcement in the plastic matrix. The polymer fiber Kevlar has been produced since the 1970s that has a specific strength above that of steel. Polyurethane is a versatile plastic: the first artificial heart was made of it; and angioplasty balloons are made of a variety of plastics, including polyethylene terephthalate (PET), nylon 11 and nylon 12.

Several additive manufacturing processes use plastic as a raw material. The first rapid prototyping technology, i.e., stereolithography, uses a photopolymer that is cured by ultraviolet radiation. Many thermosetting plastics such as ABS can be used to produce complex parts using a fused deposition modeling (FDM) process. It is interesting to note that the plastics industry is now trying to develop plastics from natural and biodegradable sources, as was the practice in the 19th century, instead of fully synthetic plastic. Polylactic acid (PLA) is a biodegradable thermoplastic produced from corn.

In earlier days, the focus of manufacturing was towards the processing of steel and iron. Later on, some nonferrous materials such as aluminum also became important. Today because of the pressure created by global competition and the requirement for sustainability, newer materials are being developed. Such development stimulates new manufacturing processes. For example, the need to process very hard materials made many nonconventional machining processes popular. Processes like electrical discharge machining (EDM), laser beam machining (LBM) and plasma arc machining (PAM) are not dependent on the hardness of the material. Lasers are used not only for machining but also for producing low formability materials. There is also an increased awareness to optimize manufacturing processes not only from the point of view of energy consumption but also by considering the influence on material properties. In the 1980–1990s several papers on the die design of extrusion processes were published, which mainly aimed at minimizing the power for extruding. Sometimes the minimization of power may lead to poor mechanical properties. Hence, it is essential to model material behavior along with an estimation of the forces derived by mechanics. Table 1.2 summarizes important events in the field of materials science.

Table 1.2 Some Important Events in the Field of Materials Science

Year	Event	Impact on Manufacturing
1709	Abraham Darby makes iron with coke.	Replacement of charcoal (a wood product) with coke (produced by burning mineral coal) improved the quality of iron produced and reduced dependency on wood. Interest in coal mining enhanced.
1740	Benjamin Huntsman developed the crucible steel technique.	Helped to control the quality of steels of different grades. The process is not in vogue now.
1839	Vulcanized (heated with sulfur) rubber invented by Charles Goodyear.	Use of rubber as an engineering material increased due to improved hot strength of rubber.
1855	The Bessemer process for mass production was patented.	This took nearly 25 years to become popular in industry, though now the process is obsolete. However, the basic oxygen process that is in use, in essence follows the similar principle.
1886	The Hall–Héroult process for producing aluminum was invented.	Price of aluminum came down.
1909	Bakelite was announced to the public by Belgian chemist Baekeland.	This was the beginning of the plastic age.
1912	Harry Brearley invented stainless steel.	Production of "rust free" steel enhanced its applicability and immediately found a place in cutlery.
1916	Jan Czochralski developed a method for growing a single crystal of a material.	This process has revolutionized the manufacturing of electronic devices.
1931	Wallace Hume Carothers developed nylon.	First used for making the bristles of toothbrushes, nylon was extensively used for making parachutes.
1980	Duplex stainless steel was developed.	High corrosion resistance and strength makes it an important engineering material.
1991	Discovery of nanotubes.	A plethora of applications of nanotubes have been found.
2004	Graphene is developed.	Perhaps the strongest material discovered by mankind, it finds application in several fields.

1.7 AUTOMATION

From time immemorial, humanity kept slaves for performing difficult, boring and repetitive tasks. And not only slaves of its own species, but other animals too. They also developed machines to minimize efforts and tried to use them as slaves too, though in fact humanity became the slave of the machine, as illustrated by the film *Modern Times* directed by Charlie Chaplin. In mythologies, there are several examples of automation. It is very difficult to ascertain how much is truth and how much is imagination in mythological stories: only imprecise references can be made from the traces of old civilizations. It has been inferred that around 5000 BC, the people of Mohenjo-Daro and Harappa in the Indus Valley made a number of toys with automatic features. Around 400 to 350 BC, Archytas of Tarentum made a steam-powered pigeon, which was perhaps supported on wires. Ctesibius (285 to 222 BC) of Alexandria, also spelled Ktesibios or Tessibius, is called the Father of Pneumatics. Among his various inventions, one is an improved clepsydra (water clock), in which water dripping at a constant rate was stored in a vessel. The raising of the water level lifted a float to which a pointer was attached. Ctesibius did not leave any written records but was referred to by Roman author Marcus Vitruvius Pollio and Greek mathematician and engineer Heron of Alexandria. Not much is known about the life of Heron; many historians believe that he lived from AD 10 to 85. He developed the first steam turbine called an "aeolipile". In his book *Automatopoietica*, Heron described several automatic devices, which closed and opened temple doors, coin-operated machines and statues that poured wine.

The modern era of automation started in the 18th century. A textile worker, Basile Bouchon, invented a method for controlling a loom with perforated tape in 1725. In 1945, Jacques Vaucanson automated it. However, the first programmable loom was developed by Joseph Marie Jacquard in 1804. This motivated the development of other types of automatic devices. In 1837, Charles Babbage began to create the first prototype of an "Analytical Engine", which was the world's first mechanical computer. He died before completing it. In the analytical engine, the input was provided through punched cards as in Jacquard's loom. Almost a hundred years after, the first general purpose computer Z3 was built. This electromechanical computer was designed by a German civil engineer, Konard Zuse. The first electronic computer, ENIAC, was developed in 1946. In 1952, IBM's first commercial computer, the IBM 701, was introduced. Meanwhile, efforts to automate machine tools had started in the 1950s. The first Numerical Control (NC) machine appeared in 1952 and the first Computer Numerical Control (CNC) machine in 1957.

There were attempts to develop robots in the 20th century, which had caught the imagination of literary and artistic people. In the 19th century (and perhaps even before) some artists had made mechanized horses. Very little is known about such inventions, because they were not replicated on a reasonable scale and so knowledge of them remained confined to the creators. In 1921–1922, Karel Capek, a Czechoslovakian, wrote a play called *Rossumovi Univerzální Roboti* (*Rossum's Universal Robot*). Isaac Asimov, a fiction writer, invented three laws of robotics in about 1940.

In the 1940s, a number of manipulator robots were developed for carrying out dangerous tasks, particularly in nuclear power plants. At the Argonne National Laboratory in the USA, Roy C. Goertz developed a tele-operated manipulator in 1944. In 1959, the Planet Corporation introduced a pick and place robot. Unimation brought out the first industrial robot in 1961. In 1971, INTEL created the first microprocessor 4004 that ran at a clock speed of 108 kHz. This prompted developments in the field of robotics and CNC technology.

The first automated guided vehicle came in the 1950s. Today unmanned vehicles and unmanned factories are becoming a reality. However, there will always be a need for human intervention. Thus, most systems are really support systems for human beings. Starting from the 1990s, applications of artificial intelligence (AI) increased. It remains to be seen what level of intelligence these systems will achieve.

1.8 THE FUTURE OF MANUFACTURING

The study of past helps to forecast the future. From time immemorial, humans have been manufacturing. Primitive humans used tools for their individual needs. As the strength and efficiency of a human being is limited but his or her desires are endless, there have been attempts to use technology for manufacturing. Science and technology have brought much progress, but they have also produced many side-effects due to the disruption of a natural lifestyle. To mitigate these adverse side-effects, solutions need to be searched for in science and technology alone.

Comparing ancient civilization with the modern era, the following salient observations can be made:

1. In the modern age, an individual's significance has been lessened in favor of the masses. Instead of producing goods according to the needs of the individual, manufacturers tried to make them for mass consumption. Hence, the product specifications were not for suiting the need of a particular customer, but were at best the optimum specifications for the masses. Even today several custom-made products are too expensive.
2. Shifting from a domestic to a factory system of production affected the family system and had an adverse impact on culture, as workers had to forsake their geographical roots. Excessive urbanization

forced people to live in nuclear families rather than joint families and to forgo their cultural identities.

3. Due to the heavy consumption of natural resources and pollution from the manufacturing sector, the environment is facing serious threat.

To combat these problems, the manufacturing sector is reforming/transforming in the following ways:

- Additive manufacturing technology is again bringing the individual to the forefront. When this technology is matured, it will be possible to produce economical custom-made products.
- Almost all factories will come to follow a lights-out manufacturing methodology, thanks to developments in the IoT and cyber-physical systems. With the help of data science and cloud computing, it will be possible to control factories remotely. Figure 1.2 depicts the concept of controlling unmanned (lights-out) factories with the help of the internet and cloud computing. Chatterjee et al. (2020) have discussed a framework for the prediction of cutting force in machining processes by collecting data from various machine tools. The approach can be extended to overall machining performance prediction. Figure 1.3 illustrates how the data from various machines can be collected and deposited in a central data repository (CDR), where they can be processed using data analytic techniques. Accordingly, instructions can be provided to various machines remotely. Work from home will become

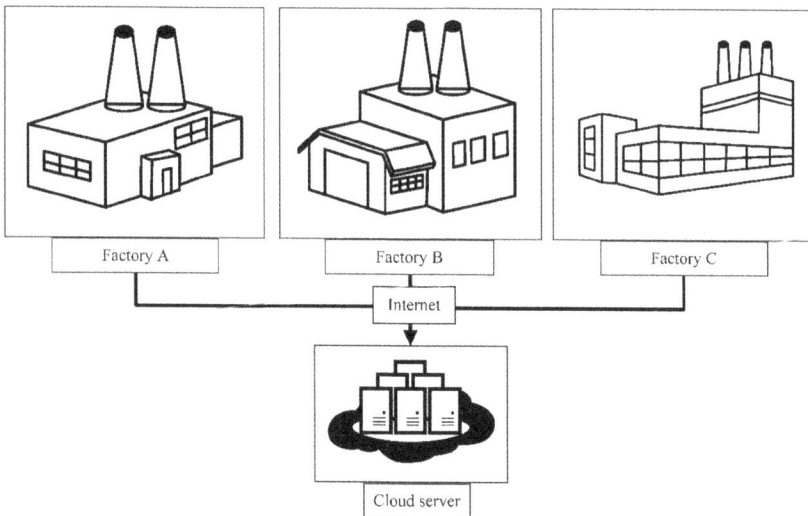

Figure 1.2 Concept of unmanned factories connected through the internet and receiving help from cloud computing (Chatterjee et al., 2020, reproduced under CC BY 3.0).

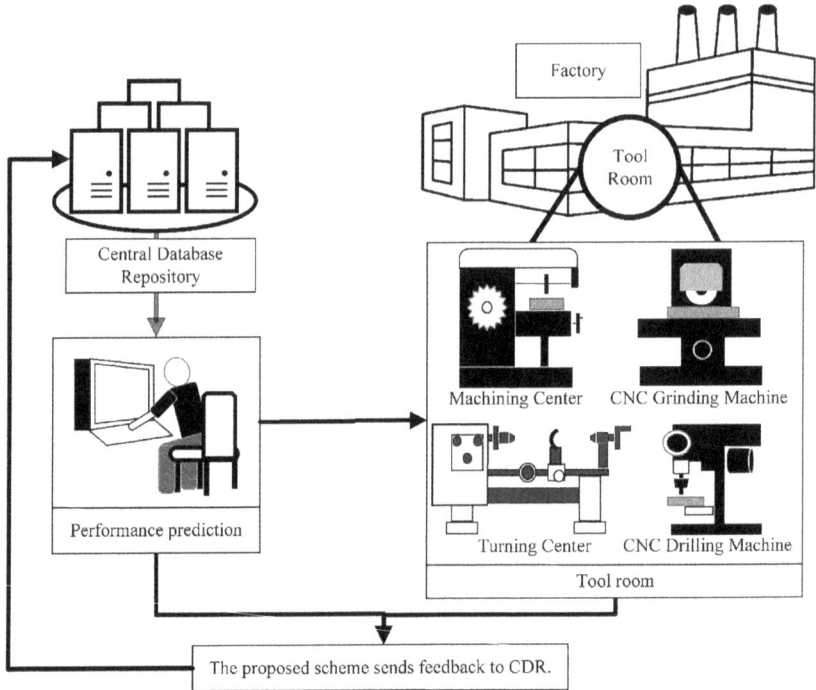

Figure 1.3 Effective control of machine tools with the help of a central database repository (CDR) (Chatterjee et al., 2020, reproduced under CC BY 3.0).

a common norm. This will also be economical as the cost of commuting to the workplace will reduce. In addition, cities will not suffer problems from traffic and pollution.

• There is increasing emphasis on environmentally friendly and sustainable manufacturing systems. New technologies like additive manufacturing are seemingly cleaner. It is true that in sustainability analysis, one has to consider the entire product life cycle. For example, while assessing the sustainability of a metal 3D printing process such as selective laser melting (SLM), the cost and effect on the environment of the metal powder production process should also be considered. It is possible that this process may not be completely benign to the environment. However, the powder production can be centralized to produce bulk quantity and the production facility can be kept away from the main part of a city. Metal powder can be distributed to different units through an unmanned transportation system.

The growth of manufacturing technology and systems can be depicted as moving in a circle with increasing tangential velocity. Figure 1.4 presents

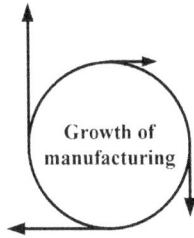

Figure 1.4 Symbolic representation of the growth of manufacturing. Researchers, scientists and practicing engineers are trying to build sustainable manufacturing systems that can bring progress without hampering the natural life style of the days of yore.

this in a simplified manner, where it can be seen that it is possible to grow in such a manner that the natural life and culture of previous times is not hampered. Working from home and green manufacturing will bring back the good old days.

1.9 CHALLENGES

There are several challenges on the path of the growth of manufacturing. Although several new technologies have been developed, the issue of overall sustainability remains. Additive manufacturing, for example, is still not cost-competitive with conventional manufacturing processes. The social aspects also need to be carefully addressed. Disruptive technologies often impact employment. Theoretically, employment opportunities should not reduce, because the desires of human beings have no end. If a certain new manufacturing process enhanced the production volume manifolds with reduced manpower, one can still employ the same number of personnel by deciding to enhance production volume. However, the personnel need to be trained to use new technology; their old skills may become obsolete. Thus, the real problem is not of employment but making people adaptable to new technology. People face employment problems only when technological development takes place at a higher pace than people can assimilate. That is why technology-induced unemployment is a temporary phenomenon. This can be easily seen by observing the impact of computers and robotics on employment. Although they created a temporary manpower reduction in industries, today these technologies are offering several attractive and interesting jobs.

The issue of security is also important. Too much dependence on the internet raises several security challenges. Valuable data can be hacked. Hence, researchers are trying to develop technologies that are cyber-secure. For example, blockchain technology, conceptualized in 2008, is supposed to be a much safer technology for data. There are challenges associated with additive manufacturing, which will enable decentralized or distributed manufacturing, which will be difficult to control. There will be a need of proper governmental policy as well as technological safeguards that will protect intellectual property rights and the production of forbidden goods.

Reducing the pollution due to manufacturing activities is also not an easy task. Nowadays several research groups are working on a circular economy, in which manufacturers design reusable products. For example, plastic products can be recycled into pellets for making new products. However, the full implementation of a circular economy is still a far goal.

Another challenge is related to manufacturing education. Modern manufacturing technologies are based on different branches of science. The boundaries between different engineering disciplines are fast vanishing. Making an optimized syllabus for engineering students with the proper blending of important scientific and technological topics is a challenge.

1.10 CONCLUSION

Manufacturing is integrally linked with human civilization. For ages, humankind has been trying to improve productivity and quality. This led to the invention of several manufacturing processes. However, those processes produced a number of side-effects. To mitigate these, engineers and scientists have been making continuous efforts, the success of which has led to sustainable development. Now, more attention is being paid to preserve ecological resources, flora and fauna, culture, health and the general happiness of the masses. This type of development is like moving on a circular track; although we are moving forward, we want to reach back to the good old days!

REFERENCES

Astakhov, V.P. (2006), An opening historical note. *International Journal of Machining and Machinability of Materials*, 1, 3–11.

Bryson, S., Kenwright, D., Cox, M., Ellsworth, D. and Haimes, R. (1999), Visually exploring gigabyte data sets in real time. *Communications of the ACM*, 42(8), 82–90. Retrieved May 3, 2021 from https://dl.acm.org/doi/pdf/10.1145/310930.310977

Chatterjee, K., Zhang, J. and Dixit, U. S. (2020), Data-driven framework for the prediction of cutting force in turning, *IET Collaborative Intelligent Manufacturing*, 2(2), 87–95. doi:10.1049/iet-cim.2019.0055.

Ernst, H. and Merchant, M. E. (1941), *Chip Formation, Friction and Finish*. Cincinnati Milling Machine Company, Cincinnati, OH.

Ghosh, A. (2021), Birth of mechanical engineering and a glimpse into the future trends, in *Mechanical Sciences: The Way Forward*, ed. U. S. Dixit and S. K. Dwivedy, Springer, Singapore, 1–28.

Guergov, S. (2018), A review and analysis of the historical development of machine tools into complex intelligent mechatronic systems, *Journal of Machine Engineering*, 18(1), 107–119. doi:10.5604/01.3001.0010.8828.

Hamaguchi, T. (1985). Prospects for self-reliance and indigenisation in automobile industry: case of Maruti-Suzuki project, *Economic and Political Weekly*, 20(35), M115–M122. Retrieved May 2, 2021, from http://www.jstor.org/stable/4374769

Jones, R., Haufe, P., Sells, E., Iravani, P., Olliver, V., Palmer, C. and Bowyer, A. (2011), RepRap–the replicating rapid prototyper. *Robotica*, 29(1), 177–191. doi:10.1017/S026357471000069X

Lazerson, M. (1995), A new phoenix?: Modern putting-out in the Modena knitwear industry. *Administrative Science Quarterly*, 40(1), 34–59. doi:10.2307/2393699.

Lee, E.H. and Shaffer, B.W. (1951), The theory of plasticity applied to problems of machining. *ASME Journal of Applied Mechanics*, 18, 405–413.

McNeil, I. (1990), Introduction: Basic tools, devices and mechanisms, in An Encyclopedia of the History of Technology (ed. by McNeil, I.), Routledge, London, pp. 1–40.

Miller, A. B., Sears, M. E., Morgan, L. L., Davis, D. L., Hardell, L., Oremus, M. and Soskolne, C. L. (2019), Risks to health and well-being from radio-frequency radiation emitted by cell phones and other wireless devices. *Frontiers in Public Health*, 7, Article 223, 10 pages. doi:10.3389/fpubh.2019.00223.

Piispanen, V. (1937). Lastunmuodostumisen teoriaa. *Teknillinen aikakauslehti*, 27(9), 315–322.

Pollard, S. (1963), Factory discipline in the industrial revolution, *The Economic History Review*, 16(2), 254–271. doi:10.2307/2598639.

Purvis, B., Mao, Y. and Robinson, D. (2019), Three pillars of sustainability: In search of conceptual origins. *Sustainability Science*, 14(3), 681–695. doi:10.1007/s11625-018-0627-5.

Putnik, G. D. and Putnik, Z. (2012), Lean vs agile in the context of complexity management in organizations, *The Learning Organization*, 19(3), 248–266. doi:10.1108/09696471211220046.

Rangwala, S. S. and Dornfeld, D. A. (1989). Learning and optimization of machining operations using computing abilities of neural networks. *IEEE Transactions on Systems, Man, and Cybernetics*, 19(2), 299–314. doi:10.1109/21.31035.

Smith, M. (1973), John H. Hall, Simeon North, and the Milling Machine: the nature of innovation among antebellum arms makers, *Technology and Culture*, 14(4), 573–591. doi:10.2307/3102444.

Thilak, V. M. M., Devadasan, S. R. and Sivaram, N. M. (2015), A literature review on the progression of agile manufacturing paradigm and its scope of application in pump industry, *The Scientific World Journal*, 2015, Article ID 297850, 9 pages. doi:10.1155/2015/297850.

Verstraeten, S. V., Aimo, L. and Oteiza, P. I. (2008), Aluminium and lead: molecular mechanisms of brain toxicity, *Archives of Toxicology*, 82(11), 789–802.

Viollet, P. L. (2017), From the water wheel to turbines and hydroelectricity. Technological evolution and revolutions, *Comptes Rendus Mécanique*, 345(8), 570–580. doi:10.1016/j.crme.2017.05.016.

Womack, J.P. and Jones, D.T. (2013), *Lean Thinking: Banish Waste and Create Wealth in Your Corporation*, Simon & Schuster, London.

Woodbury, R. S. (1960), The legend of Eli Whitney and interchangeable parts. *Technology and Culture*, 1(3), 235–253. doi:10.2307/3101392.

World Commission on Environment and Development (2013), *Our Common Future, Chapter 2: Towards Sustainable Development.* http://www.ask-force.org/web/Sustainability/Brundtland-Our-Common-Future-1987-2008.pdf, Retrieved May 3, 2021.

Zorev, N.N. (Ed.) (1966), *Metal Cutting Mechanics*, Pergamon Press, Oxford.

Zvorykin, K.A. (1896) 'On the force and energy needed to separate the chip from the workpiece (in Russian)', *Tekhicheskii Sbornik i Vestnic Promyslinosty*, 123, 57–96.

Chapter 2

Grinding and Recent Trends

Kamal Kishore and Manoj Kumar Sinha

National Institute of Technology Hamirpur, Hamirpur, India

Dinesh Setti

Indian Institute of Technology Palakkad, Palakkad, India

CONTENTS

2.1 INTRODUCTION

Manufacturing is a dynamic process, which has been continuously evolving for the last 250 years, from the steam engine to the ongoing Industry 4.0 era. Within current global competition, manufacturing is one of the most substantial contributors to the world's gross domestic product (GDP), estimated to be nearly one-third [1]. The contribution of manufacturing by the top ten countries to world GDP is shown in Figure 2.1. Among manufacturing, subtractive processes play a pioneering role in product realisation. Even

DOI: 10.1201/9781003327394-2

31

```
China  🌐                                                28.4%
United States  🏛                                16.6%
Japan  ●                       7.2%
Germany  ⬤               5.8%
South Korea  :●:       3.3%
India  ⚊          3.0%
Italy  ( )         2.3%
France  ( )       1.9%
United Kingdom  ✦      1.8%
Mexico  (·)      1.5%
```

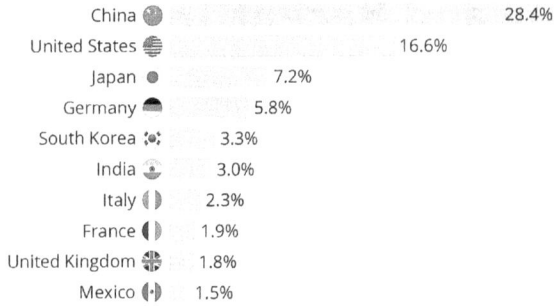

Figure 2.1 Contribution of the manufacturing sector by the top ten countries to world GDP [1].

today, the post-processing of products manufactured by additive manufacturing technologies is done using subtractive processes [2].

Grinding is one of the most widely used subtractive manufacturing processes, mainly to machine high-strength materials with a high surface finish and close tolerances. This process has covered a unique journey from a finishing process to a bulk material removal process. This process covers nearly 25% of conventional machining processes, which is a crucial indicator of its versatility. With developments in various high strength materials, such as superalloys, composites and ceramics, the grinding process has revamped itself as a most useful process in machining these high strength materials [3]. Grinding is a multifaceted, dynamic and stochastic process where material removal occurs by the interaction of numerous hard abrasives with the work surface in the form of multiple scratches. The grinding process has evolved with the adoption of new concepts, technological advancements and market demands [4]. The application of sustainable technologies, hybrid grinding, micro/nano-grinding, high-speed grinding, textured grinding wheels, 3D printed grinding wheels, the integration of artificial/intelligent systems and precision-shaped grain wheels are some of the key advancements in the grinding process. These recent advancements have entirely revamped the application domain of the grinding process. These potential advancements are briefly described in the following sections.

2.2 SUSTAINABLE MACHINING TECHNIQUES

In grinding, a lot of heat generation takes place due to its inherent material removal characteristics. The heat generated induces thermal defects in the workpiece, such as grinding burn, residual stress generation and microstructure change, which adversely affect the ground components' performance. Grinding performance has been improved using suitable metalworking fluids (MWFs) in wet cooling mode. However, the cost of MWF and,

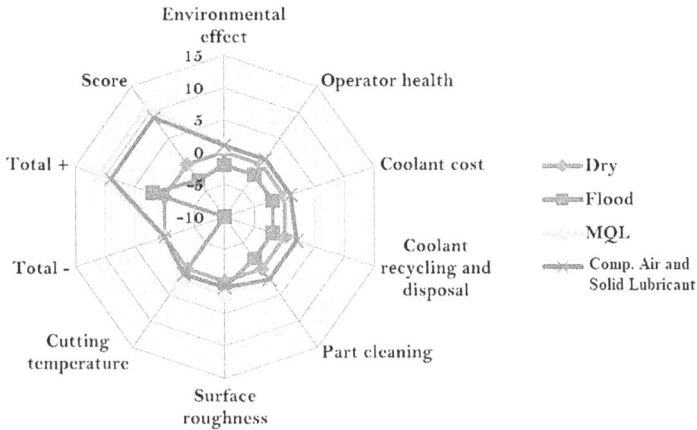

Figure 2.2 The impact of different grinding conditions on sustainability [8].

more importantly, the associated threats to people, machines and materials have raised a question mark over its usage. Also, the prolonged application of water-based MWFs may cause staining, corrosion and produce micro-organisms, which cause severe irritation and diseases like dermatitis and asthma to the operator [5]. Debnath et al. [6] note that about 80% of industry operators suffer from one or more diseases due to the harmful effect of MWFs. With strict legislation aimed at controlling health hazards and environmental pollution, manufacturing industries are supposed to adhere to ISO 14000 standards and zero-waste provision [7]. To counter these adverse implications of MWFs along with the socio-economic aspects, there is an urgent need to replace such techniques with safer and sustainable practices. This situation has led to sustainable cooling and lubrication techniques, such as minimum quantity lubrication (MQL), cryogenic cooling techniques, hybrid techniques and allied techniques for machining high strength materials [5]. Figure 2.2 summaries the impact of different grinding environments on various vital pillars of sustainability.

2.2.1 Minimum Quantity Lubrication (MQL)

As the name suggests, MQL aims to minimise cutting-fluid consumption and achieve near-to-dry machining conditions. In MQL, the mixture of compressed air and coolant is supplied to the machining zone through a nozzle in an atomised form in order to cover a sizeable grinding area. There are two types of MQL systems: one is MQL with an external aerosol supply and the other is MQL with an internal aerosol supply. In the external one, aerosol formation occurs either in the outer chamber or in the nozzle itself and supplies it to the machining zone as in grinding. In the internal one, the prepared aerosol is delivered to the machining zone through the cutting tool's spindle and internal channel. Sinha et al. [4] has proposed and experimentally validated a unique technique to optimise various MQL parameters while grinding an Inconel 718 superalloy. Silva et al. [9] performed cylindrical grinding of AISI 4340 hardened steel in standard wet cooling and MQL with a self-developed rectangular nozzle. They concluded that MQL has better results than traditional cooling in terms of surface integrity and diametric wheel wear. Although MQL shows promising results, the parameters, like the aerosol concentration and droplet size, are still a matter of concern. However, clogging of the grinding wheel and reduced heat transfer are two significant challenges associated with MQL techniques. Sato et al. [10] explained the behaviour of clogging in MQL and experimentally showed that water-based MQL reduced the clogging of the wheel, surface roughness and roundness error by 40, 51 and 50%, respectively. Further, such issues have led to the development of nanofluid MQL (nMQL) and allied cooling techniques.

2.2.2 Nanofluid MQL (nMQL)

Nanofluids are engineered colloidal fluid in which nanoparticles (NPs) are suspended. The thermal conductivity, heat transfer coefficient, thermal diffusivity and the viscosity of the prepared nanofluids (NFs) are much superior to the base fluid [11]. The base fluids widely used for making NFs are oil, water, glycol and other organic and inorganic liquids. The common types of NPs generally used for making nanofluids are [12]:

1. Stable metals: silver, gold, copper, nickel, etc..
2. Oxide, nitride and the carbides of metals: SiO_2, CuO, Fe_2O_3, TiO_2, ZrO_2, Al_2O_3, SiC, etc.
3. Carbon and its various forms, including graphene platelets, carbon nanotubes CNTs) and carbon nanohorns.

There are two types of mechanisms involved in heat transfer through NFs:

1. Structural mechanism: The mechanism primarily consists of liquid layering and particle aggregation. Liquid layering refers to a process in which a liquid layer surrounds the solid particles and is responsible for increasing the overall conductivity of NFs. In particle aggregation, NPs form chain-like structures by aggregating, which eventually helps in heat conduction (see Figure 2.3).
2. Dynamic mechanism: This mechanism is based on the Brownian motion of the NPs. Robert Brown first observed the zigzag motion of solute particles in a colloidal suspension. In NFs, it has a significant effect during heat transfer.

Ibrahim et al. [16] developed an ecofriendly mixture of palm oil and graphene nanoplatelets (GNPs) for utilisation in the MQL grinding of

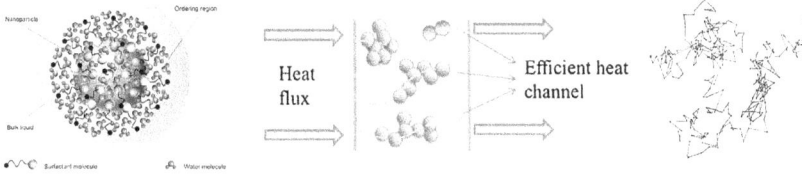

Figure 2.3 The mechanism of liquid layering, particle aggregation and Brownian motion in nanofluids [13–15].

Ti-6Al-4V. The obtained results were compared with dry grinding and commercially available Acculub LB2000. Findings revealed that the developed nanofluid reduced the specific grinding energy by 91.78 and 80.25% compared to dry grinding and grinding with Acculub LB2000, respectively, at 0.1wt% of GNPs.

Nowadays, the use of hybrid NFs, that is NFs having more than one NP, is increasing so as to employ both NPs' desirable behaviour. Zhang et al. [17] performed surface grinding on GH4169 (an Ni-based alloy) to analyse hybrid nanofluid behaviour in MQL. MoS_2 NPs with adequate lubrication capacity and carbon nanotubes with high heat conductive capacity was used for this purpose. The study revealed that the hybrid mixture showed better grinding performance than a single NP fluid due to the physical synergistic effect. NPs help in heat extraction and assist in machining to obtain good surface texture and dimensional accuracy. A few established mechanisms of NP-based grinding is depicted in Figure 2.4. The possible mechanisms of NF-induced anti-wear and anti-friction properties while machining can be summarised as [18–20]:

1. A nanofluid forms a tribofilm layer on the surface that allows for mass loss and improves the anti-wear property. This effect is called the mending effect.
2. Nanotubes and spherical NPs generally roll between the surfaces and change the sliding friction into mixed rolling and sliding friction.

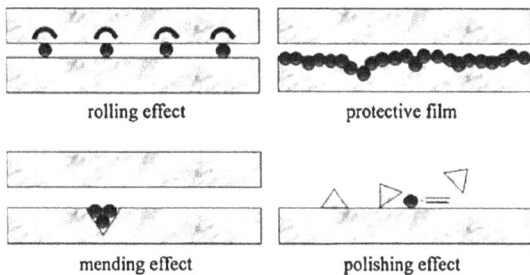

Figure 2.4 Different functions of NPs in machining [19,20].

3. Smaller NPs interact with the surface quickly and hence form a protective layer.
4. High compressive stress concentration associated with tools is reduced by NPs, which bear it.

2.2.3 Cryogenic Cooling

The word "kryos" originates from Greek, which means frost or cold [21]. According to the Cryogenic Society of America, a temperature below $-150\ °C$ is considered cryogenic [22]. Gases like nitrogen, helium, hydrogen, neon and oxygen are widely used for achieving a cryogenic environment. In machining, liquid nitrogen (LN_2) and liquid CO_2 are commonly used [23]. In 1953, Bartley was the first to perform cryogenic machining using liquid CO_2 as a coolant [24]. After that, many researchers reportedly used cryogenic machining. Strict environmental laws aim to achieve sustainability and have diverted recent research towards cryogenics from conventional cooling. The cryogens that are used in machining vaporise automatically without leaving any negative impact on the environment. Also, cryogenic cooling reduces specific energy, cutting forces and surface defects, helping to achieve a better surface finish and dimensional accuracy. Elanchezhian and Pradeep Kumar [25] studied the effect of the nozzle angle and depth of cut in the cryogenic grinding of Ti-6Al-4V, investigated in terms of normal force (F_n), tangential force (F_t), surface roughness (R_a), grinding temperature (GT) and specific energy. The findings revealed a reduction of 21, 90, 48 and 33% in Ft, Fn, GT and Ra, respectively, compared to conventional cooling at a 45° nozzle angle and an infeed of 30 μm. Amini et al. [26] compared the different cooling techniques on the grinding performance of MO40 steel. The study's outcome suggested that the cutting velocity and depth of cut are significant while assessing surface roughness and hardness in cryogenic cooling.

2.2.4 Hybrid Cooling Methods

Hybrid cooling techniques are the combination of two processes for utilising the benefits of both approaches simultaneously. For instance, the combination of cryogenic cooling and MQL, called N_2MQL, is simultaneously referred to in much of the literature [27]. Nadolny and Kieraś [28] implemented the hybrid cooling and lubrication technique utilising a solid-lubricant-impregnated grinding wheel and a cold air gun to achieve sustainability. The internal cylindrical grinding of a 100Cr6 bearing was performed. The experiment revealed that the grinding wheel wear reduced significantly with a four-fold increase in the grinding ratio. Lopes et al. [29] investigated AISI 4340 steel's grinding performance in external cylindrical plunge grinding. The two hybrid techniques, MQL with cooled air and MQL with wheel jet cleaning, were compared with traditional MQL. The findings revealed that both methods outperformed traditional MQL in terms of grinding power, wheel wear, surface roughness and specific

Table 2.1 Summary of Reviewed Works in the Field of Grinding with
Sustainable Cooling Techniques

Year	Authors	Response studied	Workpiece	C/L technique	Country
2021	Esmaeili et al. [30]	SR, SSD	Inconel 718	MQL	Iran
2021	Sato et al. [31]	SR, RE, GP, GR, MH	AISI 4340	MQL	Brazil
2021	Moraes et al. [32]	SR, RE, DWE, GR, GP, MH	AISI 4340	Hybrid	Brazil
2021	Bensaid et al. [33]	MH, RS, GWW, SR	Hardox 500	Cryogenic	Tunisia
2021	Arafat et al. [34]	GF, SR, CP, GWC	DIN100Cr6	Hybrid	Germany
2021	Dang et al. [35]	SM, MH, RS, MH, MS	300N steel	Cryogenic	China
2021	Peng et al. [36]	GT, SR, MS, RS	Inconel 718	nMQL	China
2021	Sui et al. [37]	SGF, GT, GR	YG8	nMQL	China
2021	Rodriguiz et al. [38]	SR, MH, GP, AE, GR	AISI 4340	Hybrid	Brazil
2020	Moretti et al. [39]	SR, RE, DWE, CP, MH, AE	DINGGG70	MQL	Brazil
2020	Gao et al. [40]	SR, MC, FD, GT	CFRP	nMQL	China
2020	Virdi et al. [41]	GR, GE, GT, SR	Inconel 718	MQL	India
2019	Kumar et al. [42]	SGF, SGE, GFR, SR, MH	Si₃N₄		India
2019	Awale et al. [43]	SGE, GF, GFR, SR	H13 die steel	Cryogenic	India
2021	Naskar et al. [44]	TCF, GP, RS	Ti-6Al-4V	Hybrid	India

Notes: SSD sub-surface damage, SR surface roughness, RE roundness error, DWE diametral wheel wear, CP cutting power, MH microhardness, AE acoustic emission, GP grinding power, GR grinding ratio, TCF tangential cutting force, GWW grinding wheel wear, GF grinding force, MS microstructure, MC micro-fractal characteristics, FD fractal dimension, GT grinding temperature, SGF specific grinding force, SGE specific grinding energy, GE grinding energy.

grinding energy. Further, the result of MQL with the wheel jet was compared with flood cooling. Table 2.1 summarises the recent work on sustainable cooling and lubrication (C/L) techniques used in the grinding of different engineering materials.

2.3 HYBRID GRINDING TECHNIQUES

Today, the development of hybrid machining is increasing rapidly. This combines two or more machining principles/techniques used simultaneously to achieve the desired requirements. The two crucial hybrid types of grinding processes are discussed below.

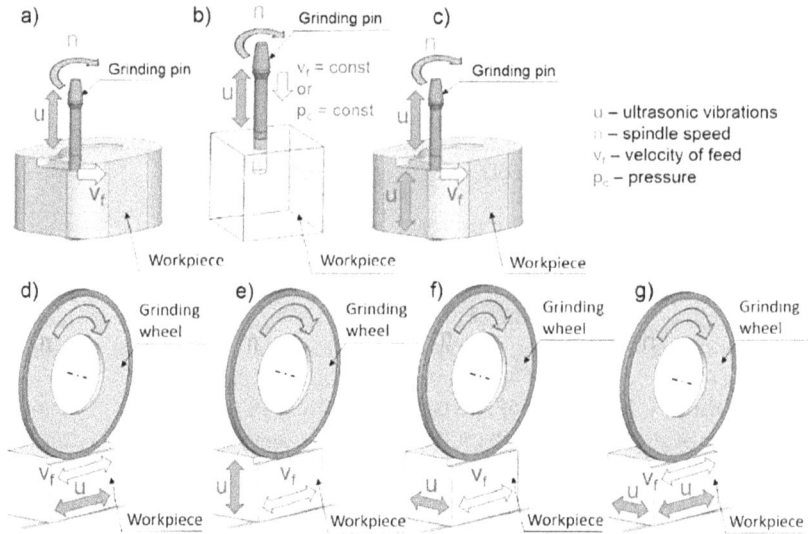

Figure 2.5 Possible variations in different types of 1D and 2D UAGs [48].

2.3.1 Ultrasonic Assisted Grinding (UAG)

UAG was introduced in the 1950s; however, by the 1980s, it had gained popularity due to its overwhelming results while machining difficult-to-machine (DTM) materials. UAG is the integration of an ultrasonic system into that of a conventional grinding technique. This integration combines the impact-based material removal process with the grinding mechanism [45]. The introduction of ultrasonic vibrations in the grinding changes the nature of the tool–workpiece interaction. The high frequency and small amplitudes of vibration enhance the ground surface quality, decrease the grinding forces, reduce tool wear and help in the self-sharpening of abrasive grits due to intermittent impact action. 1D-UAG and 2D-UAG are the two essential types of widely used UAGs shown in Figure 2.5 [46]. In 1D-UAG, the ultrasonic vibrations are applied either along the wheel's rotational axis or perpendicular to it. There are two types of 1D-UAG: the first is 1D axial UAG (1D-AUAG) where a vibration is applied along the rotational axis; the other is called 1D vertical UAG (1D-VUAG) when the pulse is applied perpendicular to the rotational axis [47]. In 2D-UAG, the vibration is applied to the workpiece in two axes, simultaneously creating elliptical or circular geometries. Hence it is sometimes also referred to as elliptical UAG. In general, 2D-UAG is better than 1D-UAG as it produces a better surface finish and tolerance.

2.3.2 Laser-Assisted Grinding (LAG)

In LAG, the laser beam is focused onto the workpiece, but only in the region just ahead of the grinding wheel. Local heating occurs on the workpiece where the laser falls and the material becomes soft. In this process, the laser's intensity is kept low so that the material does not evaporate. Generally, a CO_2 laser and an Nd: YAG laser are widely used in LAG [49]. LAG allows the grit to penetrate the workpiece more efficiently, which eventually reduces the specific energy and grinding forces, and increases process performance. Ma et al. [50] studied the effect of LAG on the surface roughness, surface morphology and subsurface damage of zirconia. A ductile fracture in LAG was visible compared to the brittle fracture in traditional grinding, resulting in better grinding efficacy and reduced surface roughness. Li et al. [51] investigated the material removal mechanism, specific grinding energy and grinding force ratio during the LAG of RB-SiC ceramics. Finite element analysis (FEA) simulation was also carried out to predict the temperature gradient induced by laser radiation. Better surface integrity was obtained through LAG when the temperature gradient in the laser radiation was near to 15%. LAG is generally performed in dry conditions; hence it is much closer to being a sustainable machining process.

2.4 MICRO-GRINDING

The continued development in grinding technology and the trending demand to make things compact and precise for industries like optics, biomedicine, telecommunications and electronics form the basis of the emergence of micro-machining [52]. This process also finds wide acceptance in the fabrication of microactuators, microsensors and fluidic devices. Micro-grinding is one of the micro-machining processes that can rapidly be used in precision and ultra-precision manufacturing. It is also adopted as the final process for fixing the quality of products. A few of the key differences between macro-grinding and micro-grinding processes are listed in Table 2.2 [52]. Sometimes micro-grinding is also known as ultra-precision grinding due to its low depth of cut and material removal rate (MRR) [53].

Micro-grinding is generally classified into two types based on the interaction between the tools and workpiece: micro-slot grinding and micro-surface grinding. The tools typically used in micro-grinding are electroplated or sintered chemical vapours deposited on a super-abrasive substance like cubic

Table 2.2 Representation of Macro-Grinding and Micro-Grinding Differences [52]

Remark	Macro-Grinding	Micro-Grinding
Ratio of the depth of cut to the grit radius	50–100	0.1–1
Ploughing effect	Not significant ($\approx$ 0%)	Significant ($\approx$ 20–30%)
Material removal rate	$10^n \sim 10^{-1}$ mm³/mm·s	$10^{-1} \sim 10^{-3}$ mm³/mm·s

boron nitride (cBN) and diamond for the fabrication of micro-featured parts. Kadivar et al. [54] investigated the effect of cutting speed and feed rate to a depth of the cut ratio on residual stress and specific micro-grinding energy when grinding Ti-6Al-4V. They found that residual stress is directly related to feed rate to the depth of the cut ratio and inversely to the cutting speed. The specific micro-grinding energy didn't significantly affect the variation of the feed rate to the depth of the cut ratio while it had an inverse relation with the cutting speed.

2.5 HIGH-SPEED GRINDING (HSG)

As technological advancement occurs, the trend for high-speed grinding increases rapidly to reduce machining time, surface roughness and increase the MRR. Figure 2.6 shows the effect of high speed on various grinding parameters. Better surface integrity, better accuracy, less time consumption, lower cost and productivity increase are a few of the significant benefits of the HSG process [55]. A few of the effective HSG processes are described briefly in the following sections.

2.5.1 Creep Feed Grinding

Creep feed grinding is an abrasive machining process characterised by low feed and high infeed. It is placed in the category of established techniques which attain high accuracy in the grinding of superalloys [56,57]. Zang et al. [58] experimentally performed creep feed grinding on IC10 (an Ni_3Al based superalloy) to investigate the effect on the material's surface integrity and fatigue life. The effect of the grinding forces and temperature were considered in these experiments. Moreover, the influence of surface roughness

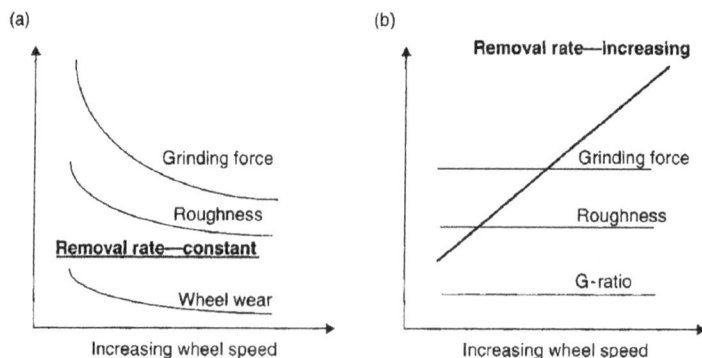

Figure 2.6 Effect of increasing wheel speed (a) at a constant MRR and (b) at an increasing MRR [4].

and surface hardening was also considered to identify their impact on the specimen's fatigue life. They concluded that the infeed was the prime factor influencing the grinding forces and temperature whereas the surface roughness showed a positive impact on fatigue life.

2.5.2 High-Efficiency Deep Grinding (HEDG)

HEDG is a novel abrasive machining process used to ultra-finish operations to a moderate accuracy level. It is characterised by high MRR, high wheel speed, low work speed and considerable infeed. It utilises the combined mechanics of creep feed grinding with efficient high-rate grinding [59,60]. Batako et al. [61] performed HEDG to achieve a high MRR. They developed and used a unique and in-house setup, an aluminium oxide wheel instead of a superabrasive wheel, for this grinding. They concluded that with an increase in MRR, the specific energy decreases. Further, due to the considerable depth of cut and contact length, the coolant could not penetrate the grinding zone despite the grinding wheel's porous nature.

2.5.3 Speed Stroke Grinding (SSG)

This is defined as a surface grinding process in which the table speed is higher than in standard conventional surface grinding. In this grinding, bigger chips are formed, which decreases the specific energy of the grinding [62]. Zeppenfeld [63] investigated chip formation and wear mechanism for SSG while grinding γ-titanium aluminide. The bigger chips result in lower heat transfer to the workpiece, which improves the product quality. Linke et al. [64] showed how the combination of speed stroke grinding and high-speed grinding could enhance the product quality and performance parameters. Further, they developed the FEA model to predict the temperature and residual stress generated in the workpiece. The validation experiment showed that a decrease in grinding power, energy and tool wear was obtained through this combination while it worsens the surface roughness.

2.6 TEXTURED GRINDING WHEEL (TGW)

Grinding performance is significantly affected by the wheel's topography, morphology and its grit sharpness. They are the critical factors in any grinding experiment. The stochastic nature, uneven protrusion and random orientation of grits in the conventional grinding wheel lead to low heat dissipation [65]. According to Li et al. [66], a maximum of 40% of cutting fluid can penetrate the grinding zone. The low dissipation of heat from the grinding area is responsible for defects like grinding burn, unwanted thermal residual stress and change in workpiece microstructures. To overcome these defects, the surface texturing of the wheel is one possible solution.

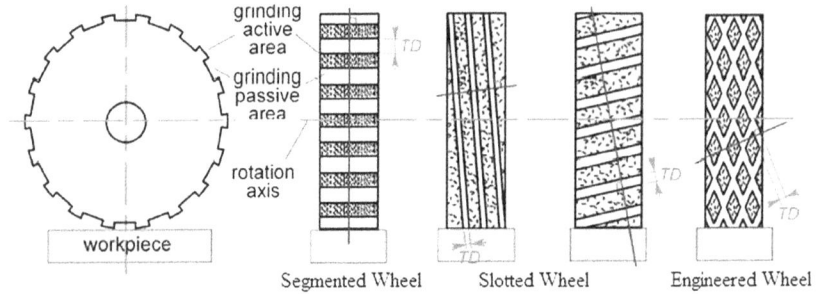

Figure 2.7 Different types of textured wheels [68].

According to Irani et al. [67], the textured wheel's grooves improved the grinding performance in the super-abrasive wheel. These grooves helped in penetrating the coolant and lubrication to the grinding zone in a convenient way. Textured grinding wheels are specially engineered grinding wheels that have either a pattern or slot over the periphery. These wheels have two types of parts on their periphery, called active and passive. The part of the textured wheel actively participating in grinding is the active area and the remaining area is known as the passive area. The passive part of the wheel provides the channel for coolants to penetrate the grinding zone. Some examples of textured wheels are shown in Figure 2.7. The term "texture dimension" (TD) is purposely defined by Li and Axinte [68] to describe the minimum gap between the repeated patterns. Based on the different design types, the textured wheel is broadly classified into three categories: the slotted grinding wheel, segmented grinding wheel and engineered/patterned grinding wheel.

Rodriguez et al. [69] studied the interrupted grinding process's performance using a specially designed grooved cBN grinding wheel. Their findings revealed that the grinding performance increased significantly. The coolant quickly reached the grinding zone compared to the conventional grinding wheel, where the superabrasive wheel's porosity was a problem. Further, surface roughness increased significantly as compared to traditional grinding due to grooves on the periphery. Denkena et al. [70] performed a grinding experiment with hardened tool steel using a pattern grinding wheel to investigate its overall performance. The outcome of the study suggested that there was a reduction in grinding forces and grinding burn defect.

2.7 3D PRINTED GRINDING WHEELS

Additive manufacturing (AM), also known as 3D printing, will be a demand of future manufacturing. This manufacturing technique works on an almost zero wastage principle. Therefore, researchers are endeavouring to utilise the benefits of AM in nearly every sector. In grinding, too, 3D printing technologies are used to print precise grinding wheels. These wheels have

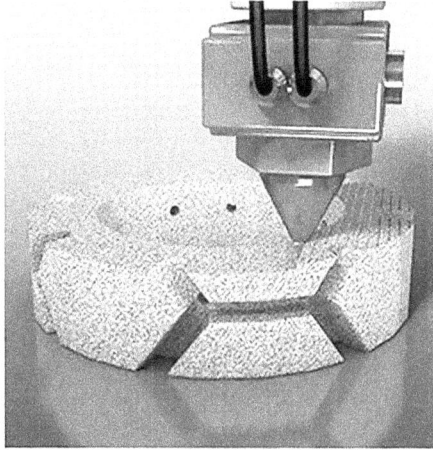

Figure 2.8 3D printed grinding wheel [71].

flexible capabilities which help in the enhancement of the grinding performance. Figure 2.8 shows a typical 3D printed grinding wheel.

Denkena et al. [72] studied the possibilities of printing a grinding wheel using a laser powder bed fusion technique. Their main aim was to evaluate the performance of an NiTi alloy as a metal bond material. They considered a pure NiTi alloy and pre-alloyed NiTi as a bonding material, and diamond as an abrasive for printing a grinding wheel. The experiment was performed and it was concluded that pre-alloyed NiTi bonding offered better performance than a pure NiTi alloy as the bonding material. As metal bonds have high thermal conductivity and good grain retention, they are considered a good option as a bonding material for superabrasive wheels. But their dense structure limits the number of pores in the grinding wheel, which hinders the coolant supply in the machining zone. To overcome this drawback, Li et al. [73] proposed a novel design and fabrication method using selective laser melting for the printing of porous metal bonded grinding wheels. The grinding performance of the 3D printed grinding wheels was evaluated and compared with the electroplated grinding wheel. The outcome of the experiment showed a significant reduction in the grinding forces and the specific grinding energy when utilising a 3D printed grinding wheel. Likewise, Tian et al. [74] compared the performance of three different kinds (an octahedron structured wheel, honeycomb structured wheel and solid structured wheel) of 3D printed grinding wheels to that of the electroplated wheel. The grinding performance of these wheels was evaluated in terms of grinding forces, surface roughness, MRR and hardness. They concluded that MRR was high for the solid structure, whereas the smallest friction force and highest cutting force were obtained in the octahedron structure. Better surface roughness was obtained using a 3D printed grinding wheel rather than the electroplated wheel.

2.8 ARTIFICIAL INTELLIGENCE (AI) IN GRINDING

The development of AI in recent years is one of the most critical events in human history. The development of the Internet of Things, robotics and intelligent systems has revamped the way of manufacturing. The integration of these systems helps to enhance the quality of the manufactured product and helps achieve sustainability. In grinding, the integration of AI is mainly done in the development of simulation-based tools and the integration of a closed feedback system with self-optimising controller programs. The features of AI for grinding are shown in Figure 2.9 [75].

In grinding, various sensors with other auxiliary devices such as cameras and microphones are used in association with smart algorithms for the online monitoring of grinding burn, wheel position and selection of wheels. Buchmeister et al. [76] developed a program that automatically selects the grinding wheel and its work to grind the helical flute to manufacture solid rotatory cutting tools. Lee et al. [77] developed an auditory-based system integrated with deep learning to monitor the grinding wheel. This system includes a microphone and acoustic emission sensors for collecting audio data which were analysed using spectrum analysis. The data obtained from the spectrum analysis were fed into deep convolution neural networks to evaluate the wheel's condition. To validate the system, 820 experiments were carried out with 410 sharp wheels and 410 worn-out wheels; it was concluded that the system achieved an accuracy of 97.44% with a precision of 98.26%.

Integrated AI applied to grinding generates a lot of data, which are essential for future use. All generated data fall under the umbrella of "Big Data". These stored data were later used as resourceful mining information. Guo et al. [78] developed a grinding circuit based on data-driven reinforcement

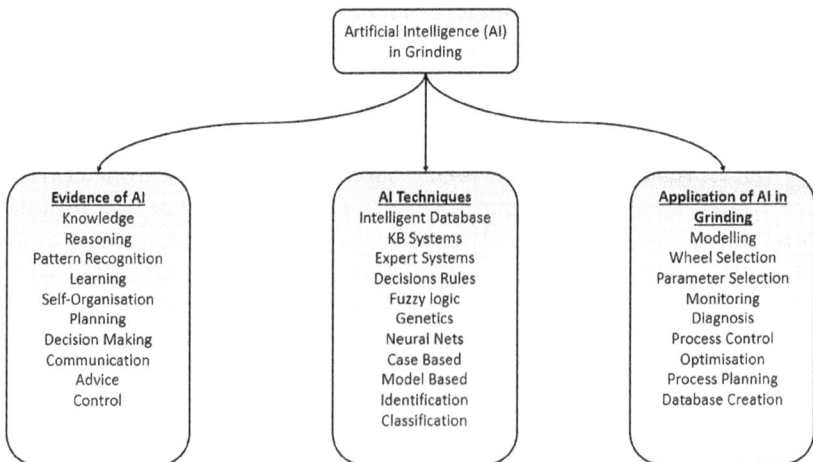

Figure 2.9 AI techniques applied in grinding [13].

learning to control the grinding process and grinding quality of ores in the mineral industry.

2.9 PRECISION SHAPED GRITS

In abrasive machining, grinding is the oldest known process. In grinding, material removal takes place with the help of abrasive particles called grits. These grits have an undefined geometry, which results in high heat generation, variation in the dynamically cutting force and a greater grinding power requirement. In recent years, to minimise these negative impacts, several steps have been taken in terms of technological advancement, like the development of the textured wheel, the optimisation of machining parameters and the introduction of sustainable cooling and lubrication techniques. In all these technological advancements, the focus is given on improvement in machine tool technologies rather than grinding wheels. Precision shaped grains are the new type of abrasive grain that are being developed by 3M. These grains have a well defined geometrical shape and hence seem to behave like cutting edges. To ascertain the grinding performance of these grains, various experiments were performed by the developer. They concluded that the grinding wheel made by these grains outperformed the conventional grinding wheel in terms of MRR, wheel life and reduced thermal damage [79].

2.10 SUMMARY

Grinding is perhaps one of the oldest known machining processes. Its application covers a wide range, from the grinding of spices in households to the industry-based machining of alloys. Still, this process is unceasingly evolving due to the incorporation of newer concepts and techniques. For instance: stone and stone-bed-based grinding in homes has been replaced with compact grinding machines. Similar is the case for industries where new grinding methods and integration with smart controllers are implemented for large production. The emergence of a new techno-economic theme with the ambition of high production has triggered the desire to inculcate sustainable grinding by ignoring the utilisation of harmful cutting fluids. At the same time, machining needs to be addressed to achieve sustainability in grinding. To overcome this situation the MQL technique, cryogenic cooling and allied methods of C/L have been widely adopted by researchers and scientists. Other grinding techniques like UAG and LAG are also considered to achieve high productivity in a sustainable way. The incorporation of AI in grinding has enabled the process with simulation tools and self-optimising programs, which minimises human interference significantly. The emergence of the latest technologies like 3-D printed grinding wheels and precision-shaped abrasive grits have opened a new realm for grinding techniques. The cyclic trend to introduce new concepts and techniques is always in fashion to make the grinding process easier and more sophisticated and precise.

REFERENCES

[1] Statista, 2018, "*China Is the World's Manufacturing Superpower*" [Online]. Available: www.statista.com/chart/20858/top-10-countries-by-share-of-global-manufacturing-output.

[2] Kishore, K., and Sinha, M. K., 2021, "A State-of-the-Art Review on Fused Deposition Modelling Process," *Adv. Manuf. Ind. Eng.*, pp. 855–864.

[3] Cheng, K., 2008, *Machining Dynamics: Fundamentals, Applications and Practices*, Springer Science & Business Media.

[4] Sinha, M. K., Setti, D., Ghosh, S. and Rao, P. V., 2016. "An Investigation on Surface Burn During Grinding of Inconel 718", *J. Manuf. Proc.*, **21**, pp. 124–133.

[5] Setti, D., Sinha, M. K., Ghosh, S., and Rao, P. V., 2015, "Performance Evaluation of Ti–6Al–4V Grinding Using Chip Formation and Coefficient of Friction Under the Influence of Nanofluids", *Int. J. Mach. Tools & Manuf.*, **88**, pp. 237–248.

[6] Debnath, S., Reddy, M. M., and Yi, Q. S., 2014, "Environmental Friendly Cutting Fluids and Cooling Techniques in Machining: A Review," *J. Clean. Prod.*, **83**, pp. 33–47.

[7] Ghosh, S., and Rao P. V., 2015, "Application of Sustainable Techniques in Metal Cutting for Enhanced Machinability: A Review," *J. Clean. Prod.*, **100**, pp. 17–34.

[8] Mia, M., Gupta, M. K., Singh, G., Królczyk, G., and Pimenov, D. Y., 2018, "An Approach to Cleaner Production for Machining Hardened Steel Using Different Cooling-Lubrication Conditions," *J. Clean. Prod.*, **187**, pp. 1069–1081.

[9] Silva, L. R., Corrêa, E. C. S., Brandão, J. R., and de Ávila, R. F., 2020, "Environmentally Friendly Manufacturing: Behavior Analysis of Minimum Quantity of Lubricant - MQL in Grinding Process," *J. Clean. Prod.*, **256**, p. 103287.

[10] Sato, B. K., Lopes, J. C., Diniz, A. E., Rodrigues, A. R., de Mello, H. J., Sanchez, L. E. A., Aguiar, P. R., and Bianchi, E. C., 2020, "Toward Sustainable Grinding Using Minimum Quantity Lubrication Technique with Diluted Oil and Simultaneous Wheel Cleaning," *Tribol. Int.*, **147**, p. 106276.

[11] Ganvir, R. B., Walke, P. V., and Kriplani, V. M., 2017, "Heat Transfer Characteristics in Nanofluid—A Review," *Renew. Sustain. Energy Rev.*, 75(October 2016), pp. 451–460.

[12] Sundar, L. S., Sharma, K. V., Singh, M. K., and Sousa, A. C. M., 2017, "Hybrid Nanofluids Preparation, Thermal Properties, Heat Transfer and Friction Factor – A Review," *Renew. Sustain. Energy Rev.*, 68(October 2016), pp. 185–198.

[13] Cardellini, A., Fasano, M., Bozorg Bigdeli, M., Chiavazzo, E., and Asinari, P., 2016, "Thermal Transport Phenomena in Nanoparticle Suspensions," *J. Phys. Condens. Matter*, 28(48), p. 483003.

[14] Wong, K. V., and Castillo, M. J., 2010, "Heat Transfer Mechanisms and Clustering in Nanofluids," *Adv. Mech. Eng.*, **2010**, pp. 49–71.

[15] Song, D., Zhou, J., Wang, Y., and Jing, D., 2016, "Choice of Appropriate Aggregation Radius for the Descriptions of Different Properties of the Nanofluids," *Appl. Therm. Eng.*, **103**, pp. 92–101.

[16] Ibrahim, A. M. M., Li, W., Xiao, H., Zeng, Z., Ren, Y., and Alsoufi, M. S., 2020, "Energy Conservation and Environmental Sustainability during Grinding Operation of Ti–6Al–4V Alloys via Eco-Friendly Oil/Graphene Nano Additive and Minimum Quantity Lubrication," *Tribol. Int.*, **150**(April) p. 106387.

[17] Zhang, Y., Li, C., Jia, D., Zhang, D., and Zhang, X., 2015, "Experimental Evaluation of the Lubrication Performance of MoS2/CNT Nanofluid for Minimal Quantity Lubrication in Ni-Based Alloy Grinding," *Int. J. Mach. Tools Manuf.*, **99**, pp. 19–33.

[18] Peng, D. X., Kang, Y., Hwang, R. M., Shyr, S. S., and Chang, Y. P., 2009, "Tribological Properties of Diamond and SiO2 Nanoparticles Added in Paraffin," *Tribol. Int.*, **42**(6), pp. 911–917.

[19] Lee, K., Hwang, Y., Cheong, S., Choi, Y., Kwon, L., Lee, J., and Kim, S. H., 2009, "Understanding the Role of Nanoparticles in Nano-Oil Lubrication," *Tribol. Lett.*, **35**(2), pp. 127–131.

[20] Hemmat Esfe, M., Bahiraei, M., and Mir, A., 2020, "Application of Conventional and Hybrid Nanofluids in Different Machining Processes: A Critical Review," *Adv. Colloid Interface Sci.*, **282**, p. 102199.

[21] Junghare, H., Hamjade, M., Patil, C. K., Girase, S. B., and Lele, M. M., 2017, "A Review on Cryogenic Grinding," *Int. J. Curr. Eng. Technol.*, **7**(7), pp. 420–423.

[22] Jawahir, I. S., Attia, H., Biermann, D., Duflou, J., Klocke, F., Meyer, D., Newman, S. T., Pusavec, F., Putz, M., Rech, J., Schulze, V., and Umbrello, D., 2016, "Cryogenic Manufacturing Processes," *CIRP Ann. - Manuf. Technol.*, **65**(2), pp. 713–736.

[23] Ravi, S., and Gurusamy, P., 2020, "Role of Cryogenic Machining: A Sustainable Manufacturing Process," *Mater. Today Proc.*, (xxxx).

[24] Chattopadhyay, A. B., Bose, A., and Chattopdhyay, A. K., 1985, "Improvements in Grinding Steels by Cryogenic Cooling," *Precis. Eng.*, **7**(2), pp. 93–98.

[25] Elanchezhian, J., and Pradeep Kumar, M., 2018, "Effect of Nozzle Angle and Depth of Cut on Grinding Titanium under Cryogenic CO_2," *Mater. Manuf. Process.*, **33**(13), pp. 1466–1470.

[26] Amini, S., Baraheni, M., and Esmaeili, S. J., 2019, "Experimental Comparison of MO40 Steel Surface Grinding Performance under Different Cooling Techniques," *Int. J. Light. Mater. Manuf.*, **2**(4), pp. 330–337.

[27] Gupta, M. K., Mia, M., Singh, G. R., Pimenov, D. Y., Sarikaya, M., and Sharma, V. S., 2019, "Hybrid Cooling-Lubrication Strategies to Improve Surface Topography and Tool Wear in Sustainable Turning of Al 7075-T6 Alloy," *Int. J. Adv. Manuf. Technol.*, **101**(1–4), pp. 55–69.

[28] Nadolny, K., and Kieraś, S., 2020, "New Approach for Cooling and Lubrication in Dry Machining on the Example of Internal Cylindrical Grinding of Bearing Rings," *Sustain. Mater. Technol.*, **24**, p. e00166.

[29] Lopes, J. C., Garcia, M. V., Valentim, M., Javaroni, R. L., Ribeiro, F. S. F., de Angelo Sanchez, L. E., de Mello, H. J., Aguiar, P. R., and Bianchi, E. C., 2019, "Grinding Performance Using Variants of the MQL Technique: MQL with Cooled Air and MQL Simultaneous to the Wheel Cleaning Jet," *Int. J. Adv. Manuf. Technol.*, **105**(10), pp. 4429–4442.

[30] Esmaeili, H., Adibi, H., and Rezaei, S. M., 2021, "Study on Surface Integrity and Material Removal Mechanism in Eco-Friendly Grinding of Inconel 718 Using Numerical and Experimental Investigations," *Int. J. Adv. Manuf. Technol.*, **112**(5–6), pp. 1797–1818.

[31] Sato, B. K., Lopes, J. C., Rodriguez, R. L., Garcia, M. V., Ribeiro, F. S. F., Aguiar, P. R., and Bianchi, E. C., 2021, "Eco-Friendly Manufacturing towards the Industry of the Future with a Focus on Less Cutting Fluid and High Workpiece Quality Applied to the Grinding Process," *Int. J. Adv. Manuf. Technol.*, **113** (3), pp. 1163–1172.

[32] de Moraes, D. L., Lopes, J. C., Andrioli, B. V., Moretti, G. B., da Silva, A. E., da Silva, J. M. M., Ribeiro, F. S. F., de Aguiar, P. R., and Bianchi, E. C., 2021. Advances in Precision Manufacturing Towards Eco-Friendly Grinding Process by Applying MQL with Cold Air Compared with Cooled Wheel Cleaning Jet. *Int. J. Adv. Manuf. Technol.*, **113**(11), pp. 3329–3342

[33] Bensaid, K., Dhiflaoui, H., Bouzaiene, H., Yahyaoui, H., and Fredj, N.B., 2021. Effects of the cooling mode on the integrity and the multi-pass micro-scratching wear resistance of Hardox 500 ground surfaces. *Int. J. Adv. Manuf. Technol.*, **113**(9), pp. 2865–2882.

[34] Arafat, R., Madanchi, N., Thiede, S., Herrmann, C., and Skerlos, S. J., 2021, "Supercritical Carbon Dioxide and Minimum Quantity Lubrication in Pendular Surface Grinding - A Feasibility Study," *J. Clean. Prod.*, **296**, p. 126560.

[35] Dang, J., Zhang, H., An, Q., Ming, W., and Chen, M., 2021, "Surface Modification of Ultrahigh Strength 300M Steel under Supercritical Carbon Dioxide (ScCO2)-Assisted Grinding Process," *J. Manuf. Process.*, **61**(October 2020), pp. 1–14.

[36] Peng, R., He, X., Tong, J., Tang, X., and Wu, Y., 2021, "Application of a Tailored Eco-Friendly Nanofluid in Pressurized Internal-Cooling Grinding of Inconel 718," *J. Clean. Prod.*, **278**, p. 123498.

[37] Sui, M., Li, C., Wu, W., Yang, M., Ali, H. M., Zhang, Y., Jia, D., Hou, Y., Li, R., and Cao, H., 2021, "Temperature of Grinding Carbide with Castor Oil-Based MoS2 Nanofluid Minimum Quantity Lubrication," *J. Therm. Sci. Eng. Appl.*, **13**(October), pp. 1–30.

[38] Rodriguez, R. L., Lopes, J. C., Garcia, M. V., Ribeiro, F. S. F., Diniz, A. E., Eduardo de Ângelo Sanchez, L., José de Mello, H., de Aguiar, P. R., and Bianchi, E. C., 2021, "Application of Hybrid Eco-Friendly MQL+WCJ Technique in AISI 4340 Steel Grinding for Cleaner and Greener Production," *J. Clean. Prod.*, **283**, p. 124670.

[39] Moretti, G. B., de Moraes, D. L., Garcia, M. V., Lopes, J. C., Ribeiro, F. S. F., Foschini, C. R., de Mello, H. J., Sanchez, L. E. D. A., Aguiar, P. R., and Bianchi, E. C., 2020, "Grinding Behavior of Austempered Ductile Iron: A Study about the Effect of Pure and Diluted MQL Technique Applying Different Friability Wheels," *Int. J. Adv. Manuf. Technol.*, **108**(11–12), pp. 3661–3673.

[40] Gao, T., Li, C., Jia, D., Zhang, Y., Yang, M., Wang, X., Cao, H., Li, R., Ali, H. M., and Xu, X., 2020, "Surface Morphology Assessment of CFRP Transverse Grinding Using CNT Nanofluid Minimum Quantity Lubrication," *J. Clean. Prod.*, **277**, p. 123328.

[41] Virdi, R. L., Chatha, S. S., and Singh, H., 2021, "Experimental Investigations on the Tribological and Lubrication Behaviour of Minimum Quantity Lubrication Technique in Grinding of Inconel 718 Alloy," *Tribol. Int.*, **153**(July 2020), p. 106581.

[42] Kumar, A., Ghosh, S., and Aravindan, S., 2019, "Experimental Investigations on Surface Grinding of Silicon Nitride Subjected to Mono and Hybrid Nanofluids," *Ceram. Int.*, **45**(14), pp. 17447–17466.

[43] Awale, A., Srivastava, A., Vashista, M., and Yusufzai, M. Z. K., 2019, "Surface Integrity Characterisation of Ground Hardened H13 Hot Die Steel Using Different Lubrication Environments," *Mater. Res. Express*, **6**(2), pp. 983–997.

[44] Naskar, A., Choudhary, A. and Paul, S., 2021. Surface Generation in Ultrasonic-Assisted High-Speed Superabrasive Grinding Under Minimum Quantity Cooling Lubrication with Various Fluids. *Tribology International*, 156, p. 106815.

[45] Das, S., and Pandivelan, C., 2020, "Grinding Characteristics During Ultrasonic Vibration Assisted Grinding of Alumina Ceramic in Selected Dry and MQL Conditions," *Mater. Res. Express*, 7(8), p. 085404.

[46] Dambatta, Y. S., Sarhan, A. A. D., Sayuti, M., and Hamdi, M., 2017, "Ultrasonic Assisted Grinding of Advanced Materials for Biomedical and Aerospace Applications — a Review," 92(9), pp. 3825–3858.

[47] Singh, A. K., Kumar, A., Sharma, V., and Kala, P., 2020, "Sustainable Techniques in Grinding: State of the Art Review," *J. Clean. Prod.*, **269**, p. 121876.

[48] Wdowik, R., 2018, "Measurements of Surface Texture Parameters after Ultrasonic Assisted and Conventional Grinding of Carbide and Ceramic Samples in Selected Machining Conditions," *Procedia CIRP*, **78**, pp. 329–334.

[49] Chryssolouris, G., Anifantis, N., and Karagiannis, S., 1997, "Laser Assisted Machining: An Overview," *J. Manuf. Sci. Eng. Trans. ASME*, **119**(4B), pp. 766–769.

[50] Ma, Z., Wang, Z., Wang, X., and Yu, T., 2020, "Effects of Laser-Assisted Grinding on Surface Integrity of Zirconia Ceramic," *Ceram. Int.*, **46**(1), pp. 921–929.

[51] Li, Z., Zhang, F., Luo, X., Chang, W., Cai, Y., Zhong, W., and Ding, F., 2019, "Material Removal Mechanism of Laser-Assisted Grinding of RB-SiC Ceramics and Process Optimization," *J. Eur. Ceram. Soc.*, **39**(4), pp. 705–717.

[52] Park, H. W., 2008, *Development of Micro-Grinding Mechanics and Machine Tools*, Georgia Institute of Technology.

[53] Hof, L. A., 2018, High-precision micro-machining of glass for mass-personalization (Doctoral dissertation, Concordia University)..

[54] Kadivar, M., Azarhoushang, B., Daneshi, A., and Krajnik, P., 2020, "Surface Integrity in Micro-Grinding of Ti6Al4V Considering the Specific Micro-Grinding Energy," *Procedia CIRP*, **87**, pp. 181–185.

[55] Schulz, H., and Moriwaki Toshimichi, 1992, "High-Speed Machining," *Compr. Mater. Process.*, **41**(2), pp. 637–643.

[56] Maksoud, T. M. A., 2005, "Heat Transfer Model for Creep-Feed Grinding," *J. Mater. Process. Technol.*, **168**(3), pp. 448–463.

[57] Ortega, N., Bravo, H., Pombo, I., Sánchez, J. A., and Vidal, G., 2015, "Thermal Analysis of Creep Feed Grinding," *Procedia Eng.*, **132**, pp. 1061–1068.

[58] Zhang, S., Yang, Z., Jiang, R., Jin, Q., Zhang, Q., and Wang, W., 2020, "Effect of Creep Feed Grinding on Surface Integrity and Fatigue Life of Ni3Al Based Superalloy IC10," *Chinese J. Aeronaut.*, (July), **34**(1), pp. 1–11.

[59] Ghosh, S., Chattopadhyay, A. B., and Paul, S., 2008, "Modelling of Specific Energy Requirement during High-Efficiency Deep Grinding," *Int. J. Mach. Tools Manuf.*, **48**(11), pp. 1242–1253.

[60] Bell, A., Jin, T., and Stephenson, D. J., 2011, "Burn Threshold Prediction for High Efficiency Deep Grinding," *Int. J. Mach. Tools Manuf.*, **51**(6), pp. 433–438.

[61] Batako, A. D. L., Morgan, M. N., and Rowe, B. W., 2013, "High Efficiency Deep Grinding with Very High Removal Rates," *Int. J. Adv. Manuf. Technol.*, **66**(9–12), pp. 1367–1377.

[62] Weiß, M., Klocke, F., and Wegner, H., 2013. "Process Machine Interaction in Pendulum and Speed-Stroke Grinding". In *Process Machine Interactions* (pp. 101–119). Springer, Berlin.

[63] Zeppenfeld, C., 2006, "Speed Stroke Grinding of γ-Titanium Aluminides," *CIRP Ann. - Manuf. Technol.*, **55**(1), pp. 333–338.

[64] Linke, B., Duscha, M., Klocke, F., and Dornfeld, D., 2011. Combination of speed stroke grinding and high speed grinding with regard to sustainability. *CIRP ICMS*, (1). p. 7.

[65] Kishore, K., Sinha, M. K., Singh, A., Archana Gupta, M. K. and Korkmaz, M. E., 2022. A comprehensive review on the grinding process: Advancements, applications and challenges. *Proc. Inst. Mech. Eng., Part C: Journal of Mechanical Engineering Science*, p. 09544062221110782.

[66] Li, C., Zhang, Q., Wang, S., Jia, D., Zhang, D., Zhang, Y., and Zhang, X., 2015, "Useful Fluid Flow and Flow Rate in Grinding: An Experimental Verification," *Int. J. Adv. Manuf. Technol.*, **81**(5–8), pp. 785–794.

[67] Irani, R. A., Bauer, R. J., and Warkentin, A., 2005, "A Review of Cutting Fluid Application in the Grinding Process," *Int. J. Mach. Tools Manuf.*, **45**(15), pp. 1696–1705.

[68] Li, H. N., and Axinte, D., 2016, "Textured Grinding Wheels: A Review," *Int. J. Mach. Tools Manuf.*, **109**, pp. 8–35.

[69] Rodriguez, R. L., Lopes, J. C., Garcia, M. V., Tarrento, G. E., Rodrigues, A. R., de Ângelo Sanchez, L. E., de Mello, H. J., de Aguiar, P. R., and Bianchi, E. C., 2020, "Grinding Process Applied to Workpieces with Different Geometries Interrupted Using CBN Wheel," *Int. J. Adv. Manuf. Technol.*, **107**(3–4), pp. 1265–1275.

[70] Denkena, B., Grove, T., and Göttsching, T., 2015, "Grinding with Patterned Grinding Wheels," *CIRP J. Manuf. Sci. Technol.*, **8**, pp. 12–21.

[71] 3M, 2021, "*Precision Structured Wheels Overview*". Available: https://www.3m.com/3M/en_US/metalworking-us/applications/precision-grinding/technology/superabrasives/precision-structured-wheels/.

[72] Denkena, B., Krödel, A., Harmes, J., Kempf, F., Griemsmann, T., Hoff, C., Hermsdorf, J., and Kaierle, S., 2020, "Additive Manufacturing of Metal-Bonded Grinding Tools," *Int. J. Adv. Manuf. Technol.*, **107**(5–6), pp. 2387–2395.

[73] Li, X., Wang, C., Tian, C., Fu, S., Rong, Y., and Wang, L., 2021, "Digital Design and Performance Evaluation of Porous Metal-Bonded Grinding Wheels Based on Minimal Surface and 3D Printing," *Mater. Des.*, **203**, p. 109556.

[74] Tian, C., Li, X., Zhang, S., Guo, G., Wang, L., and Rong, Y., 2018, "Study on Design and Performance of Metal-Bonded Diamond Grinding Wheels Fabricated by Selective Laser Melting (SLM)," *Mater. Des.*, **156**, pp. 52–61.

[75] Rowe, W. B., Li, Y., Mills, B., and Allanson, D. R., 1996, "Application of Intelligent CNC in Grinding," *Comput. Ind.*, **31**(1), pp. 45–60.

[76] Buchmeister, B., Palcic, I., and Ojstersek, R., 2019, *Artificial Intelligence in Manufacturing Companies and Broader: An Overview*, DAAAM International.

[77] Lee, C. H., Jwo, J. S., Hsieh, H. Y., and Lin, C. S., 2020, "An Intelligent System for Grinding Wheel Condition Monitoring Based on Machining Sound and Deep Learning," *IEEE Access*, **8**, pp. 58279–58289.

[78] Guo, L., Wang, H., and Zhang, J., 2019, "Data-Driven Grinding Control Using Reinforcement Learning," *Proc. - 21st IEEE Int. Conf. High Perform. Comput. Commun. 17th IEEE Int. Conf. Smart City 5th IEEE Int. Conf. Data Sci. Syst. HPCC/SmartCity/DSS 2019*, pp. 2817–2824.

[79] Graf, W., 2014, "*Cubitron II: Precision-Shaped Grain Turns the Concept of Gear Grinding Upside Down New Abrasive Grains Developed by 3M are Redefining the Process of Gear Grinding*," pp. 1–15.

Chapter 3

Recent Advances in Ultrasonic Manufacturing and Its Industrial Applications

Ravinder P. Singh
Maharishi Markandeshwar (Deemed to be University), Mullana, India

Vishal Gupta
Thapar Institute of Engineering & Technology
(Deemed to be University), Patiala, India

Girish C. Verma
Indian Institute of Technology Indore, Indore, India

Pulak Mohan Pandey
Indian Institute of Technology Delhi, Delhi, India

Uday Shanker Dixit
Indian Institute of Technology Guwahati, Guwahati, India

CONTENTS

DOI: 10.1201/9781003327394-3

3.1 INTRODUCTION

The dimensional and geometrical accuracy of any product is one of the bottlenecks of present manufacturing industries. This is primarily due to the dependence of various functional responses on geometric and dimensional accuracy. To meet these tough demands of high accuracy and precision, industries are looking into advanced manufacturing techniques, primarily due to the variance present in conventionally machined products, especially in the case of tougher and harder materials. The dimensional and geometric accuracy and precision of a part produced through any machining process is primarily dependent on the process parameters [1].

High cutting temperature and force generation during the conventional machining process are mainly responsible for inaccurate machining. Current manufacturing strategies involve the application of a coolant to deal with this problem. However, the use of conventional coolants makes the machining process ecologically harmful and costly [2,3]. Researchers have found that the application of cryogenic cooling during the machining operation also results in improved responses which further leads to better product quality. Experimental results [4,5] suggest that due to cryogenic cooling the length of the secondary shear zone is shortened, further resulting in a reduced temperature rise and cutting forces. The minimum quantity lubrication (MQL) approach was also found to be efficacious in lowering the cutting force and temperature [6]. Experimental findings [7] suggest that the application of MQL with nanoparticle additives further increases cooling and lubrication efficacy, which is inversely related to the roughness and cutting forces.

Recent studies in the area of ultrasonic assisted machining (UAMc) have drawn the attention of several researchers. In the UAMc process, a high frequency vibration (≥ 20 kHz) is superimposed with a conventional cutting motion [8]. Based on ease of application, vibration assistance can be given to the cutting tool or to the workpiece. Experimental investigations on ultrasonic assisted machining processes show an improvement in the process responses and product quality [8].

It has been reported [9] that ultrasonic vibration assistance during the milling of composite material shows better accuracy and improved machinability. Researchers have also concluded that the cutting energy spent decreases due to the existence of vibrations [9]. Recent studies [10] have shown that the effect of tool wear on cutting forces also decreases with the assistance of ultrasonic vibration. Additionally, the increment in surface roughness due to flank wear also diminishes with the assistance of ultrasonic vibration. Researchers have concluded that most of these effects are linked with the ultrasonic vibration energy provided during the process [10]. The influence of this ultrasonic energy also affects the residual stresses generated during the machining process [11]. A few researchers have also found significant (due to ultrasonic vibration assistance) reduction in surface roughness while machining additively manufactured parts [12].

Recently introduced ultrasonic assisted drilling (UAD) has shown potential over conventional drilling (CD) in terms of cutting force and temperature produced [13]. Many prior studies [14–17] reported that providing ultrasonic vibration to the rotary drilling tool resulted in a lower temperature rise in comparison to CD. Alam et al. [15] reported that UAD produced a lower temperature rise compared to CD in the drilling of bovine bone. In another study, Wang et al.[18] reported that UAD produces less mechanical damage to the bone surface during drilling and generated fewer micro-cracks. Recently Gupta and Pandey [19] introduced a novel drilling technique which is a combination of rotary ultrasonic machining and orthopedic drilling; namely the rotary ultrasonic bone drilling process. The authors reported that rotary ultrasonic bone drilling (RUBD) produced a lower cutting force and temperature rise, and comparatively lesser micro-cracks [20] in comparison to CD. Prior studies [13,18–20] showed that UAD produced better results in comparison to the CD process in biomedical applications.

3.2 BASIC CONCEPT

The UAMc process is an established method that improves machinability and machining responses, especially in the case of several airspace grade alloys with high toughness and strength such as Ti6Al4V and Al7075. The influence of vibration assistance on machining responses was first assessed on the turning process in the late 1950s [8,21,22]. After that, researchers performed several experimental studies to explore the potential of different manufacturing processes. The results of experimental investigations have shown that UAMc of different classes of materials, such as ductile, brittle and soft ones, resulted in improved machinability and tool life [8]. Researchers credited this enhancement in machinability to the intermittent cutting phenomenon/effect. This effect is basically a periodic disengagement and engagement of tool and workpiece which occurs at very high frequency (equal to the applied vibrational frequency). Experimental and theoretical studies have shown that the presence of an intermittent cutting effect in UAMc processes affects the mechanics of cutting and workpiece material properties.

3.2.1 Mechanics of the Cutting UAMc Process

It has already been stated that most research has suggested that the improvement in responses is mainly due to the occurrence of the intermittent cutting phenomenon/effect. This conclusion is primarily based on the fast fourier transform (FFT) analysis outcome of the force data. However, in a few experimental studies [23], researchers have also analyzed chip morphology to assess the effect of ultrasonic vibration on the cutting mechanics in UAMc turning process. Their findings also confirmed the effect of intermittent cutting which produced a reduced formation of built-up edges and resulted in thinner and uniform chip formation [23]. A few researchers [24] have also reported that the decrease in cutting force is also due to low chip–tool interface friction. Therefore, it can be concluded that, due to the direct or indirect effect of intermittent cutting, the responses improve.

Among most of the reported literature, researchers have suggested that intermittent cutting is only effective in a specific range of process parameters. Additionally, a few studies [25] have also reported a detrimental effect of intermittent cutting on tool health. However, they have reported that this detrimental effect was observed in a very narrow range of process parameters. Researchers have concluded that this is primarily due to the shockwave generated at the tool tip due to the abrupt interaction between the tooltip and workpiece.

3.2.2 Influence on the Cutting Mechanism

Due to the superposition of ultrasonic vibration at the tip of the cutting tool an additional sinusoidal motion is added to the existing cutting motion. Due to this an extra velocity V_{ul} (sinusoidal in nature) is superimposed at the tool tip which can be calculated at any given time 't' by [8]:

$$V_{ul} = 2\pi af \cos(2\pi ft) \qquad (3.1)$$

where a and f are the amplitude and frequency of vibration respectively. The presence of the ultrasonic velocity V_{ul} alters the magnitude and direction of the total/actual cutting velocity ($V_T = V_c + V_{ul}$) at the cutting tool tip. The periodic disengagement between the tip of the cutting tool and the machining zone occurs due to the change in the amplitude direction of V_T at $V_{ul} \gg V_c$. Figure 3.1 shows the intermittent cutting effect. It can be seen from Figure 3.1 that the cutting occurs with a total velocity of $V_c + V_{ul}$ when V_{ul} acts along with V_c, but, as V_{ul} changes its direction, the cutting velocity reduces, and in a small time period the tool tip starts separating from the machining zone (with total velocity, $V_{ul} - V_c$).

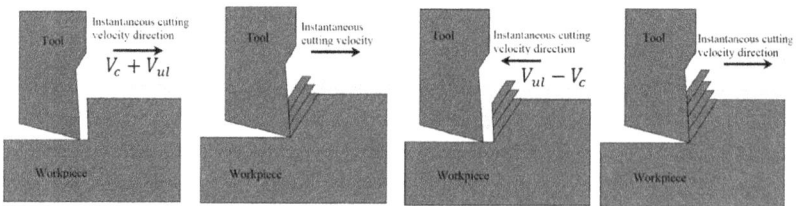

Figure 3.1 Schematic depiction of the intermittent cutting effect in the UAMc process [26].

As the velocity is sinusoidal in nature this phenomenon keeps repeating and is termed "intermittent cutting". As V_T depends on f, a and V_c, these parameters also influence the intermittent cutting effect. Experimental results have also verified the influence of vibration parameters on the intermittent cutting effect. Researchers have reported that as the frequency or amplitude of the vibration increases the machining force decreases due to an increase in the intermittent cutting effect [8, 27, 28].

Researchers have suggested that this is mainly due to the lower value of the contact ratio [23, 29], which is the ratio of the time period during which the cutting tip performs a cutting operation in one oscillation cycle (of the applied vibration). Theoretical studies [30, 31] have proposed that the contact ratio depends on the process parameter and may vary from 1.0 to 0.5. These theoretical studies also suggested that as the contact ratio increases and approaches 1.0 the intermittent cutting effect reduces and the UAMc is converted to a conventional process. Experimental studies [32–34] have also verified the proposed diminishing theory of the intermittent cutting effect, as similar cutting force trends were observed in the proposed range of cutting velocity ($V_{ul} > V_c$). Apart from evaluating the ultrasonic vibration effect, researchers [35–37] have also experimented with a frequency less than 20 kHz, that is with subsonic vibrations. However, in the case of subsonic vibration assistance, the machining forces were higher compared to a conventional machining process.

Apart from the reduction of the average cutting force, improvement in other responses has also been observed during UAMc. Tool deflection is also reduced due to the lower machining forces, which further results in higher machining accuracy. Due to this the process finds its application in the area of precision machining. However, this additional energy expended for ultrasonic vibration generation makes these processes costly for mass production. Therefore, it is considered suitable for manufacturing of precision parts.

3.3 MECHATRONICS INVOLVED IN UAMc

Generally the vibration applied in UAMc has a very low amplitude and a high frequency. For generating the vibrations, commonly two types of transducer materials are used: piezoelectric or magnetostrictive. These transducer materials convert the electrical energy into mechanical vibrations. Magnetostrictive materials generate vibrations by applying an oscillating magnetic field; piezoelectric transducers do the same thing with a high frequency AC supply. Piezoelectric materials are considered better than magnetostrictive ones, as they have a higher efficiency, durability and wider working range [38].

The amplitude of vibration generated by any transducer material depends on the thickness of the element (across which the potential was applied);

however, this is generally very low. Therefore, in order to amplify the vibration, amplitude horns are used [39]. These horns are made of specific alloys (with a low attenuation property) and designed through simulation (based on harmonic and modal analysis) [40]. Additionally, a high frequency AC power supply of 20 kHz is used to supply the required electric power to the transducer. In most of the prior studies the ultrasonic effect was mostly applied to the stationary component due to its ease of application [8]. However, in a few studies [41,42] on milling and drilling, the assistance of vibration was also provided to the moving component (tool).

3.4 THE ECONOMIC ASPECT OF UAMc

As discussed in prior sections, the assistance of ultrasonic vibration has several positive effects on process responses. However, to assess the economic aspect of ultrasonic assistance different responses need to be considered, such as the machining cost or post-processing cost. Experimental studies have shown that the UAMc process improves product quality and reduces burr formation by up to 64%. Because of this the cost of post-processing is reduced by up to 30% [43]; however, due to ultrasonic vibration application a significant amount of cost is added, which supersedes the reduction due to reduced post-processing. Additionally, due to the low post-processing requirement, the overall production time is reduced by 5–10% [43]. This reduces the total production cost primarily for precision components. A few researchers [8] have also assessed the impact of the ultrasonic effect on macro-scale machining and found it to be insignificant. Due to this the UAMc process becomes suitable only for precision manufacturing. As no research has studied the economical aspect of the UAMc processes, there is a need to carry out a comprehensive commercial investigation. At the same time, focus should also be towards the development of an economical technology which can be used for producing ultrasonic vibrations.

3.5 INFLUENCE OF THE ULTRASONIC EFFECT ON VARIOUS MACHINING PROCESSES

Experimental studies on the ultrasonic assisted turning (UAT) process attracted several researchers to assess the effect of ultrasonic vibrations on various manufacturing techniques [8]. Prior investigations [28] have shown the improvement of various output responses to the incorporation of ultrasonic vibrations in traditional and non-traditional processes. However, in order to keep this chapter concise only three UAMc processes – namely drilling, turning and milling – will be discussed, though, in order to show the applicability of UAMc processes, two case studies will be presented.

3.5.1 UAT Process

Due to the ease in evaluating cutting mechanics, most studies are performed in the area of the UAT process. From reported experimental studies [8] it can be seen that the direction of applied ultrasonic vibration influences the output responses (like surface roughness and forces) in dissimilar manners. UAT with vibration in the direction of feed has resulted in very low surface roughness [44]. This is mainly due to the rubbing motion between workpiece and tool whose effect is similar to a burnishing operation. However, the assistance of vibration in a transverse direction results in a surge in surface roughness. This is primarily because the impact happens between the cutting tip and the machined surface. Due to this impact the cutting tip intrudes into the machined surface, resulting in a dimpled textured surface [45]. Despite an increase in roughness, researchers have found that these dimpled textures enhance the tribological properties and wettability of the machined surface [46,47].

Researchers [23,48] have also suggested that the improvement in tool life can also be credited to an improved cooling and lubrication effect. This is mainly because of the periodic separation which allows better coolant flow between workpiece and tool [49]. However, in a few studies [50] a minor rise in temperature was reported in the UAT of Inconel 718. Researchers have suggested high ultrasonic energy attenuation and an insignificant acoustic softening effect is the reason for such a rise in temperature.

Researchers have also tried to assess the influence of vibration assistance simultaneously in multiple directions [8]. In all the reported work on two dimensional (2D) UAT, it has been suggested that, due to the superposition of two ultrasonic velocities in perpendicular directions, there was a change in the trajectory of the tool tip. In previous studies it was also concluded that the assistance of 2D vibration has a better effect on responses as compared to UAT with vibration assistance in one direction (1D) [51]. It has been suggested that in 2D UAT the improvement of responses is primarily due to a change in the magnitude and direction of the friction force. Furthermore, a few researchers have also investigated the influence of ultrasonic vibration in all three directions in the turning process [45]. It has been reported that 2D assistance results in a comparatively higher surface finish than 1D and 3D UAT.

In a few experimental investigations [45] the influence of the ultrasonic effect on grain deformation behavior was also analyzed in UAT. From experimental results it has been found that the grain size of deformation after UAT was bigger in comparison to traditional turning. Researchers have proposed that the extra energy transmitted due to vibration is stored in the working material which further enhances the dynamic recrystallization, resulting in a larger grain size [45].

Theoretical studies [8] have also reported on the modelling of responses to the UAT process. In the case of UAT force modelling, most researchers incorporated the instantaneous cutting speed and uncut chip thickness for incorporating the influence of ultrasonic vibration assistance on cutting mechanics. They also incorporated the intermittent cutting effect to predict the lowering of average UAT forces [23]. For modeling surface roughness after the UAT process, the mathematical model for the tool path was considered, which was further used to predict the generated texture pattern [52]. Further they increased the prediction accuracy by incorporating the material's spring-back and ploughing effects.

3.5.2 Ultrasonic Assisted Milling (UAM) Process

Researchers have also assessed the influence of ultrasonic vibration assistance on responses in the milling process. Prior investigation [27] has revealed that, like UAT, the assistance of vibration in the milling process also improves the responses. In the UAM process, the effect of ultrasonic vibration was assessed by applying it to the tool and also in some cases to the work specimen. In UAM, the tool can only be vibrated in the axial direction; however, the assistance of vibration can be given to the work piece in any direction (axial, cross-feed and feed). Like UAT, the assistance of vibration in the milling process (of different types of materials, like alloys with high toughness, strength and brittleness, and composites) also resulted in improved process responses [27].

In the literature most investigation [42] was focused on assessing the influence of ultrasonic vibration assistance in the feed direction. However, a few researchers have also assessed the effect of the vibrations in the axial direction, that is along the axis of rotation. The results have shown that axial ultrasonic vibration during milling is beneficial, as it offers the flexibility to perform machining in any direction with a uniform and better surface finish [44,53–55].

Researchers [42] have reported that the reduced machining force in UAM is due to the intermittent cutting mechanism during material removal, which is similar to that of UAT. The influence of the intermittent cutting effect becomes very significant in UAM when the vibrations are applied in the direction of the feed of the cutting tool. This is mainly because of the collinearity between the cutting velocity and oscillating ultrasonic velocity. However, in the case of axial vibration assisted UAM, the helical profile of the milling tooth is responsible for the intermittent cutting [30,56].

Researchers [42] have also investigated the influence of ultrasonic vibration at the bottom surface and the side walls of material in UAM. From experimental investigations [57] on UAM it was found that due to the ironing effect the surface roughness of side walls reduces. However, roughness at the bottom surface increases due to the impingement of the tool tip into the machined surface; it was also found that the engraved dimple marks were generated at the bottom surface [57].

Studies also confirmed that UAM helps in significantly improving machining accuracy. Experimental findings suggest that reduced cutting forces and tool deviations in UAM result in higher machining accuracy. In most of the studies improvement in machining accuracy was assessed through slot dimensional comparison. However, in a few studies [58,59], the improvement in UAM was also verified on thin-walled structures.

Analytical studies [42] were also performed to model the responses of the UAM process. Similar to the UAT process, in most of the theoretical studies the effect of ultrasonic vibration assistance on chip thickness and cutting speed was incorporated. These considerations resulted in adequate prediction accuracy, mainly for UAM with vibration assistance in a transverse direction (feed direction). A few researchers [30, 31] have also incorporated the acoustic softening effect which further results in the higher prediction accuracy of cutting forces and temperature prediction [31].

3.5.3 Ultrasonic Assisted Drilling

The ultrasonic assisted drilling (UAD) process has also been reported to be very efficacious in improving process responses. The UAD process has almost a similar setup and tool motion to the rotary ultrasonic drilling (RUD) process. One of the major differences between the RUD and UAD processes is that the drilling tool is used in both operations (see Figure 3.2). In RUD, abrasive coated tools are used, but in UAD conventional drilling tools are used, due to which the cutting mechanics are different in both operations. UAD finds application in industries such as aviation and biomedical equipment. In the UAD process the ultrasonic effect can be provided to the drilling tool or to the workpiece; however, it can only be provided in the axial direction because of dimension related constraints.

Experimental studies [13,61–63] have shown that the assistance of ultrasonic vibration in the axis of rotation of the cutting tool during the drilling process not only reduces the cutting forces, but also improves the surface quality and dimensional accuracy. As in UAT and UAM researchers have

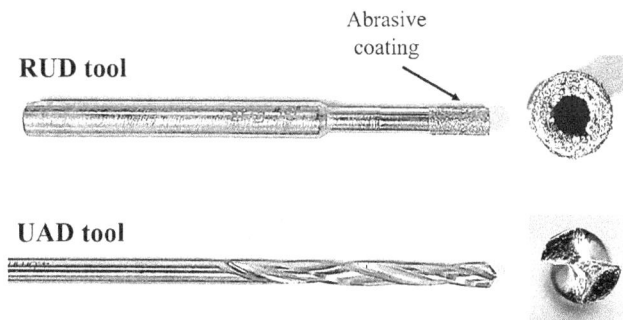

Figure 3.2 Drilling tool for the RUD and UAD processes [60].

credited the intermittent cutting effect to the decrease in the drilling force [13,61–63]. However, in a few studies, it was proposed that the lower cutting force is due to reduction in the frictional force, which is basically the result of the oscillating cutting tip. Furthermore, it was found that the temperature in the UAD process was higher in comparison to the traditional drilling process [16,62,64,65], which was not the case in UAM and UAT. It has been proposed that the higher heat generation in UAD is mainly due to the impact between workpiece and tool.

A few researchers [66] have also reported an increase in the ploughing effect in UAD. They have suggested that this is primarily due to higher heat generation during UAD, which softens the material. Due to this the ploughing effect increases, which further results in an increase in burr height. The applicability of UAD was also tested in the biomedical domain by assessing the influence of ultrasonic vibration during machining on the cutting force and temperature [13,16]. It has been found that in a specific range of drilling parameters, the assistance of ultrasonic vibration reduces heat generation which further results in a reduced chance of thermal necrosis (cell damage due to high temperature). However, in the UAD of metals, most researchers have reported a higher temperature as compared to CD. Therefore, in order to establish any relation between the process parameter and temperature rise, further experimental studies are required.

Analytical studies [61,64,67] have also been carried out for modeling the responses of the UAD process. Similar to UAM and UAT force models, most researchers have incorporated the instantaneous chip thickness and cutting speed for modeling UAD forces. In order to improve accuracy in a few studies [36], the ploughing effect during machining was also considered. Simulation results [61,68] have also verified the occurrence of intermittent cutting which results in reduced drilling force.

3.6 INDUSTRIAL APPLICATION OF THE UAM AND RUD PROCESSES

As discussed in the preceding sections, most of the experimental studies on UAT are mainly focused on understanding the cutting mechanics of the UAM process. However, the UAM and UAD processes have been evaluated on critical components/materials to check their industrial applicability. Two such experimental studies (one on UAD and one on UAM) have been discussed in this chapter to show the improvement in the outcome. Two distinct areas of application will now be discussed which will give a broader view of the applicability of both processes.

3.6.1 Case Study of the UAM Process

The accuracy of the machined product is a significant challenge faced by modern manufacturing industries, in particular, the machining of thin-walled structures, which generally has lower rigidity. Obtaining accuracy in the machining of thin-walled components like turbine blades and cooling fins is very important as it affects performance and life. Recent studies suggest that the assistance of ultrasonic vibration in the milling process also improves the machining accuracy of thin-walled structures.

Verma et al. [30] compared the influence of ultrasonic vibration assistance on machining accuracy during the milling of thin-walled structures. In their experiment this influence was compared by milling the thin-walled structure of two different shapes namely, straight and curved (see Figures 3.3a and 3.3b). After machining, the dimensional deviation of the machined thin walls was measured using reverse engineering methodology. Reverse engineering involves a scanner or a device to generate the CAD model of an existing body by obtaining the surface information data (or point cloud data). The obtained dimensional deviation of the developed part can be used to infer the information regarding machining accuracy in the UAM and conventional machining (CM) processes.

The comparisons between the dimensional deviation in the machined thin walls (by UAM and CM) are presented in Table 3.1. The experimental results show that the deviation in the machined thin wall is higher at the top and lower at the base. It has been suggested that this is mainly due to the bending of machined walls, which results in higher deformity at the upper surface as compared to the bottom [69]. From the dimensional measurement, it was also found that the deviation at the top of the thin wall was significantly higher in comparison to the deviation at the lower surface of the structure (for both UAM and CM). It was suggested that the bending at the top was higher than the lower section due to the lower rigidity of the structure.

From the results it can be seen that the dimensional deviations in the case of UAM are significantly less than in the conventional milling process. This was due to the reduced machining force in the UAM process, which is directly related to the dimensional accuracy of the structure. As has been discussed previously, the lower average machining force in UAM is primarily due to the intermittent contact of the tool during the cutting mechanism [8,28]. The experimental results show that the dimensional accuracy improves by 33% in UAM due to ultrasonic vibration assistance.

In order to visualize the dimensional deviation in the thin walls, the CAD were mapped (Figures 3.3c, 3.3d, 3.3e and 3.3f) with deviations highlighted. The maximum deviations at the edge and middle portion (for straight and curved thin wall) are presented in Figures 3.3g, 3.3h, 3.3k and 3.3m for UAM and in Figures 3.3h, 3.3j, 3.3l and 3.3n for CM. It was also revealed that in both cases the deviations are higher at the edge. It can be seen that a maximum deviation of 93 and 56 μm was obtained in the case of straight and curved thin walls machined with CM. However, in the case of UAM a maximum deviation of 67 μm for a straight thin wall and 38 μm for a curved thin wall were observed. These deviations at the edges for both straight and curved thin walls are higher compared to the middle portion.

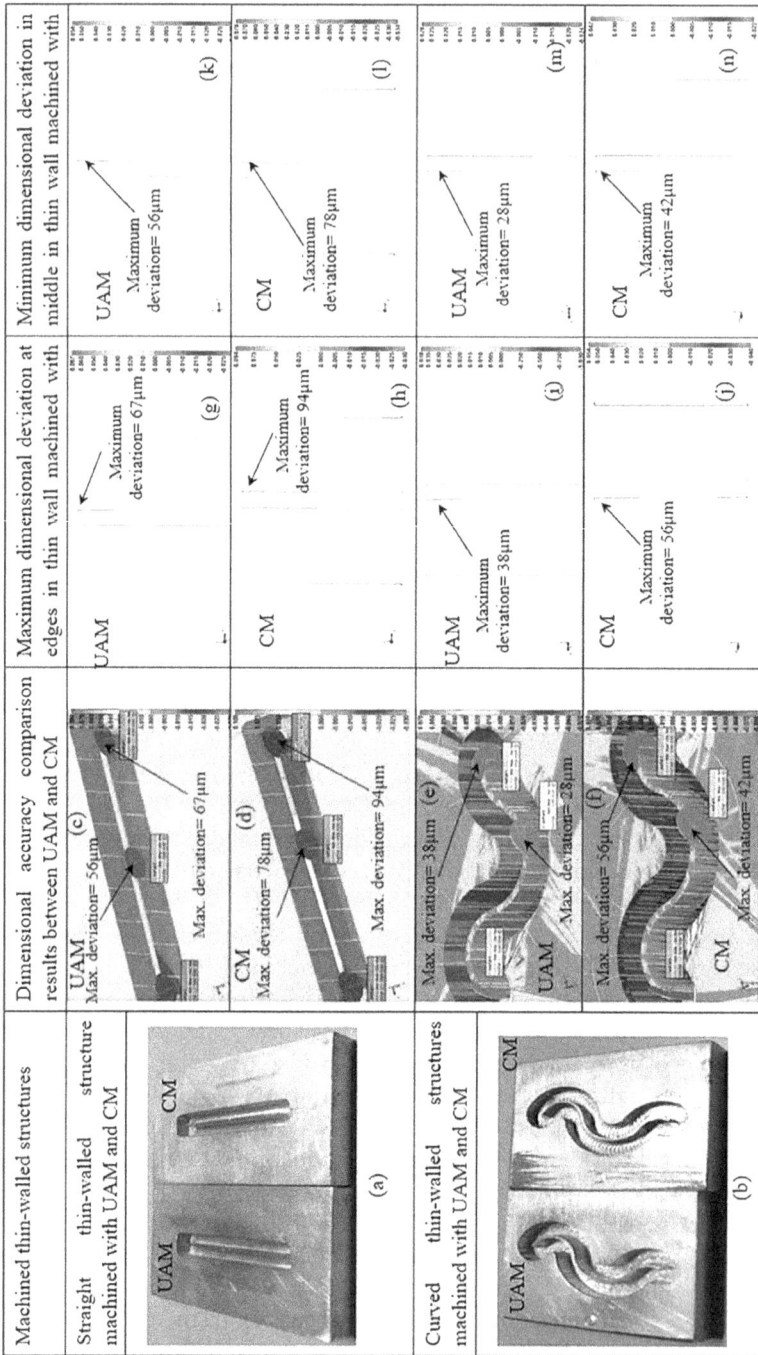

Figure 3.3 Comparison results for thin-walled structures machined with the UAM process and the conventional process [58].

Table 3.1 Comparison of Thickness at the Top and Bottom Portions of Machined Thin Walls (mm) [58]

	Straight (UAM)	Straight (CM)	Curved (UAM)	Curved (CM)
CAD Dimension	0.8	0.8	0.8	0.8
Maximum thickness of thin wall at bottom	0.790	0.788	0.786	0.785
Minimum thickness of thin wall at bottom	0.793	0.792	0.790	0.788
Average thickness of thin wall at bottom	0.792	0.790	0.788	0.787
Maximum thickness of thin wall at top	0.867	0.893	0.838	0.856
Minimum thickness of thin wall at top	0.856	0.878	0.828	0.842
Average thickness of thin wall at top	0.862	0.885	0.832	0.848

The variable stiffness of the thin wall at the middle and edge portion may be the reason for the low and high deviations at different sections. From these results it can be concluded that for achieving lower deviation in thin walls UAM can be used; however, for attaining a consistent depth of cut it must be reduced during machining at the edges.

3.6.2 Case Study of RUD in Biomedical Applications

Heat generation during drilling in orthopedic surgery plays a vital role in the success of the osteosynthesis process. High temperature at the drilling site causes problems like necrosis, that is permanent death to the bone cells when exposed to a temperature above a certain threshold. Recently, Gupta et al. [17,19,70] introduced a novel bone drilling process, namely RUBD, which is a combination of rotary ultrasonic machining and orthopedic drilling, to overcome the problem of high temperature generation. The RUBD process uses a diamond abrasive coated hollow tool; th inear ultrasonic vibrations are provided to the tool during drilling. The experimental setup for RUBD was developed [19] on a CNC milling station and the experimental investigations were performed on the porcine femur bones.

Gupta et al. [17] performed experimental investigations to study the effect of the drilling parameters on the temperature generated during drilling by conventional surgical bone drilling (CSBD) and RUBD. The authors also studied the influence of irrigation on temperature rise while drilling with the RUBD tool. Temperatures during the drilling were recorded by using K-type thermocouples which were inserted in the pre-drilled holes near the periphery of the hole to be drilled. It was also found that the temperature produced during the CSBD technique was higher than that produced by the RUBD process (see Figures 3.4a, 3.4b and 3.4c). This was due to the intermittent contact between the drill tool and the bone surface during drilling due to the ultrasonic vibration and the reduced contact ratio resulting from the abrasive contact. Comparative results show that the temperature increased with

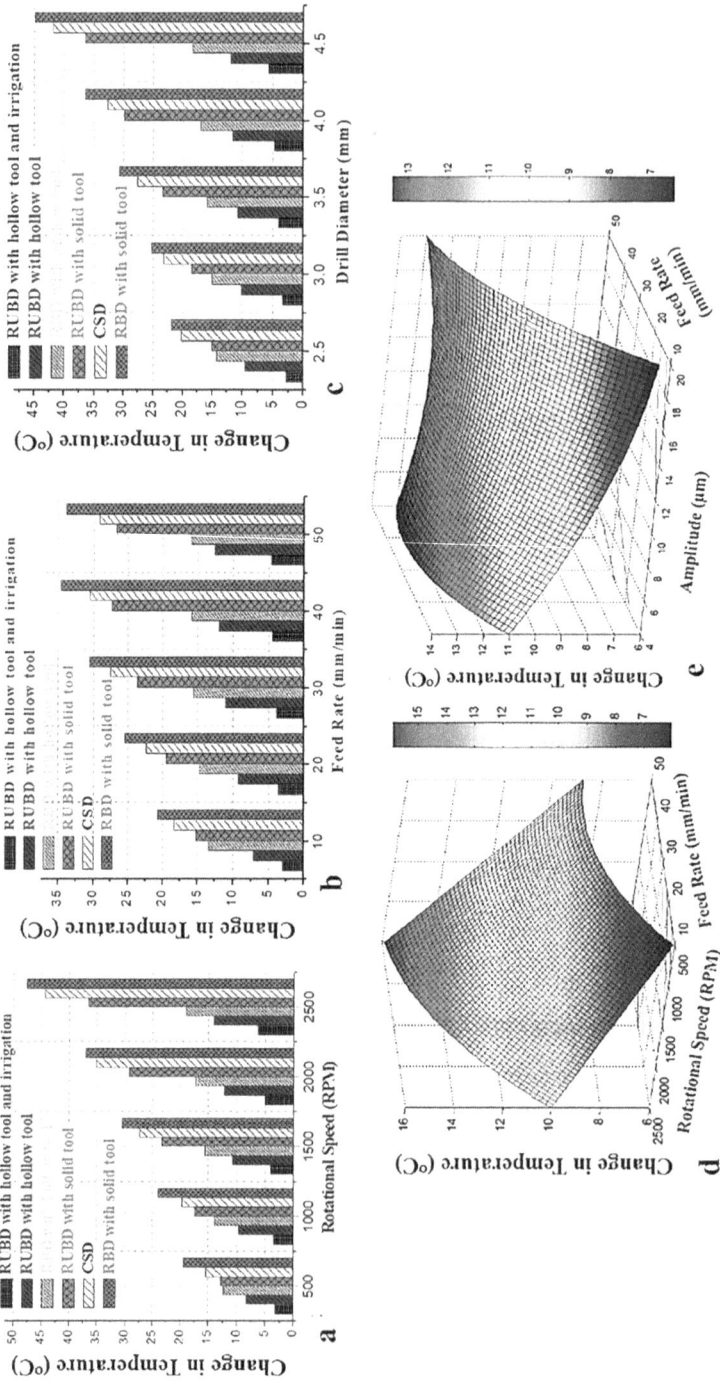

Figure 3.4 Effect of different drilling techniques on the change in temperature during drilling at varying (a) rotational speed, (b) feed rate and (c) drill diameter [17], and during significant interaction between (d) rotational speed and feed rate, and (e) amplitude and feed rate in the RUBD technique [70].

the increase in the feed rate, rotational speed and the diameter of drill tool, for all types of drilling techniques and tools considered in the study. In another statistical study [70] experimental results revealed that rotational speed, feed rate, drill diameter and vibrational amplitude significantly affected the temperature generated during drilling. The temperature increased with the increase in the feed rate, rotational speed and drill diameter. Figures 3.4d and 3.4e show the significant interaction of process parameters on the temperature generated in the RUBD technique.

Recently, Singh et al. [71–73] developed an operation-theater-compatible, light weight RUBD machine, working on the principle of the rotary ultrasonic machining (RUM) process. The experimental investigations were performed on three different human bones [74]—femur, tibia and fibula—to study the effect of drilling parameters and techniques on the temperature generated. A comparative study between the CSBD and RUBD techniques was also performed (see Figures 3.5a and 3.5b). Temperatures during the experiments were recorded by inserting the K-type thermocouple in the predrilled hole, at a distance of 0.5 mm and 1.0 mm from the periphery of the hole to be drilled. The study revealed that the temperature generated in CSBD was higher than the RUBD technique for all three human bones,

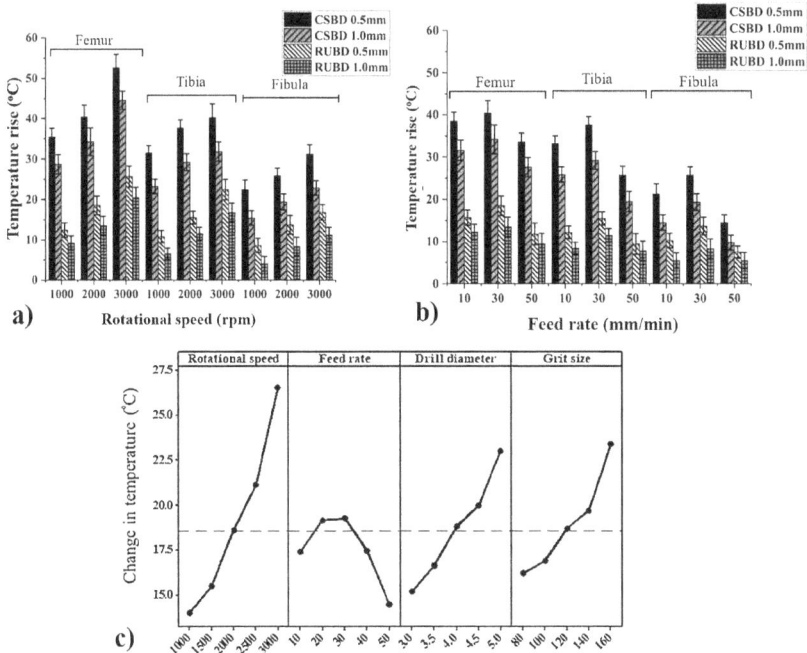

Figure 3.5 Comparative results of temperature rise while drilling different human bones by the CSBD and RUBD techniques at varying (a) rotational speed [74] and (b) feed rate [74], and (c) the effect of drilling parameters on the change in temperature in the RUBD process [60].

which was due to the reduced area of contact and the reduced time duration of contact between the tool and workpiece in RUBD. Also, the temperature produced in drilling hard bones like the femur and tibia was greater than for the fibula bone, for both drilling techniques. The reason for this was due to the hardness of the bone. The femur is the most compact bone and the hardest, and has more strength than the tibia and fibula.

Singh et al. [60] developed a statistical model and it was observed (see Figure 3.5c) that while drilling the human femur by the RUBD technique the temperature increased with an increase in rotational speed from 1000 to 3000 rpm due to the increase in shearing energy. It was found that the temperature rise increased with the increase in the feed rate from 10 to 30 mm/min due to the increase in the shear zone but a further increase in the feed rate decreased the drilling time thus decreasing the exposure time between the rotating tool and workpiece thereby resulting in a lower temperature rise. The increase in the area of contact by increasing the drill diameter resulted in a rise of temperature. The increase in grit size (i.e. the smaller size of diamond abrasives) of the abrasives was associated with a greater number of particles coming in contact with the bone surface thus increasing the area of contact and resulting in a rise of temperature.

3.6.3 Histopathological Observation to Study Thermal Necrosis

Gupta et al. [75] performed a comparative histopathological study of the RUBD and CSBD techniques to examine the damage produced to bone cells due to the heat generated while drilling in porcine bone. During the process the samples were preserved in a formaline solution and the decalcification of drilled samples were performed using 10% ethylene diamine tetra acetic acid (EDTA) chelating. Hematoxylin and eosin (H&E) staining was performed for the microscopic examination. The results of the study revealed (see Figure 3.6a and 3.6b) that high temperature in CSBD resulted in empty lacunae near the drilling site, but the reduced temperature generated in RUBD drilling resulted in the presence of osteocytes near the drilling site. Thus RUBD resulted in the elimination of thermal necrosis.

Recently, Singh et al. [77] performed similar investigations on the human femur; the experiments were performed on an indigenously developed RUBD machine. The results of the study were in line with the previous study on porcine bone [75] and it was found that RUBD produced intact bone cells near the drilling site (see Figure 3.6c), whereas the higher temperature generated in the CSBD process resulted in necrosis and empty lacunae were found near the circumference of the drilled hole (Figure 3.6d).

Singh et al. [77] also performed an ultrastructural study to examine thermal damage. Transmission electron microscopic (TEM) images revealed complete or severe damage of bone cells. The integrity of the cell membrane and the subcellular cytoplasmic and nuclear structures of osteocytes were

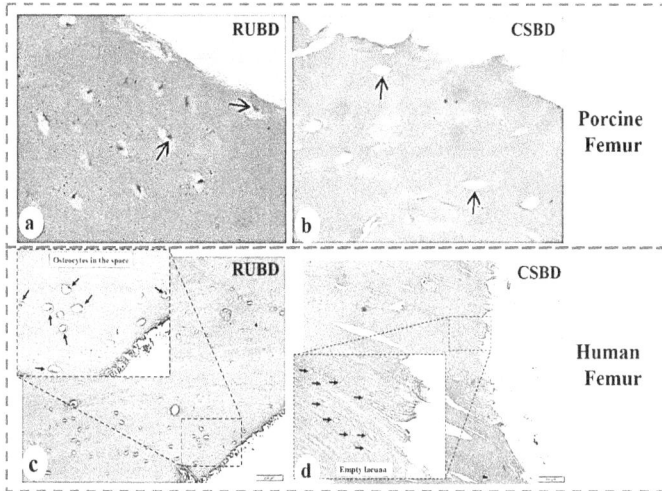

Figure 3.6 Histopathological examination of a porcine femur (a, b) [75] and human femur (c, d) [76] bone tissues drilled by the CSBD and RUBD techniques using H&E staining.

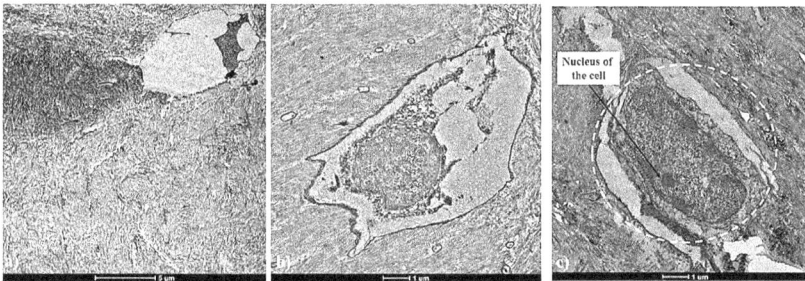

Figure 3.7 (a) and (b) show TEM images of CSBD specimens with severe damage of the cytoplasm, nucleus and cell membrane of osteocytes. (c) RUBD specimen showing the presence of intact cells within the lacunae without discernible damage to the osteocyte [76].

lost (Figure 3.7a and 3.7b). In RUBD the temperature during drilling was much below the threshold value. The cellular damage was minimal or absent and this was also demonstrated by TEM. Intact healthy osteocytes were present within the lacunae (Figure 3.7c).

3.7 CONCLUSION

From the present investigation it can be concluded that the assistance of ultrasonic vibration improves the process responses as well as the machining accuracy. However, due to the high running cost the UAM process is only considered suitable for precision manufacturing. From the present study the following conclusions can be drawn:

1. The assistance of ultrasonic vibration during conventional machining processes significantly decreases the machining forces (by 50 %). The intermittent cutting effect can be considered the primary reason of such a reduction; however, it only occurs in a specific range of process parameters.
2. Surface roughness also reduces due to ultrasonic vibration assistance in the feed direction. However, in a transverse direction the surface roughness increases the generation of texture.
3. The introduction of ultrasonic vibrations during the milling of a thin-walled structure is efficacious in improving machining accuracy. The dimensional accuracy improved by 33% in the machining of a curved thin-walled structure. Therefore ultrasonic assisted milling can be used for fabricating curved thin-walled structures like turbine blades and heat exchanger fins.
4. Ultrasonic drilling also shows potential in biomedical application, where it was found that rotary ultrasonic drilling causes a reduced temperature during orthopedic drilling and causes minimal damage to the bone cells.

REFERENCES

[1] P. Wang, S. Zhang, Z. Li, and J. Li, "Tool path planning and milling surface simulation for vehicle rear bumper mold," *Adv. Mech. Eng.*, vol. 8, no. 3, pp. 1–10, 2016, doi:10.1177/1687814016641569.

[2] M. Soković and K. Mijanović, "Ecological aspects of the cutting fluids and its influence on quantifiable parameters of the cutting processes," *J. Mater. Process. Technol.*, vol. 109, no. 1–2, pp. 181–189, 2001, doi:10.1016/S0924-0136(00)00794-9.

[3] U. S. Dixit, D. K. Sarma, and J. P. Davim, *Environmentally Friendly Machining.* Springer, New York, 2012.

[4] Z. Y. Wang and K. P. Rajurkar, "Cryogenic machining of hard-to-cut materials," *Wear*, vol. 239, no. 2, pp. 168–175, 2000, doi:10.1016/S0043-1648(99)00361-0.

[5] X. Huang, X. Zhang, H. Mou, X. Zhang, and H. Ding, "The influence of cryogenic cooling on milling stability," *J. Mater. Process. Technol.*, vol. 214, no. 12, pp. 3169–3178, 2014, doi:10.1016/j.jmatprotec.2014.07.023.

[6] V. S. Sharma, G. Singh, and K. Sorby, "A review on minimum quantity lubrication for machining processes," *Mater. Manuf. Process.*, vol. 30, no. 8, pp. 935–953, 2015, doi:10.1080/10426914.2014.994759.

[7] P. H. Lee, J. S. Nam, C. Li, and S. W. Lee, "An experimental study on micro-grinding process with nanofluid minimum quantity lubrication (MQL)," *Int. J. Precis. Eng. Manuf.*, vol. 13, no. 3, pp. 331–338, 2012, doi:10.1007/s12541-012-0042-2.

[8] D. E. Brehl and T. A. Dow, "Review of vibration-assisted machining," *Precis. Eng.*, vol. 32, no. 3, pp. 153–172, 2008, doi:10.1016/j.precisioneng.2007.08.003.

[9] L. Yang, L. Zhibing, W. Xibin, and H. Tao, "Experimental study on cutting force and surface quality in ultrasonic vibration-assisted milling of C/SiC composites," *Int. J. Adv. Manuf. Technol.*, vol. 112, no. 7–8, pp. 2003–2014, 2021, doi:10.1007/s00170-020-06355-x.

[10] G. Gao, Z. Xia, T. Su, D. Xiang, and B. Zhao, "Cutting force model of longitudinal-torsional ultrasonic-assisted milling Ti-6Al-4V based on tool flank wear," *J. Mater. Process. Technol.*, vol. 291, pp. 1–16, 2021, doi:10.1016/j.jmatprotec.2021.117042.

[11] N. Ying, J. Feng, Z. Bo, G. Guofu, and N. Jing-Jing, "Theoretical investigation of machining-induced residual stresses in longitudinal torsional ultrasonic–assisted milling," *Int. J. Adv. Manuf. Technol.*, vol. 108, no. 11–12, pp. 3689–3705, 2020, doi:10.1007/s00170-020-05495-4.

[12] Y. Bai, Z. Shi, Y. J. Lee, and H. Wang, "Optical surface generation on additively manufactured AlSiMg0.75 alloys with ultrasonic vibration-assisted machining," *J. Mater. Process. Technol.*, vol. 280, 2020, doi:10.1016/j.jmatprotec.2020.116597.

[13] K. Alam, A. V. Mitrofanov, and V. V. Silberschmidt, "Experimental investigations of forces and torque in conventional and ultrasonically-assisted drilling of cortical bone," *Med. Eng. Phys.*, vol. 33, no. 2, pp. 234–239, 2011, doi:10.1016/j.medengphy.2010.10.003.

[14] K. Alam and V. V. Silberschmidt, "Analysis of temperature in conventional and ultrasonically-assisted drilling of cortical bone with infrared thermography.," *Technol. Heal. care Off. J. Eur. Soc. Eng. Med.*, vol. 22, no. 2, pp. 243–252, 2014.

[15] K. Alam, E. Hassan, and I. Bahadur, "Experimental measurements of temperatures in ultrasonically assisted drilling of cortical bone," *Biotechnol. Biotechnol. Equip.*, vol. 29, no. 4, pp. 753–757, 2015, doi:10.1080/13102818.2015.1034176.

[16] E. Shakouri, M. H. Sadeghi, M. R. Karafi, M. Maerefat, and M. Farzin, "An in vitro study of thermal necrosis in ultrasonic-assisted drilling of bone," *Proc. Inst. Mech. Eng. Part H J. Eng. Med.*, vol. 229, no. 2, pp. 137–149, 2015, doi:10.1177/0954411915573064.

[17] V. Gupta, R. P. Singh, P. M. Pandey, and R. Gupta, "In vitro comparison of conventional surgical and rotary ultrasonic bone drilling techniques," *Proc. Inst. Mech. Eng. Part H J. Eng. Med.*, vol. 234, no. 4, pp. 398–411, 2020, doi:10.1177/0954411919898301.

[18] Y. Wang, M. Cao, Y. Zhao, G. Zhou, W. Liu, and D. Li, "Experimental investigations on microcracks in vibrational and conventional drilling of cortical bone," *J. Nanomater.*, vol. 2013, pp. 1–5, 2013, doi:10.1155/2013/845205.

[19] V. Gupta and P. M. Pandey, "An in-vitro study of cutting force and torque during rotary ultrasonic bone drilling," *Proc. Inst. Mech. Eng. Part B J. Eng. Manuf.*, vol. 232, no. 9, pp. 1549–1560, 2018, doi:10.1177/0954405416673115.

[20] V. Gupta, P. M. Pandey, and V. V. Silberschmidt, "Rotary ultrasonic bone drilling: Improved pullout strength and re duce d damage," *Med. Eng. Phys.*, vol. 41, pp. 1–8, 2017, doi:10.1016/j.medengphy.2016.11.004.

[21] M. Xiao, K. Sato, S. Karube, and T. Soutome, "The effect of tool nose radius in ultrasonic vibration cutting of hard metal," *Int. J. Mach. Tools Manuf.*, vol. 43, no. 13, pp. 1375–1382, 2003, doi:10.1016/S0890-6955(03)00129-9.

[22] R. C. Skelton, "Turning with an oscillating tool," *Int. J. Mach. Tool Des. Res.*, vol. 8, no. 4, pp. 239–259, 1968, doi:10.1016/0020-7357(68)90014-0.

[23] C. Nath and M. Rahman, "Effect of machining parameters in ultrasonic vibration cutting," *Int. J. Mach. Tools Manuf.*, vol. 48, no. 9, pp. 965–974, 2008, doi:10.1016/j.ijmachtools.2008.01.013.

[24] A. V. Mitrofanov, N. Ahmed, V. I. Babitsky, and V. V. Silberschmidt, "Effect of lubrication and cutting parameters on ultrasonically assisted turning of Inconel 718," *J. Mater. Process. Technol.*, vol. 162–163, pp. 649–654, 2005, doi:10.1016/j.jmatprotec.2005.02.170.

[25] K.-M. Li and S.-L. Wang, "Effect of tool wear in ultrasonic vibration-assisted micro-milling," *Proc. Inst. Mech. Eng. Part B J. Eng. Manuf.*, vol. 228, no. 1, pp. 847–855, 2013, doi:10.1177/0954405413510514.

[26] U. S. Dixit, P. M. Pandey, and G. C. Verma, "Ultrasonic-assisted machining processes: a review," *Int. J. Mechatronics Manuf. Syst.*, vol. 12, no. 3/4, pp. 227–254, 2019.

[27] W. Chen, D. Huo, J. Hale, and H. Ding, "Kinematics and tool-workpiece separation analysis of vibration assisted milling," *Int. J. Mech. Sci.*, vol. 136, no. December 2017, pp. 169–178, 2018, doi:10.1016/j.ijmecsci.2017.12.037.

[28] M. N. Kumar, S. Kanmani Subbu, P. Vamsi Krishna, and A. Venugopal, "Vibration assisted conventional and advanced machining: A review," *Procedia Eng.*, vol. 97, pp. 1577–1586, 2014, doi:10.1016/j.proeng.2014.12.441.

[29] S. Patil, S. Joshi, A. Tewari, and S. S. Joshi, "Modelling and simulation of effect of ultrasonic vibrations on machining of Ti6Al4V," *Ultrasonics*, vol. 54, no. 2, pp. 694–705, 2014, doi:10.1016/j.ultras.2013.09.010.

[30] G. Chandra, P. Mohan, and U. Shanker, "Estimation of workpiece-temperature during ultrasonic-vibration assisted milling considering acoustic softening," *Int. J. Mech. Sci.*, vol. 140, pp. 547–556, 2018, doi:10.1016/j.ijmecsci.2018.03.034.

[31] G. Chandra, P. Mohan, and U. Shanker, "Modeling of static machining force in axial ultrasonic-vibration assisted milling considering acoustic softening," *Int. J. Mech. Sci.*, vol. 136, pp. 1–16, 2018, doi:10.1016/j.ijmecsci.2017.11.048.

[32] M. R. Razfar, P. Sarvi, and M. M. A. Zarchi, "Experimental investigation of the surface roughness in ultrasonic-assisted milling," *Proc. Inst. Mech. Eng. Part B J. Eng. Manuf.*, vol. 225, no. 9, pp. 1615–1620, 2011, doi:10.1177/0954405 411399331.

[33] X. H. Shen, J. Zhang, D. X. Xing, and Y. Zhao, "A study of surface roughness variation in ultrasonic vibration-assisted milling," *Int. J. Adv. Manuf. Technol.*, vol. 58, no. 5–8, pp. 553–561, 2012, doi:10.1007/s00170-011-3399-y.

[34] M. M. Abootorabi Zarchi, M. R. Razfar, and A. Abdullah, "Investigation of the effect of cutting speed and vibration amplitude on cutting forces in ultrasonic-assisted milling," *Proc. Inst. Mech. Eng. Part B J. Eng. Manuf.*, vol. 226, no. 7, pp. 1185–1191, 2012, doi:10.1177/0954405412439666.

[35] H. G. Toews, W. D. Compton, and S. Chandrasekar, "A study of the influence of superimposed low-frequency modulation on the drilling process," *Precis. Eng.*, vol. 22, no. 1, pp. 1–9, 1998.

[36] S. S. F. Chang and G. M. Bone, "Thrust force model for vibration-assisted drilling of aluminum 6061-T6," *Int. J. Mach. Tools Manuf.*, vol. 49, no. 14, pp. 1070–1076, 2009, doi:10.1016/j.ijmachtools.2009.07.011.

[37] A. S. Adnan and S. Subbiah, "Experimental investigation of transverse vibration-assisted orthogonal cutting of AL-2024," *Int. J. Mach. Tools Manuf.*, vol. 50, no. 3, pp. 294–302, 2010, doi:10.1016/j.ijmachtools.2009.11.004.

[38] F. Claeyssen, "Magnetostrictive actuators compared to piezoelectric actuators," *Proc. SPIE Eur. Work. Smart Struct. Eng. Technol.*, vol. 4763, pp. 194–200, 2003, doi:10.1117/12.508734.

[39] R. Singh and J. S. Khamba, "Ultrasonic machining of titanium and its alloys: A review," *J. Mater. Process. Technol.*, vol. 173, no. 2, pp. 125–135, 2006, doi:10.1016/j.jmatprotec.2005.10.027.

[40] D. A. Wang, W. Y. Chuang, K. Hsu, and H. T. Pham, "Design of a Bézier-profile horn for high displacement amplification," *Ultrasonics*, vol. 51, no. 2, pp. 148–156, 2011, doi:10.1016/j.ultras.2010.07.004.

[41] J. Jallageas, J. Y. K'Nevez, M. Chérif, and O. Cahuc, "Modeling and optimization of vibration-assisted drilling on positive feed drilling unit," *Int. J. Adv. Manuf. Technol.*, vol. 67, no. 5–8, pp. 1205–1216, 2013, doi:10.1007/s00170-012-4559-4.

[42] W. Chen, D. Huo, Y. Shi, and J. M. Hale, "State-of-the-art review on vibration-assisted milling: principle, system design, and application," *Int. J. Adv. Manuf. Technol.*, vol. 97, pp. 2033–2049, 2018.

[43] C. Ma, E. Shamoto, T. Moriwaki, Y. Zhang, and L. Wang, "Suppression of burrs in turning with ultrasonic elliptical vibration cutting," *Int. J. Mach. Tools Manuf.*, vol. 45, no. 11, pp. 1295–1300, 2005, doi:10.1016/j.ijmachtools.2005.01.011.

[44] A. Suárez, F. Veiga, L. N. L. de Lacalle, R. Polvorosa, S. Lutze, and A. Wretland, "Effects of Ultrasonics-Assisted Face Milling on Surface Integrity and Fatigue Life of Ni-Alloy 718," *J. Mater. Eng. Perform.*, vol. 25, no. 11, pp. 5076–5086, 2016, doi:10.1007/s11665-016-2343-6.

[45] S. A. Sajjady, H. Nouri Hossein Abadi, S. Amini, and R. Nosouhi, "Analytical and experimental study of topography of surface texture in ultrasonic vibration assisted turning," *Mater. Des.*, vol. 93, pp. 311–323, 2016, doi:10.1016/j.matdes.2015.12.119.

[46] L. Qin, P. Lin, Y. Zhang, G. Dong, and Q. Zeng, "Influence of surface wettability on the tribological properties of laser textured Co-Cr-Mo alloy in aqueous bovine serum albumin solution," *Appl. Surf. Sci.*, vol. 268, pp. 79–86, 2013, doi:10.1016/j.apsusc.2012.12.003.

[47] D. Xing, J. Zhang, X. Shen, Y. Zhao, and T. Wang, "Tribological properties of ultrasonic vibration assisted milling aluminium alloy surfaces," *Procedia CIRP*, vol. 6, pp. 539–544, 2013, doi:10.1016/j.procir.2013.03.008.

[48] C. Ni, L. Zhu, C. Liu, and Z. Yang, "Analytical modeling of tool-workpiece contact rate and experimental study in ultrasonic vibration-assisted milling of Ti–6Al–4V," *Int. J. Mech. Sci.*, vol. 142–143, pp. 97–111, 2018, doi:10.1016/j.ijmecsci.2018.04.037.

[49] M. Zhou, Y. T. Eow, B. K. A. Ngoi, and E. N. Lim, "Vibration-Assisted Precision Machining of Steel with PCD Tools," *Mater. Manuf. Process.*, vol. 18, no. 5, pp. 825–834, 2003, doi:10.1081/AMP-120024978.

[50] V. I. Babitsky, A. V. Mitrofanov, and V. V. Silberschmidt, "Ultrasonically assisted turning of aviation materials: Simulations and experimental study," *Ultrasonics*, vol. 42, no. 1–9, pp. 81–86, 2004, doi:10.1016/j.ultras.2004.02.001.

[51] C. Ma, E. Shamoto, T. Moriwaki, and L. Wang, "Study of machining accuracy in ultrasonic elliptical vibration cutting," *Int. J. Mach. Tools Manuf.*, vol. 44, no. 12–13, pp. 1305–1310, 2004, doi:10.1016/j.ijmachtools.2004.04.014.

[52] X. Zhang, A. S. Kumar, M. Rahman, and K. Liu, "Modeling of the effect of tool edge radius on surface generation in elliptical vibration cutting," *Int.*

J. Adv. Manuf. Technol., vol. 65, no. 1, pp. 35–42, Mar. 2013, doi:10.1007/ s00170-012-4146-8.

[53] P. Sarvi Hampa, M. R. Razfar, M. Malaki, and A. Maleki, "The Role of Dry Aero-acoustical Lubrication and Material Softening in Ultrasonically Assisted Milling of Difficult-to-Cut AISI 304 Steels," *Trans. Indian Inst. Met.*, vol. 68, no. 1, pp. 43–49, 2014, doi:10.1007/s12666-014-0429-0.

[54] K. Marcel, Z. Marek, and P. Jozef, "Investigation of ultrasonic assisted milling of aluminum alloy AlMg4.5Mn," *Procedia Eng.*, vol. 69, pp. 1048–1053, 2014, doi:10.1016/j.proeng.2014.03.089.

[55] E. Uhlmann, F. Protz, B. Stawiszynski, and S. Heidler, "Ultrasonic Assisted Milling of Reinforced Plastics," *Procedia CIRP*, vol. 66, pp. 164–168, 2017, doi:10.1016/j.procir.2017.03.278.

[56] G. Verma, P. M. Pandey, and U. S. Dixit, "Modeling of cutting forces in ultrasonic-vibration assisted milling," 5611725, 2017.

[57] K. Zheng, W. Liao, Q. Dong, and L. Sun, "Friction and wear on titanium alloy surface machined by ultrasonic vibration-assisted milling," *J. Brazilian Soc. Mech. Sci. Eng.*, vol. 40, no. 9, p. 411, 2018, doi:10.1007/s40430-018-1336-9.

[58] G. C. Verma, P. M. Pandey, and U. S. Dixit, "Experimental Investigations to Evaluate Machining Accuracy of Ultrasonic-Assisted Milling on Thin-Walled Structures," *Adv. Micro Nano Manuf. Surf. Eng.*, pp. 141–151, 2019.

[59] J. Tong, G. Wei, L. Zhao, X. Wang, and J. Ma, "Surface microstructure of titanium alloy thin-walled parts at ultrasonic vibration-assisted milling," *Int. J. Adv. Manuf. Technol.*, vol. 101, pp. 1007–1021, 2019, doi:10.1007/ s00170-018-3005-7.

[60] R. P. Singh, P. Mohan, and A. Ranjan, "An in-vitro study of temperature rise during rotary ultrasonic bone drilling of human bone," *Med. Eng. Phys.*, vol. 79, pp. 33–43, 2020, doi:10.1016/j.medengphy.2020.03.002.

[61] M. Lotfi and S. Amini, "Experimental and numerical study of ultrasonically-assisted drilling," *Ultrasonics*, vol. 75, pp. 185–193, 2017, doi:10.1016/j. ultras.2016.11.009.

[62] M. A. Kadivar, J. Akbari, R. Yousefi, A. Rahi, and M. G. Nick, "Investigating the effects of vibration method on ultrasonic-assisted drilling of Al/SiCp metal matrix composites," *Robot. Comput. Integr. Manuf.*, vol. 30, no. 3, pp. 344–350, 2014, doi:10.1016/j.rcim.2013.10.001.

[63] Y. S. Liao, Y. C. Chen, and H. M. Lin, "Feasibility study of the ultrasonic vibration assisted drilling of Inconel superalloy," *Int. J. Mach. Tools Manuf.*, vol. 47, no. 12–13, pp. 1988–1996, 2007, doi:10.1016/j.ijmachtools.2007.02.001.

[64] J. Pujana, A. Rivero, A. Celaya, and L. N. López de Lacalle, "Analysis of ultrasonic-assisted drilling of Ti6Al4V," *Int. J. Mach. Tools Manuf.*, vol. 49, no. 6, pp. 500–508, 2009, doi:10.1016/j.ijmachtools.2008.12.014.

[65] F. Makhdum, V. A. Phadnis, A. Roy, and V. V. Silberschmidt, "Effect of ultrasonically-assisted drilling on carbon-fibre-reinforced plastics," *J. Sound Vib.*, vol. 333, no. 23, pp. 5939–5952, 2014, doi:10.1016/j.jsv.2014.05.042.

[66] S. S. F. Chang and G. M. Bone, "Burr height model for vibration assisted drilling of aluminum 6061-T6," *Precis. Eng.*, vol. 34, no. 3, pp. 369–375, 2010, doi:10.1016/j.precisioneng.2009.09.002.

[67] B. Azarhoushang and J. Akbari, "Ultrasonic-assisted drilling of Inconel 738-LC," *Int. J. Mach. Tools Manuf.*, vol. 47, no. 7–8, pp. 1027–1033, 2007, doi:10.1016/ j.ijmachtools.2006.10.007.

[68] A. Sanda, I. Arriola, V. Garcia Navas, I. Bengoetxea, and O. Gonzalo, "Ultrasonically assisted drilling of carbon fibre reinforced plastics and Ti6Al4V," *J. Manuf. Process.*, vol. 22, pp. 169–176, 2016, doi:10.1016/j.jmapro.2016.03.003.

[69] S. Ratchev, S. Liu, W. Huang, and A. A. Becker, "Milling error prediction and compensation in machining of low-rigidity parts," *Int. J. Mach. Tools Manuf.*, vol. 44, no. 15, pp. 1629–1641, 2004, doi:10.1016/j.ijmachtools.2004.06.001.

[70] V. Gupta and P. M. Pandey, "Experimental investigation and statistical modeling of temperature rise in rotary ultrasonic bone drilling," *Med. Eng. Phys.*, vol. 38, no. 11, pp. 1330–1338, 2016, doi:10.1016/j.medengphy.2016.08.012.

[71] R. P. Singh and P. M. Pandey, "Comparison Of Conventional And Ultrasonic Drilling On Cutting Force In Porcine And Human Femur," in *Proceedings of the ASME 2020 15th International Manufacturing Science and Engineering Conference MSEC2020*, pp. 1–8, 2020.

[72] R. P. Singh, P. M. Pandey, A. R. Mridha, and T. Joshi, "Experimental investigations and statistical modeling of cutting force and torque in rotary ultrasonic bone drilling of human cadaver bone," *Proc IMechE Part H J Eng. Med.*, vol. 234, no. 2, pp. 148–162, 2020, doi:10.1177/0954411919889913.

[73] R. P. Singh, V. Gupta, P. M. Pandey, and A. R. Mridha, "Effect of Drilling Techniques on Microcracks and Pull-Out Strength of Cortical Screw Fixed in Human Tibia: An In-Vitro Study," *Ann. Biomed. Eng.*, vol. 49, pp. 382–393, 2021, doi:10.1007/s10439-020-02565-2.

[74] R. P. Singh, P. M. Pandey, C. Behera, and A. R. Mridha, "Effects of rotary ultrasonic bone drilling on cutting force and temperature in the human bones," *Proc. Inst. Mech. Eng. Part H J. Eng. Med.*, vol. 234, no. 8, pp. 829–842, 2020, doi:10.1177/0954411920925254.

[75] V. Gupta, P. M. Pandey, R. K. Gupta, and A. R. Mridha, "Rotary ultrasonic drilling on bone: A novel technique to put an end to thermal injury to bone," *Proc. Inst. Mech. Eng. Part H J. Eng. Med.*, vol. 231, no. 3, pp. 189–196, 2017, doi:10.1177/0954411916688500.

[76] R. P. Singh, *Development of Rotary Ultrasonic Bone Drilling Machine and Experimental Investigations on Human Cadaver Bone*, Indian Institute of Technology Delhi, 2020.

[77] R. P. Singh, P. M. Pandey, and A. R. Mridha, "Thermal changes during drilling in human femur by rotary ultrasonic bone drilling machine: A histologic and ultrastructural study," *J. Biomed. Res. Part B Appl. Biomater.*, vol. 110, pp. 1023–1033, 2022, doi: 10.1002/jbm.b.34975.

Chapter 4

An Environmental Sustainability Assessment of a Milling Process using Life Cycle Assessment

A Case Study of India

Nitesh Sihag, Vikrant Bhakar,
and Kuldip Singh Sangwan
Birla Institute of Technology, Pilani, India

CONTENTS

4.1 INTRODUCTION

Life cycle assessment (LCA) is a proven technique for scientifically evaluating the environmental impacts of products and processes. The assessment generally starts from raw material extraction and finishes with the final disposal of the used product [1]. Every product/process's life cycle involves the consumption of various natural resources and energy, and the emission of hazardous pollutants into the environment [2]. LCA is used to measure the environmental impacts across the product and process life cycles [3]. LCA visualizes the major hotspots throughout the life cycle stages of a product/process, thereby supporting decision makers in reducing the environmental impacts by employing alternative materials and/or processes. In this context, this chapter presents a life cycle analysis of a milling process to evaluate and quantify its environmental impacts.

DOI: 10.1201/9781003327394-4

The research community has addressed the life cycle analysis of machine tools to quantify their environmental emissions during their life cycle [4–6], but the environmental impacts of actual machining processes have not been addressed effectively. Cao et al. [4] presented a carbon efficiency approach to quantifying the carbon emissions caused by machine tools during their life cycle. It was observed that the fixed emissions can be reduced by a light weight design and remanufacturing, while the variable emissions can be reduced by improving the planning of the machining process. Diaz et al. [5] calculated the energy consumption and CO_2 emissions for two machine tools during four life cycle phases of manufacturing, transportation, use and end of life (EOL). The life cycle assessment studies for a machining process are not available. In the present study, a life cycle analysis of a milling process was conducted to evaluate the environmental impacts in the production of a sample product. The life cycle analysis of the milling process encompassed raw material (aluminum), consumables (cutting tool, coolant, lubricating oil), transportation (raw material, consumables, finished goods, waste), electricity (machining and heating, ventilation and air conditioning (HVAC)), treatment/disposal (used product, chip processing, worn out cutting tool, used coolant and lubricant) and share of machine tools and factory infrastructure. A sequence of milling operations was performed on a workpiece to produce the sample part. The case study found the major hotspots and appropriate strategies to mitigate the environmental emissions and hence save energy and natural resources. The next section discusses the LCA approach in detail.

4.2 MATERIALS AND METHOD

LCA analysis of the milling process was carried out by utilizing ISO 14040 standards [7]. LCA consists of four steps in its systematic approach as suggested by these standards—goal and scope definition, inventory analysis, impact assessment, and interpretation [8,9]. Rebitzer et al. [10] discussed the ISO 14040 series standards in detail for different applications: ISO 14040 [7] for principles and framework, ISO 14041 [11] for goal and scope definition and inventory analysis, ISO 14042 [12] for life cycle impact assessment, and ISO 14043 [13] for life cycle interpretation.

The environmental impacts of the milling process were assessed with the help of the Umberto NXT universal [14] software tool and Eco-invent dataset version 3 [15]. The well-known ReCiPe method was employed for both midpoint and endpoint assessments of inventories. The ReCiPe method is known for its harmonization at both midpoint and endpoint levels [2]. It covers a wide range of midpoint and endpoint categories, which are useful for envisaging several environmental impacts. The ReCiPe method of impact assessment is an upgrade of the eco-indicator and the CML method [16].

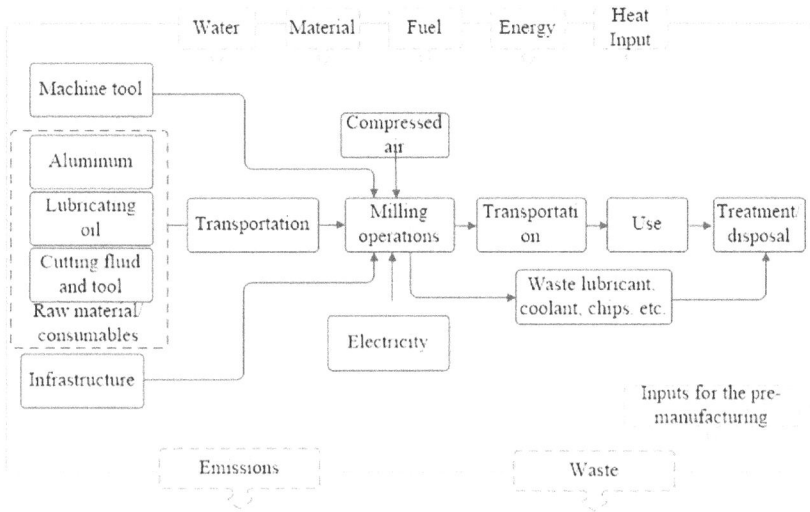

Figure 4.1 System boundary for the life cycle analysis of the milling process.

4.2.1 Goal and Scope

The goal of the study is to analyze the environmental impacts generated by a milling process for the production of a sample part.

4.2.2 Functional Unit and System Boundary

The functional unit selected for the study is the removal of 0.1 kg of aluminum as chips using end mill machining to generate the desired component shape. The system boundary for the study is shown in Figure 4.1. It consists of pre-manufacturing (raw material acquisition and transportation), manufacturing and post-manufacturing (delivery, use and EOL). The EOL of the milling process involves chip processing, treatment/disposal of the worn out cutting tool, used lubricant and coolant. The system boundary also includes the effect of infrastructure (technical building services, equipment and compressed air) and machine tool depreciation. The operational system boundary of the analysis is considered as one year.

4.2.3 Reference Factory and the HVAC System

The energy and resource requirement for the milling process to produce the sample part were identified by experimental evaluation and empirical relations. For the calculation of the HVAC system, a reference factory with a floor size of 20×50 m² for ten CNC machines was defined. Small logistics places close to the machine tools were included. The insulation technology used in the reference factory meets the current state-of-the-art standards for new non-residential buildings, which is 0.28 [W/m²*K] for opaque parts in areas with temperatures above 19 C [17].

The air conditioning needs of the reference factory was assessed based on the degree days. Degree days are widely used to assess the heating and cooling needs as these data are easily available for most of the locations in the world and further local measurements are not needed. Heating degree days (HDDs) describe how much and for how long the temperature was below a base temperature, while cooling degree days (CDDs) describe how much and for how long the temperature was above a base temperature [18]. The energy demand for cooling (Q_C) is estimated as follows [18]:

$$Q_C = \frac{\dot{m} * c * 24 * CDD}{COP} \tag{4.1}$$

where $\dot{m}$ is the mass flow rate through the cooling system, c is the heat capacity of air and COP is the coefficient of performance. In the present study, the local electricity mix was used for the cooling purpose and a base temperature of 26 °C was selected for calculation of the energy required for temperature conditioning. The energy demand by the HVAC system for the machining of 1 kg aluminum in the reference factory was calculated to be 0.955 kWh.

4.2.4 Inventory Analysis

The inventory analysis for 0.1 kg aluminum machining was carried out as described in Table 4.1. Primary data for the inventory analysis were collected using real time experimentations on a vertical milling center. Secondary process and material specific data were collected using contemporary literature and laboratory manuals. The coolant used was a mixture of mineral based soluble oil and water.

The inventory analysis, as shown in Table 4.1, was divided into basic resources (materials, energy sources and water) and related waste. The main aim of the study was to assess the environmental impacts of the milling process and therefore human and monetary resources were not included in the scope of the study. The aluminum material removed in the form of chips was considered for the calculation of raw material processing and disposal.

The inventory data shown in Table 4.1 was assumed to provide the essential process requirements to carry out the milling process and associated transportation activities. The life of the vertical milling machine tool was assumed to be 20 years for an estimation of the share of the machine tool depreciation during the milling process of the sample product. The cutting tool used for the milling process was measured to have a capability to produce ten similar components with the same sequence of operations as considered in the study, without compromising the surface characteristics. The electricity used for environmental conditioning was calculated. The

Table 4.1 Inventory Table for LCA of a Milling Process

Inventory	Unit	Quantity
Lubricating oil	kg	0.00382
Aluminum	kg	0.10
Share of metal working factory	unit	2.02E–10
Share of metal working machine	kg	0.000174
Transportation of raw material	kg-km	200
Compressed air	m³	1.28
Water cooling	m³	0.02
Mineral oil	kg	0.132
Electricity for environmental conditioning	kWh	0.01
Electricity for milling	kWh	0.57
High speed steel for cutting tool	kg	0.014
Transportation of finished part to consumer	kg-km	200
Aluminum scrap	kg	0.10
Waste mineral oil	kg	0.1352
Transportation of chips for treatment	kg-km	50
Waste water treatment	m³	0.02

electricity consumption for the study was taken to be an Indian electricity mix from the eco-invent dataset version 3 [15]. The inventory dataset for raw material acquisition (aluminum, lubricating oil, water, mineral oil, compressed air, etc.) was obtained from the eco-invent dataset contained in the Umberto NXT Universal software.

4.3 RESULTS AND DISCUSSION

The environmental impacts of the milling process were assessed using both midpoint and endpoint impact assessment methods. The well know ReCiPe method for both midpoint and endpoint impact assessment was utilized to assess the local, regional and global environmental impacts of the milling process. Local impacts majorly include: air and noise emissions, land area changes, and impacts on the local ecosystem by means of metal mining or other related activities. If the metal mining activities are carried out at far off locations rather than in the vicinity, the effect will be more regional than local. The effect of NO_x and SO_x pollutants in the environment due to the manufacturing activities may cause acid rain, which is also a regional impact. The various environmental emissions by manufacturing activities, which result in long-term distortion of the environment, are global level impacts. These environmental burdens are classified as climate change or sometimes as global warming, acidification, eutrophication, human toxicity, and so on.

The environmental impact categories selected in the endpoint assessment are: resources, human health and ecosystem quality. The same categories selected to carry out the midpoint assessment are: climate change (CC), fossil depletion (FDP), human toxicity (HTP), metal depletion (MDP), natural land transformation (NLTP), ozone depletion (ODP), particulate matter formation (PMFP) and water depletion (WDP).

4.3.1 Endpoint Assessment

The endpoint assessment results for the major activities of the milling process are shown in Figure 4.2. It is observed that the aluminum production and electricity consumption are major environmental impacting factors for the milling process, followed by compressed air and cutting fluid. The figure also shows that the highest impacts are found in the human health category, particularly from the aluminum production and electricity consumption.

4.3.2 Midpoint Assessment

The midpoint assessment results of the study show similar trends as the endpoint assessment results. The high impact activities in descending order are: aluminum production, electricity consumption, treatment and production of cutting fluid, compressed air, chip processing and cutting tool production.

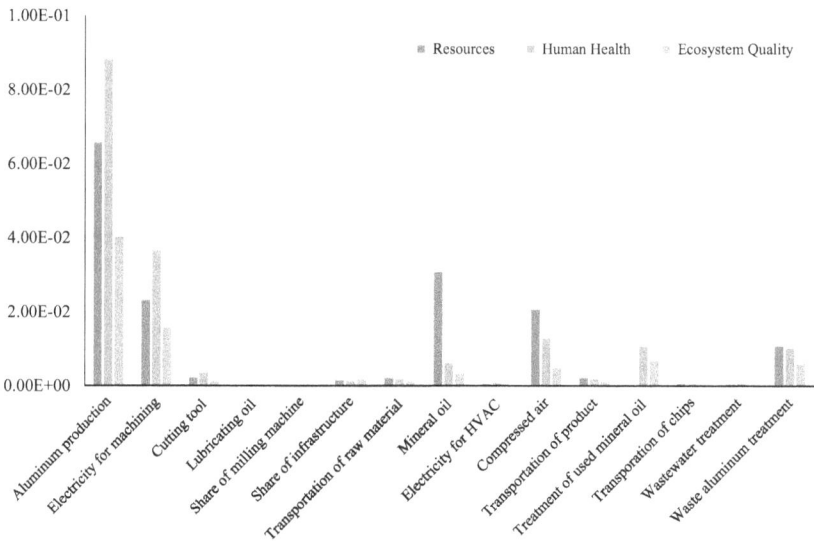

Figure 4.2 Endpoint environmental impact assessment of the three categories.

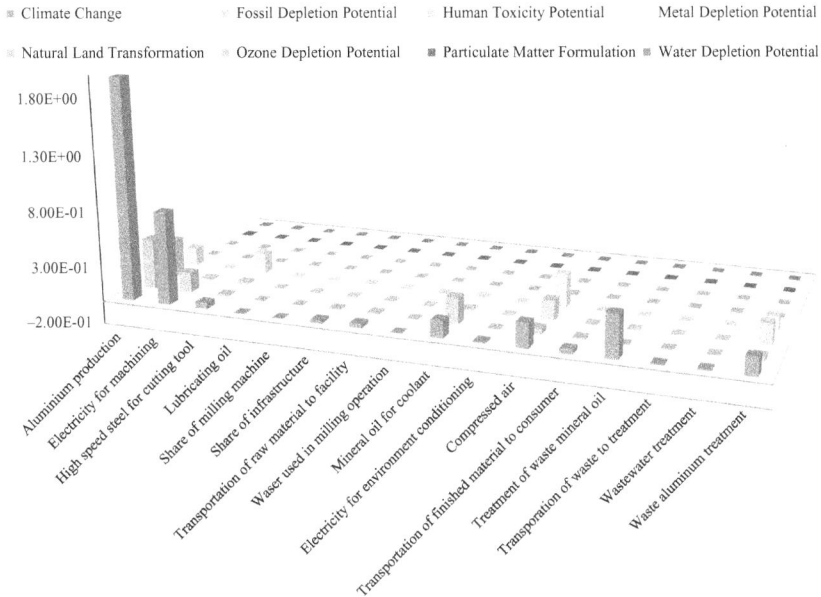

Figure 4.3 Midpoint environmental impact assessment results.

Figure 4.3 presents the midpoint impact assessment of the milling process in the selected categories (CC, FDP, HTP, MDP, NLTP, ODP, PMFP, and WDP). It is observed that in the NLTP, ODP, PMFP and WDP categories, the environmental impacts are negligible as compared to the other categories. The majority of the environmental impacts are in terms of CC followed by FDP, MDP and HTP.

4.4 PRACTICAL IMPLICATIONS AND RECOMMENDATIONS

The LCA analysis of the milling process provided a clear visualization of hotspots for environmental impacts. Aluminum production, electricity consumption, cutting fluid production and disposal, compressed air and chip treatment are found to be the major factors causing environmental impacts. It is observed from the results that the major impacts in aluminum production are due to carbon dioxide emissions, sulfur hexafluoride, methane, ethane-hexafluoro (HFC-116) and dinitrogen monoxide emissions. However, the focus of the present study has been an assessment of the environmental impacts of the milling process and therefore the material processing impacts and their reduction strategies have not been discussed in detail. It is evident from the analysis that energy consumption plays a major role in the environmental impacts in the milling process directly (for material removal) and indirectly (compressed air production and the HVAC system). Therefore, reduction in energy consumption plays a major role in reducing the environmental impacts of the milling process.

4.5 SENSITIVITY ANALYSIS

Sensitivity analysis was carried out to assess the robustness of the results obtained from the machining LCA study. In sensitivity analysis, the values of independent variables are varied and their effect on dependent variables is measured. Sensitivity analysis helps in establishing a clear interpretation of results and assessing their robustness and transparency [19]. Sensitivity analysis is carried out by varying the amount of materials and process inputs. In this study, the sensitivity analysis was performed by varying the values of seven materials and process inputs. The variation includes both increase and decrease in the values as presented in the Table 4.2. The sensitivity analysis results show that the environmental impact results obtained in the LCA analysis are robust, as shown in Figure 4.4. The error bars show the variation in CC impact, with variation in the input process parameters as shown in Table 4.2.

Table 4.2 Actual and Changed Values of Input Variables for Conducting Sensitivity Analysis

Process/Raw Material (percentage change in values)	Unit	Actual Values	Changed Values
Lubricating oil (10% increase)	kg	0.00382	0.004202
Lubricating oil (10% decrease)			0.003438
Compressed air (10% increase)	m³	1.28	1.408
Compressed air (10% decrease)			1.152
Mineral oil (10% increase)	kg	0.132	0.1452
Mineral oil (10% decrease)			0.1188
Transportation of finished part to consumer (10% increase)	kg-km	200	220
Transportation of finished part to consumer (10% decrease)			180
Waste mineral oil (10% increase)	kg	0.1352	0.1487
Waste mineral oil (10% decrease)			0.1217
Transportation of chips for treatment (10% increase)	kg-km	50	55
Transportation of chips for treatment (10% decrease)			45
Waste water treatment (10% increase)	m³	0.02	0.022
Waste water treatment (10% decrease)			0.018

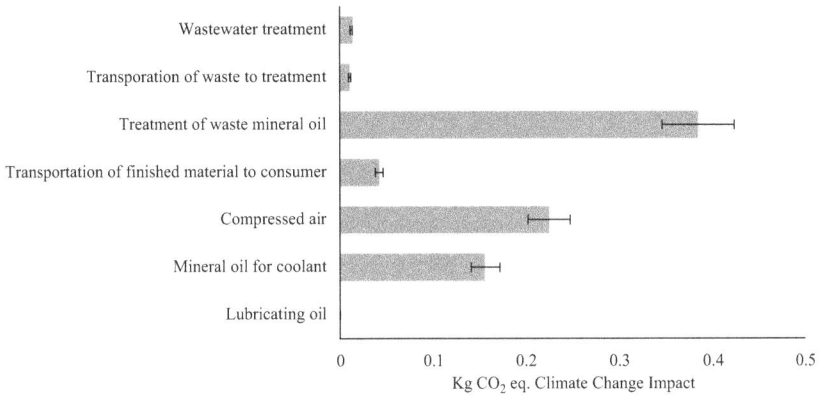

Figure 4.4 Sensitivity analysis results with actual and changed values of input variables.

4.6 SUMMARY

This chapter has presented the life cycle analysis of a milling process to visualize the major hotspots contributing to environmental impacts. The study was conducted to assess the environmental impacts of the aluminum milling process for the production of a sample part. The LCA analysis was carried out as per ISO 14040 standard by using the Umberto NXT software tool and the eco-invent dataset version 3.0. The scope of the study includes the production of raw materials and consumables (cutting tool, coolant, lubricating oil), transportation (raw material, consumables, finished goods), electricity (machining and HVAC), treatment/disposal (used product, chip processing, worn out cutting tool, waste coolant and lubricant) and share of machine tools and factory infrastructure for machining. The environmental impacts were assessed in both endpoint and midpoint impact assessment categories by using the ReCiPe method. Raw material production, electricity consumption, cutting fluid production and disposal, chip processing and compressed air production are the high impact generating activities of the milling process. The major impacts are in terms of human health and resource depletion. The effects on the climate change, fossil depletion, human toxicity and metal depletion categories are also high. This analysis can be used as a foundation for the formulation of key plans for the improvement of the environmental sustainability of machining processes.

REFERENCES

[1] Klöpffer W. Life cycle assessment: From the beginning to the current state. *Environ Sci Pollut Res Int* 1997;4:223–228. doi:10.1007/BF02986351.

[2] Huijbregts MAJ, Steinmann ZJ, Elshout PMF, Stam G, Verones F, Vieira MDM, Zijp M, Van Zelm R. *ReCiPe 2016: A Harmonized Life Cycle Impact Assessment Method at Midpoint and Enpoint Level - Report 1: Characterization.* Bilthoven; 2016.

[3] Sangwan KS. Performance value analysis for justification of green manufacturing systems. *J Adv Manuf Syst* 2006;5:59–73. doi:10.1142/S0219686706000765.

[4] Cao H, Li H, Cheng H, Luo Y, Yin R, Chen Y. A carbon efficiency approach for life-cycle carbon emission characteristics of machine tools. *J Clean Prod* 2012;37:19–28. doi:10.1016/j.jclepro.2012.06.004.

[5] Diaz N, Helu M, Jayanathan S, Chen Y, Horvath A, Dornfeld D. Environmental analysis of milling machine tool use in various manufacturing environments. In: *Proc 2010 IEEE Int Symp Sustain Syst Technol ISSST* 2010. pp. 1–6. doi:10.1109/ISSST.2010.5507763.

[6] Song S, Cao H, Li H. Evaluation method and application for carbon emissions of machine tools based on LCA. In: *International Conference on Advanced Technology of Design and Manufacture* 2010. pp. 74–78. doi:10.1049/cp.2010.1263.

[7] ISO 14040. Environmental management—life cycle assessment—principles and framework. 1997.

[8] Kellens K, Dewulf W, Overcash M, Hauschild MZ, Duflou JR. Methodology for systematic analysis and improvement of manufacturing unit process life-cycle inventory (UPLCI)—CO2PE! initiative (cooperative effort on process emissions in manufacturing). Part 1: Methodology description. *Int J Life Cycle Assess* 2012;17:69–78. doi:10.1007/s11367-011-0340-4.

[9] Sangwan KS, Choudhary K, Batra C. Environmental impact assessment of a ceramic tile supply chain – a case study. *Int J Sustain Eng* 2018;11:211–216. doi:10.1080/19397038.2017.1394398.

[10] Rebitzer G, Ekvall T, Frischknecht R, Hunkeler D, Norris G, Rydberg T, Schmidt WP, Suh S, Weidema BP, Pennington DW. Life cycle assessment Part 1: Framework, goal and scope definition, inventory analysis, and applications. *Environ Int* 2004;30:701–720. doi:10.1016/j.envint.2003.11.005.

[11] ISO 14041. Environmental management—life cycle assessment—goal and scope definition and inventory analysis. 1998.

[12] ISO 14042. Environmental management—life cycle assessment—life cycle impact assessment. 2000.

[13] ISO 14043. Environmental management—life cycle assessment—life cycle interpretation. 2000.

[14] IFU Hamburg. Umberto NXT Universal Sustainability and Productivity - One Software for Both Goals 2015. http://www.umberto.de/en/versions/umberto-nxt-universal/. (Accessed 17 September 2019).

[15] Wernet G, Bauer C, Steubing B, Reinhard J, Moreno-Ruiz E, Weidema B. The ecoinvent database version 3 (part I): overview and methodology. *Int J Life Cycle Assess* 2016;21(9):1218–1230. doi:10.1007/s11367-016-1087-8.

[16] Huijbregts, Mark AJ, Zoran JN Steinmann, Pieter MF Elshout, Gea Stam, Francesca Verones, M. D. M. Vieira, Anne Hollander, Michiel Zijp, and Rosalie van Zelm. ReCiPe 2016: a harmonized life cycle impact assessment method at midpoint and endpoint level report I: characterization. 2016.

[17] Sihag N, Leiden A, Bhakar V, Thiede S, Sangwan KS, Herrmann C. The Influence of Manufacturing Plant Site Selection on Environmental Impact of Machining Processes. *Procedia CIRP* 2019;80:186–191. doi:10.1016/j.procir.2019.01.023.

[18] CIBSE. *Degree-Days: Theory And Application - TM41*. Chartered Institution of Building Services Engineers: London; 2006.

[19] Niero M, Pizzol M, Bruun HG, Thomsen M. Comparative life cycle assessment of wastewater treatment in Denmark including sensitivity and uncertainty analysis. *J Clean Prod* 2014;68:25–35. doi:10.1016/j.jclepro.2013.12.051.

Chapter 5

Mechanically Based Non-Conventional Machining Processes

Rajesh Babbar and Aviral Misra

Dr. B. R. Ambedkar National Institute of Technology, Jalandhar, India

Girish C. Verma and Pulak Mohan Pandey

Indian Institute of Technology Delhi, New Delhi, India

CONTENTS

DOI: 10.1201/9781003327394-5

5.1 INTRODUCTION

Stiff competition in the market and the continuously rising demand for an enhanced performance of products have led to the development of newer materials like high strength alloy composites, ceramics and semiconductors. These materials have special properties such as a high strength-to-weight ratio, high hardness and high brittleness. But, the processing of these materials is a difficult proposition by the conventional method of machining. It either inflates the cost of manufacturing or results in the loss of the useful properties of the material. Also, complex shapes are very challenging to fabricate via conventional techniques. To overcome the shortcoming of conventional machining and satisfy the requirements for precision, close dimensional tolerances and miniaturization of parts require the growth of advanced machining processes. These processes use tools that are of a non-contact type and utilize different types of energy source. A classification of the various advanced machining processes with the type of energy source is presented in Table 5.1.

This chapter brings out the various non-conventional techniques that use mechanical means to remove material with the help of abrasives as the main cutting tool. The varying application of abrasive particles, such as in cutting, drilling and finishing, will be discussed, and we will discuss the details of the important developments that are taking place with various mechanically based non-conventional processes, such as abrasive jet machining, abrasive water jet machining, abrasive flow finishing and

Table 5.1 Advanced Machining Processes with Type of Energy Used and Its Source

Type of Energy Used	Name of the Process	Energy Source
Mechanical energy	AJM	Pneumatic pressure
	USM	Ultrasonic vibration
	AFM	Hydraulic pressure
	AWJM	Hydraulic pressure
	MAF	Magnetic field
Chemical	CHM	Corrosive agent
Electrochemical	ECM	Current
Thermal	LBM	Stimulated emission of radiation
	EBM	Electron beam
	PAM	Ionized beam
	EDM	Voltage

AJM: abrasive jet machining; USM: ultrasonic machining; AFM: abrasive flow machining; AWJM: abrasive water jet machining; MAF: magnetic abrasive finishing; CHM: chemical machining; ECM: electrochemical machining; LBM: laser beam machining; PAM: plasma arc machining; EDM: electro discharge machining; EBM: electron beam machining.

magnetic abrasive finishing with the process mechanics involved in material removal. The chapter also brings out details like the role of various process parameters on process characteristics.

5.2 ABRASIVE JET MACHINING

Abrasive jet machining (AJM) is a process in which abrasive particles of very small size (the order of μm) are made to impact on the surface of the workpiece at a very high velocity to remove the material from the workpiece surface due to erosive action. A highly compressed carrier gas is used (usually air) in which abrasive particles are mixed, that is allowed to pass through the nozzle. The high-pressure energy of the gas is transformed into kinetic energy with the help of the nozzle. The velocity gained by the carrier gas is also extended to the abrasive particles [1]. The set-up cost as well as the power required to operate the AJM process is relatively low compared to the other mechanical type abrasive processes.

Figure 5.1 illustrates the layout of an AJM machine. The compressed air from the compressor is first cleaned and dried to remove any foreign particles and traces of water vapor in it. The flow and pressure of air are controlled with the help of the opening valve and pressure regulator respectively. The regulated compressed air is then moved to a mixing chamber in which it is mixed with an identified quantity of abrasive particles. The high-pressure mixture of gas and abrasive particles are then forced to pass through the

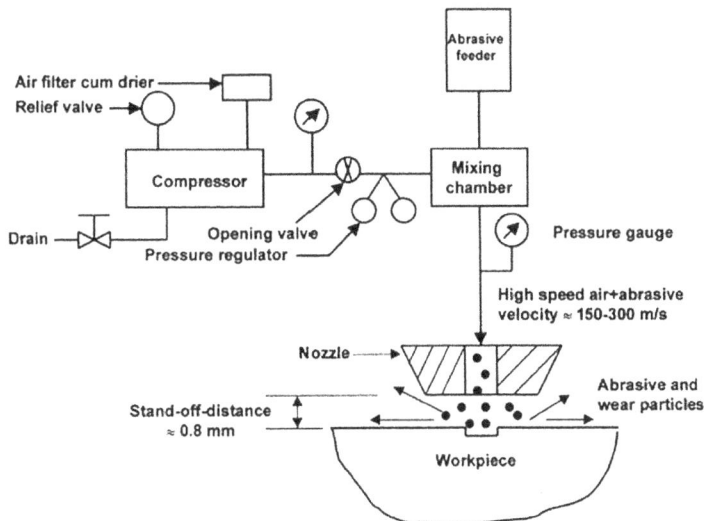

Figure 5.1 Layout of an abrasive jet machine [2].

nozzle and then allowed to impact on the workpiece surface. An abrasive particle impact causes either a tiny brittle fracture or plastic extrusion and depends upon the machining parameters and the properties of the workpiece material [3]. The chips formed due to a tiny brittle fracture are blown away from machining zone by the flow of the carrier gas.

5.2.1 Mechanism of Material Removal in AJM

The targeted flow of abrasive particles causes a tiny brittle fracture on the workpiece at the impact location in the form of chips. The cross-section of the nozzle and the direction of impact of the abrasive particles determine the machined geometrical shape on the workpiece surface due to AJM. The properties of the workpiece material and parameters of the process determine the ductile or brittle removal mechanism during erosion. The standard crack model is represented in Figure 5.2. The deformation and cracking, which occur due to the impact of abrasive particles, can be thought of in terms of a quasi-static Vickers-indentation theory [4]. The indentation of the abrasive particles creates compressive stresses in the material underneath, which forms a plastically deformed zone. If plastic deformation exceeds the fracture threshold, a radial crack propagates perpendicularly to the surface in the downward direction. The unloading phase sees a lateral crack at the bottom of the plastically deformed zone which spreads out parallel to the workpiece surface [5]. The crack which forms in the radial direction does not affect the formation of the chip, though it degrades the integrity of the surface. Generally, it has been accepted that the lateral crack regulates the volume of the removed material. Likewise, the lateral crack size and the depth of its initiation is contemplated as a depiction of the erosion phenomenon.

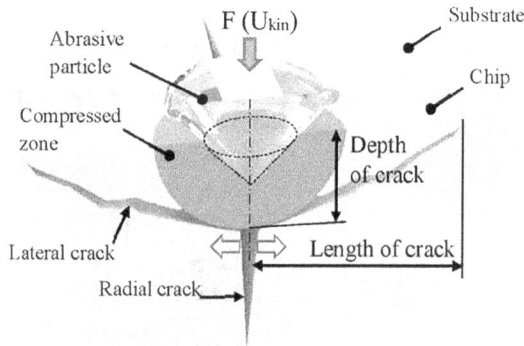

Figure 5.2 Schematic of indentation and crack propagation due to the impact of an abrasive particle [3].

The AJM process is found to be suitable for brittle materials such as ceramics and glass. The process is found to be useful for thin section parts but unsuitable for the fabrication of features in the parts that have very sharp edges.

Finnie [6] worked with the equations of motion, considering particles of angular shape for machining the workpiece surface of a ductile material. He found that the volumetric material removal Q can be related to the mass M of the abrasive particle for a nozzle having a fixed geometry as:

$$Q = \frac{Cf(\theta)MV^n}{\sigma} \tag{5.1}$$

where n and C are constants, depending on the process parameters and workpiece properties, σ is the minimum flow stress (related to the workpiece hardness), V is the velocity of the impacting particles and θ is the angle of impingement of the abrasive.

5.2.2 Process Parameters of AJM

AJM process performance is assessed in terms of the rate of material removal (MRR), the accuracy of the geometry of the part and its finish, and the nozzle wear rate. The AJM process parameters are:

- Type of abrasive;
- Size of abrasive;
- Mixing ratio;
- Jet velocity;
- Nozzle size, shape and wear properties;
- Stand-off-distance (SOD).

The abrasive particles commonly used in AJM are silicon carbide (SiC), alumina (Al_2O_3), glass beads, sodium bicarbonate and dolomite. SiC and Al_2O_3 are utilized for cutting whereas glass beads, sodium bicarbonate and dolomite are applied for cleaning, deburring or etching [2]. The volumetric material removal rate (VMRR) and shape of the cavity formed are primarily dependent on the SOD. Figure 5.3a shows that the volumetric rate increases linearly with SOD up to a certain limit and then decreases. It also indicates that VMRR increases with the mixing ratio. There is an optimum penetration rate that can be obtained for a set of SOD and mixing ratios as illustrated in Figure 5.3b; it also indicates that an increased mixing ratio for a given SOD tends to increase the penetration rate.

Figure 5.4a shows the variation of cavity diameter at the top with a change in SOD for different mixing ratios; it also reveals that an increase in SOD tends to increase the top diameter of the cavity. This is reinforced by the photograph of the machined cavity shown in Figure 5.4b.

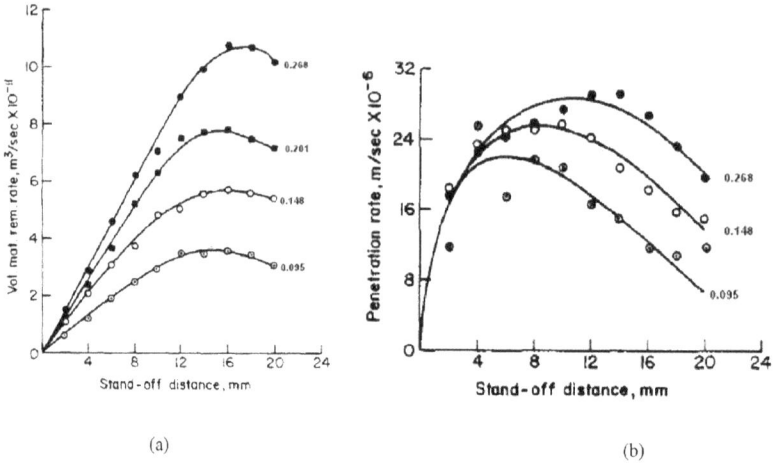

(a)

(b)

Figure 5.3 Variation of (a) volumetric material rate (b) penetration rate with SOD for varying mixing ratios [7].

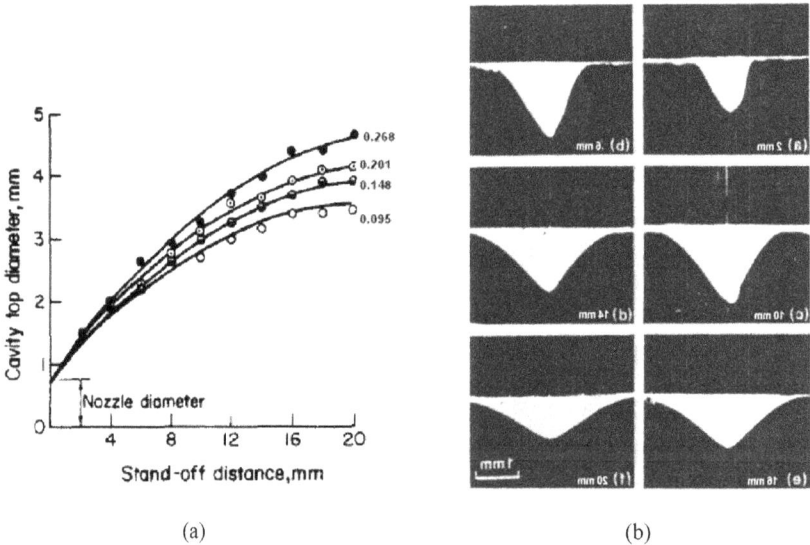

(a)

(b)

Figure 5.4 (a) Variation of machined cavity diameter at the top with SOD for varying mixing ratios; (b) photograph of cavity profile machined at different stand-off distances [7].

5.2.3 Applications of AJM

AJM is applied to the drilling, cutting and engraving of brittle materials such as marble and glass. It is also used for surface preparation, erosion testing and shot peening [8]. In the context of manufacturing, the AJM process is

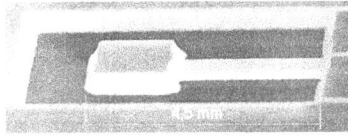

Figure 5.5 Cantilever beam fabricated by the AJM process [12].

employed for cutting, drilling, polishing and surface texturing. In the context of metallurgy, the abrasive jet particles can be employed for the cleaning of casting. The micro-blasting can be utilized for manufacturing microfluidic devices [9].

Surface patterning or texturing is a modification of the topography of the surface to create a controlled geometry of uniformly distributed asperities and depressions [10]. In the year 2000, the blasting of fine abrasive particles onto a surface was demonstrated as a high-quality mechanical etching technique [9], since then texturing through AJM has shown a constant positive trend. In 2005, a micro-scale patterning resolution was achieved. Slikkerveer et al. [11] fabricated channels in a glass material of 500–100 μm wide using 23 μm abrasive particles with a variety of erosion depth to within 5%. Figure 5.5 illustrates a typical cantilever beam fabricated by AJM for an inertial sensor application in a Pyrex glass wafer.

5.3 ABRASIVE WATER JET MACHINING

Abrasive water jet machining (AWJM) is an advanced machining process that is environmentally friendly. It removes material due to the erosive action of the abrasive laden water jet. AWJM find applications in drilling, cutting, cleaning or descaling of different materials. It can machine a variety of materials that include copper and its alloys, aluminum, steel, lead, titanium, tungsten carbide, composites, acrylic, ceramics, rocks, concrete, silica glass and graphite. The most auspicious application of AWJM in the aerospace industry comprises the machining of honeycomb sandwich structures [2]. The AWJM process performs a cutting operation by applying a high-velocity water jet containing abrasive particles using a setup presented in Figure 5.6. The abrasive particles are mixed with the water jet such that the momentum of the jet is partially transmitted to the abrasive particles. The primary aim of the water is to produce a highly coherent jet and also to accelerate the mixed abrasive particles to a high velocity [14]. The AWJM system contains several elements to form an efficient set-up, that is a pumping system that produces very high pressure, a water jet nozzle to convert pressurized water to a high velocity jet, an abrasive feed system to deliver a precisely regulated quantity of abrasive particles and which offers the efficient mixing of abrasive particles with a water jet, and a catcher that damps the energy of the water jet after machining.

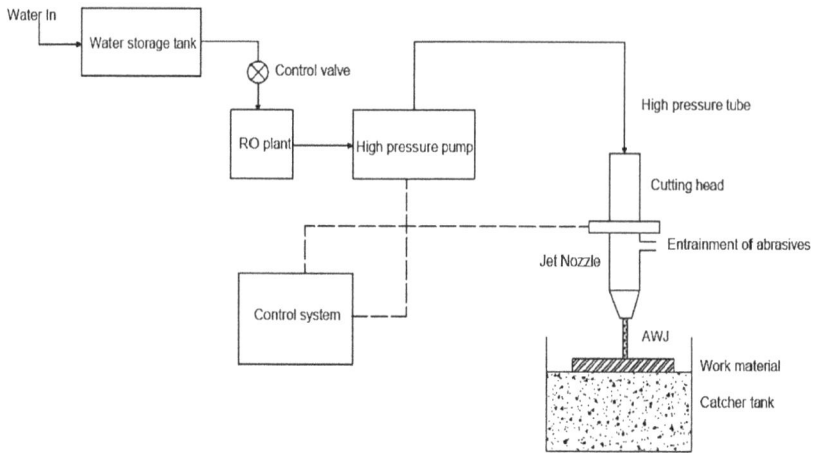

(a)

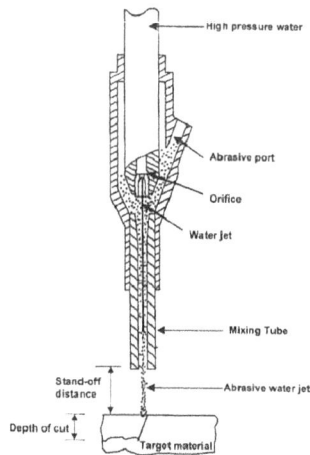

(b)

Figure 5.6 Basic scheme of (a) the set-up [13] and (b) the nozzle [2] for AWJM.

5.3.1 Material Removal Mechanism in AWJM

The mechanism of material removal during machining with a water jet and AWJM is rather complex. During machining, AWJM removes the material mainly owing to the impact of abrasive particles that leads to ploughing and micro-cutting at a low angle of impact. Initially Finnie [6] studied this process in detail. Further considering a higher angle of impact angle, it was observed that the removal of material occurs due to the plastic failure of the material at the impacting site.

5.3.2 Process Parameters of AWJM

5.3.2.1 Water Jet Pressure

The magnitude of the pressure of water determines the kinetic energy of the impacting abrasive particles. If the pressure is below a critical value, then the removal of material does not take place. At the same time, the pressure should lie in the critical range for an efficient cutting since, beyond this range, the machining process becomes ineffective [13]. The water jet pressure has a direct relation with the MRR and the penetration depth. It also has an influence on the abrasive particles in the jet.

5.3.2.2 Traverse Rate

The speed of the nozzle relative to the workpiece is termed the "traverse rate". The traverse rate of the water jet determines the quality of the surface produced by AWJM. A lower value of traverse rate is preferred as it provides a good quality of machined surface because of the large number of abrasive particles impacted on the workpiece. But, it increases the machining time.

5.3.2.3 Abrasives

Different types of artificial abrasive particles (SiC, Al_2O_3) and natural garnets can be used in this process. The performance of the AWJM process is highly dependent on the shape, size and hardness of the abrasive particles. An increased abrasive particle size tends to increase the penetration depth and MRR to a limit. Thereafter, if the abrasive particle size further increases it tends to reduce both the penetration depth and MRR.

5.3.2.4 Abrasive Mass Flow Rate

An optimum supply of abrasive particles ensures an improved surface finish with a higher machining efficiency. The optimum abrasive particle flow rate is determined by the diameter of the focusing nozzle. To ensure that an effective acceleration is gained by the abrasive particles, transmission of the jet's momentum is required, so the optimum supply of abrasive particles is ensured in the jet.

5.3.2.5 SOD

SOD is the gap between the workpiece surface and the tip of the nozzle. SOD has a large influence on the kerf profile generated during the AWJM process. Figure 5.7 illustrates different kerf profiles for varying SOD.

5.3.2.6 Jet Impingement Angle (θ)

The orientation of the abrasive water jet governs the jet impingement angle. This is the inclination between the workpiece surface and direction of

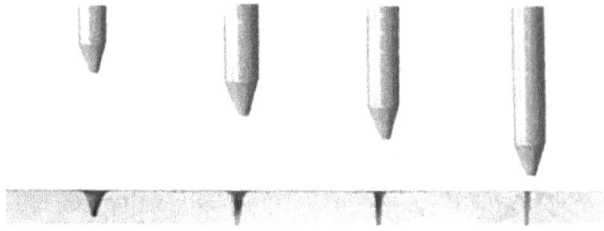

Figure 5.7 Variation of kerf profiles at different SOD [13].

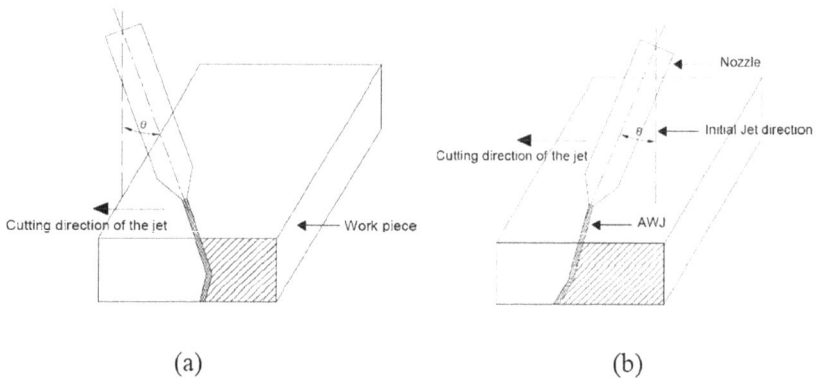

(a)   (b)

Figure 5.8 Illustration of jet impingement angle (θ) in an (a) backward and a (b) forward
direction [14].

the abrasive water jet without any deviation. The jet impingement angle
(θ) determines the angle at which the abrasive strikes the surface of the
workpiece, which further affects the erosion phenomenon. Jet impingement
angles used in the AWJM process are of two types: backward and forward
directions as shown in Figure 5.8.

5.3.3 Cutting Geometry in AWJM

AWJM can achieve high MRR; however, it does have accompanying
defects. As a water jet comes out from the nozzle, it starts diverging until
it contacts the work surface. This tends to decrease the kinetic energy of
the jet for the initial impact. The jet erodes the material until it completely
forms a through-hole. As the jet penetrates, there occurs a continuous loss
of kinetic energy. This results in a reduction in the width of the machined
material, known as kerf. There occurs the formation of a conical frustum
for a drilled hole, and a trapezium for a machined slot. The cutting geometry associated with AWJM is termed "taper cutting". Figure 5.9 illustrates
the geometry of a tapered cut as top width (W_t), bottom width (W_b) and
taper angle (θ).

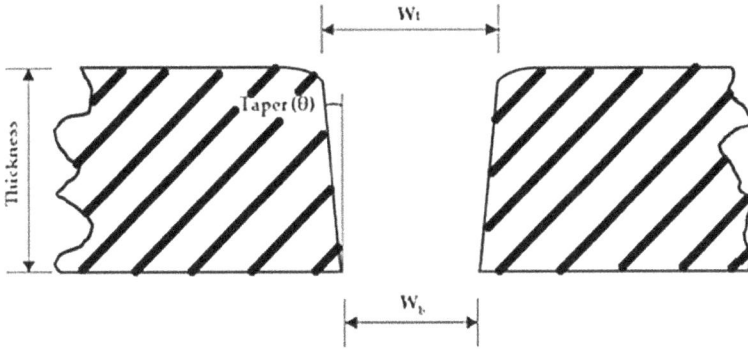

Figure 5.9 Illustration of taper defect generated in AWJM.

An estimation of the cutting geometry can be made by using the ratio $\dfrac{W_t}{W_b}$; for an improved cutting geometry, the ratio should be close to 1. Mathematically, the accuracy of the cutting geometry or kerf can be determined by the taper angle (θ) as:

$$\theta = \tan^{-1}\left(\frac{W_t - W_b}{2h}\right) \tag{5.2}$$

where h represents the thickness.

Another problem during the transverse movement of the abrasive water jet is striation formation. Figure 5.10a shows the deflection of the jet during the cutting action with a transverse feed. Most of the kinetic energy of the abrasive water jet is utilized in removing the top layer of the workpiece. An increase in the depth of penetration leads to a reduction in the quality of the surface formation known as striation formation; see Figure 5.10b.

Earlier it was discussed that using a single pass cutting in AWJM was unable to penetrate the workpiece material efficiently, so the kerf has a

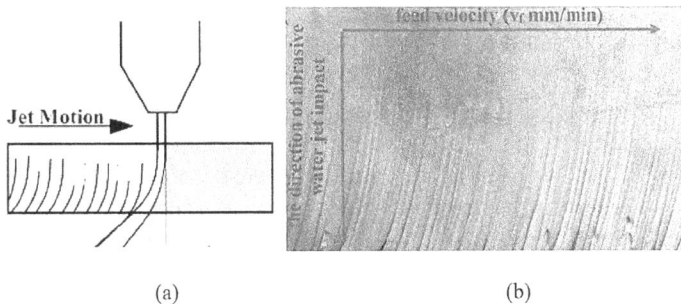

(a)   (b)

Figure 5.10 Striation formation: (a) schematic [13] and (b) observed in an AZ91D alloy [15].

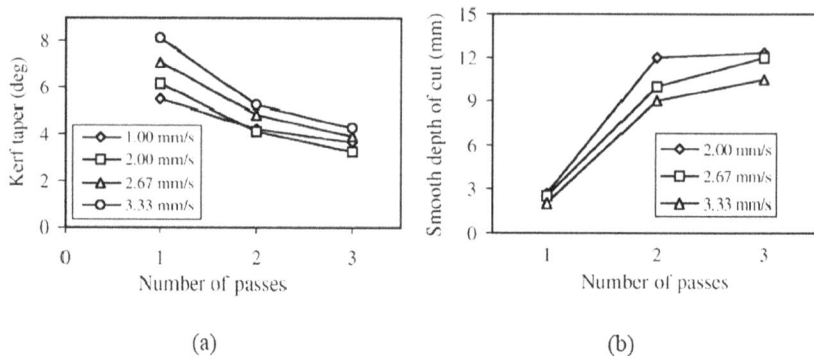

Figure 5.11 Influence of the number of passes on (a) the kerf taper and (b) on the smooth cut depth at different jet traverse speeds [16].

wider top opening but decreases gradually, having a large pocket at the bottom due to the upward deflection of the jet. Hence for a large depth multipass the material must be cut through. Experimentally it has been found that a large number of passes reduces the kerf taper. It was also found that the kerf taper angle also reduces as the transverse feed reduces; see Figure 5.11a. From Figure 5.11b it can be observed that the smoothness of the surface increases with the number of passes, and decreases with the traversed feed of the nozzle.

5.3.4 Applications of AWJM

AWJM has several advantages, including omnidirectional cutting with minimal thermal damage. The part produced has no burrs and cutting occurs with delamination, thus a high cutting speed and efficiency is obtained. AWJM, due to its several advantages, is widely used in a large number of industries from aerospace, to automotive, to mining. AWJM can be widely used for brittle materials [17]. Figure 5.12 illustrates the parts fabricated by abrasive water jet machining.

5.4 MAGNETIC ABRASIVE FINISHING (MAF)

MAF is an advanced finishing process, in which the finishing forces are manipulated by changing the magnetic field in the working gap. Magnetic abrasive particles (MAPs) fill the gap between the magnet (either electromagnet or permanent magnet) and the workpiece material [19].

5.4.1 Material Removal Mechanism in the MAF Process

MAPs utilized in MAF are classified in two types: bonded and unbonded. Figure 5.13 illustrates unbonded and bonded MAPs. Unbonded MAPs are

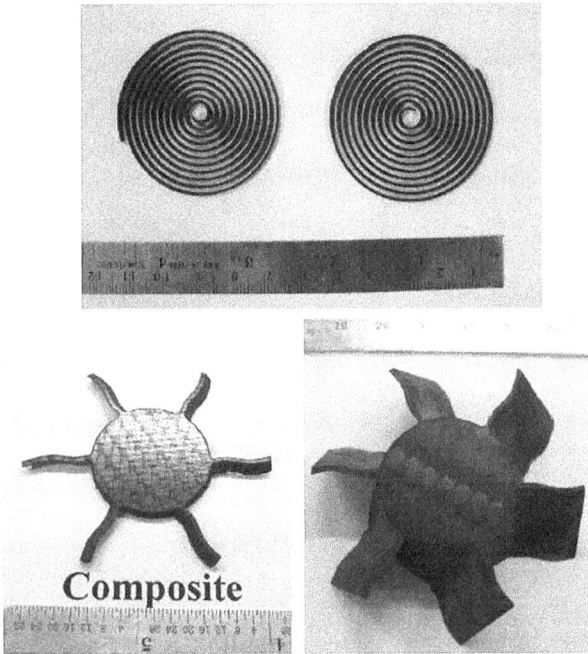

Figure 5.12 Illustration of parts fabricated by abrasive water jet machining [18].

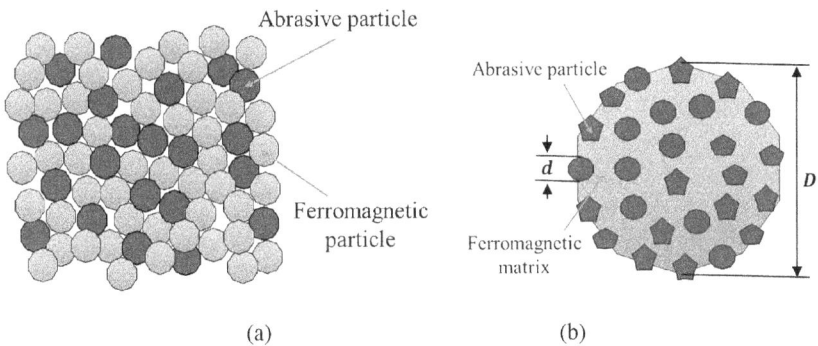

(a)   (b)

Figure 5.13 Schematic illustration of (a) unbonded and (b) bonded MAPs.

the mechanical mixture of ferromagnetic and non-magnetic abrasive particles. Bonded MAPs are fabricated by sintering a mixture of ferromagnetic particles with abrasive particles.

Under the influence of a magnetic field, MAPs develop into a flexible magnetic abrasive brush (FMAB) which is used as a finishing tool. The bonded MAPs line up with the magnetic force lines. But, in unbonded MAPs, the ferromagnetic particles line up with the magnetic force lines under the influence of the magnetic field and the non-magnetic abrasive particles take a

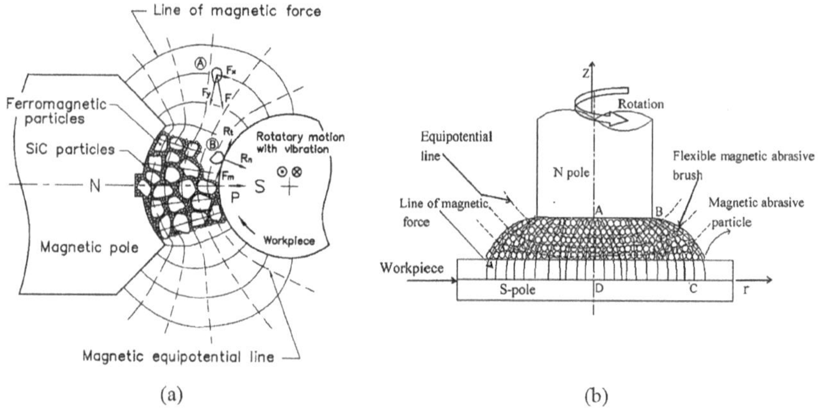

Figure 5.14 Illustration of magnetic field distribution and line of magnetic forces for (a) cylindrical [20] and (b) plane MAF [21].

position between these chains. Figure 5.14 illustrates the formation of an FMAB and the magnetic force lines for cylindrical and plane surfaces.

The abrasives that actually come in direct contact with the surface of the workpiece and are directly involved in surface finishing are designated "active abrasive particles". A magnetic levitation force, due to the presence of ferromagnetic particles, acts on these active abrasive particles to push them to indent the surface of the workpiece. Due to the relative motion provided between the magnet (either electromagnet or permanent) and the surface of the workpiece, the indented abrasive particles produce scratching/abrasion. The FMAB that comes in contact with the surface of the workpiece induces contact stresses at the interaction points, and if this produced stress surpasses a threshold value, the surface irregularities present on the workpiece will be sheared off. Hence, the removal of material in the MAF process occurs due to nano-scratching and micro-chipping [22]. The magnetic field distribution in the working gap influences the rigidity and configuration of the FMAB, and therefore considerably affects the finishing characteristics [23]. The strength of the FMAB is the principal characteristic of the process; a higher strength generates a higher indentation force, which results in a higher MRR during finishing. During MAF the material removal depends on machining parameters, like the working gap, rotational speed of the magnet, voltage, mesh size of the abrasive particles and the mass fraction of the abrasive [24]. Also, the instantaneous MRR at any time is governed by the volume of surface irregularities existing on the workpiece surface at that instant.

5.4.2 Process Parameters of MAF

The performance of MAF is evaluated by the response change in surface roughness (ΔR_a). Figure 5.15 shows the variation (ΔR_a) with process

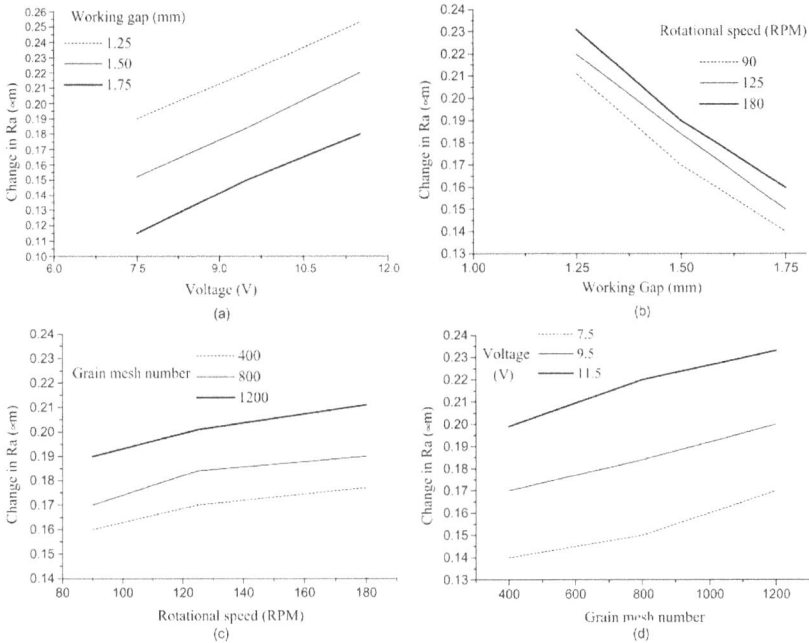

Figure 5.15 Effects of process parameter (a) voltage, (b) working gap, (c) rotational speed and (d) grain mesh number on the change in surface roughness (ΔR_a) [25].

parameters, that is the working gap, voltage, rotational speed and mesh number of the abrasive particles. The supply voltage is the primary source for creating the magnetic field in the working gap; the higher the voltage, the higher is the magnetic field, hence the higher the change in the roughness value during finishing. As the working gap increases it leads to a decrease in the value of the surface roughness (R_a). The rotational speed provides the relative motion to the FMAB; the higher the rotational speed, the higher is the improvement in the value of the surface roughness. The mesh size determines the abrasive particle size; the higher is the mesh number (smaller abrasive particle size), the higher is the ΔR_a.

5.4.3 Advances and Application of MAF

MAF has various advantages over other processes, but in order to enhance the process performance various researchers have proposed compounds or hybridizations of MAF by integrating two or more processes for various applications. Some of the variants of the MAF process involve vibration assistance to the workpiece material [26], ultrasonic-assisted MAF [27], MAF combined with an electrolytic process [28], and so on. MAF can produce electronic and mechanical components having hardly any kind of surface defect with a low R_a value on the finished surface. The MAF process can also be used for micro-deburring application with the help of a permanent

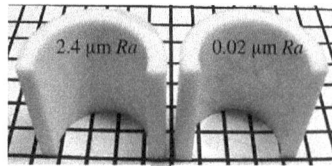

Figure 5.16 Finishing capabilities of MAF for alumina ceramic tubes [29].

magnet instead of an electromagnet [21]. Figure 5.16 illustrates the capabilities of MAF for finishing of inner surface of alumina ceramic tube. The roughness value decreases from 2.4 μm to 0.02 μm in 20 min of finishing.

5.4.4 Future Scope of MAF

In MAF the finishing rate is still a concern to the researcher, especially regarding the surfaces made of high strength alloys. Many researchers in the past had found an enhancement in the finishing rate by hybridizing MAF with different processes, though still there is scope for future investigation. MAF needs to be explored in future to overcome limitations. It is also crucial to investigate the preparation of suitable magnetic abrasive particles regarding high saturation magnetization, shear capacity and dispersibility. At present, an offline inspection of the processed workpiece can be carried out to verify the finishing quality, which affects the surface finish of the workpiece and the finishing efficiency to some extent. So, improved intelligent MAF equipment with a feedback function and online inspection could also be implemented in the future, which would help in the realization of the computerization of finishing. Enhancements in automation could bring in the MAF process as a component of manufacturing in Industry 4.0.

5.5 ABRASIVE FLOW MACHINING

Abrasive flow machining (AFM) is a fine finishing process, having the capability for finishing curved and intricate geometrical parts. AFM also has the capability to finish those regions which are unapproachable by conventional finishing processes, lapping, honing or grinding. The ability of the AFM process to produce a fine surface finish with close tolerances make it popular and also provide an upper hand in the manufacturing industry.

In AFM, the abrasive particles are suspended in a polymeric-based semi-solid medium which is forced to pass through or across the workpiece surface under high pressure. The high pressure is developed when the abrasive particles in a semi-solid deformable medium are extruded into a constricted area. If the medium experiences any restriction in motion it behaves like a deformable tool. A specially designed fixture is usually needed to form a constricting passage that conforms with the surface of the part to be finished for the movement of the medium [30].

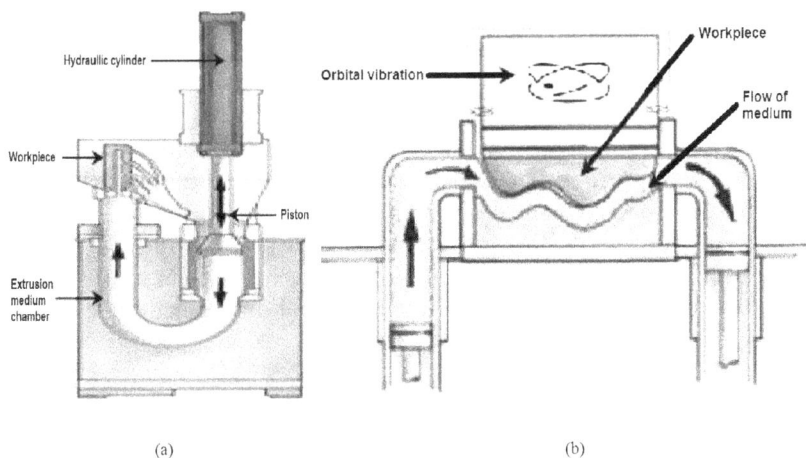

(a)   (b)

Figure 5.17 Schematic of (a) one-way and (b) orbital AFM [31].

Different types of AFM process arrangement are shown in Figures 5.17 and 5.18. AFM generally are of different types, but based on the type of relative motion obtained it can be categorized as a one-way, two-way or orbital AFM. The one-way AFM process (Figure 5.17a) has a unidirectional extrusion of the medium. As the extrusion proceeds, abrasive particles in the medium shear off the peaks of the workpiece asperities, thus reducing the roughness value. At the end of the extrusion, the medium is again recirculated back. The one-way AFM process is useful for components having contours that assist the flow in the forward direction but resist in the backward flow. Figure 5.17b is a schematic of an orbital AFM; the complex-shaped part does not have a through passage. In this illustration, the abrasive

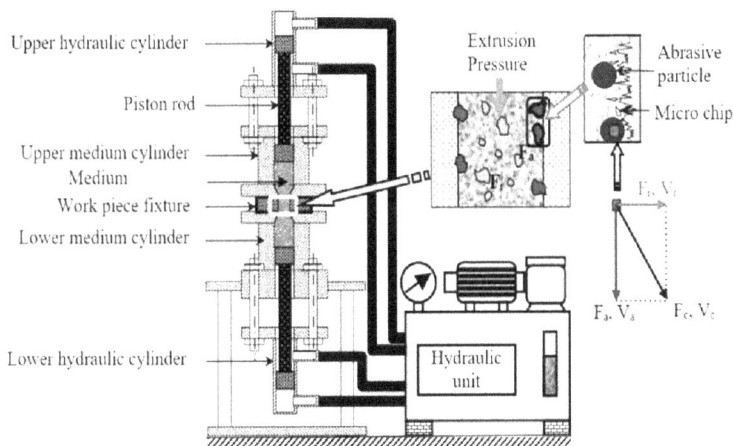

Figure 5.18 Schematic of two-way AFM [32].

medium is allowed to flow between the gap of the die and the workpiece surface. The pressure on the die is used to apply a normal force to the medium. The part having shallow cavities, such as in dies or coins, is generally best finished using orbital AFM.

Two-way AFM, popularly termed AFM only, is shown in Figure 5.18. It has two pistons and a cylinder arrangement and a constriction passage in between the pistons where the part is placed. In this way, the abrasive medium extrudes to and fro in the constriction between the two pistons. Two-way AFM is generally preferred for the finishing of cylindrical surfaces, both internal and external. The complex shape of the through-hole or cavities can also be finished using this process [33].

5.5.1 Mechanism of Material Removal in AFM

In AFM, the abrasion phenomenon induced by the abrasive particles onto the surface of the workpiece is responsible for material removal. The forces generated during finishing are low, thus the finished surface is free from residual stress due to the interaction of the abrasive. As shown in Figure 5.19c the radial force F_r acting on the abrasive particles causes indentation on the surface of the workpiece. The axial force F_a provides a relative motion to the abrasive particles in the axial direction and the material in front of the indented portion of the abrasive is removed in the form of μ-chips as shown in Figure 5.19a.

In AFM, material removal takes place following any one of three conditions [33]. Figure 5.19 shows these three conditions that can occur in material removal. In Figure 5.19a, h is the indentation depth of an abrasive particle, $h' > h$ is the increased indention depth on account of the increased extrusion pressure (Figure 5.19b), whereas $h'' < h'$ is the reduced indentation depth due to the rotation of the abrasive particles as shown in Figure 5.19c. The force F_{req} required for removal of the work material can be calculated as:

$$F_{req} = \tau_s A_p \tag{5.3}$$

where τ_s is the shear strength of work material and A_p is the area of projection for the indented portion of the abrasive particle. If $F_{req} < F_a$, then the removal of material takes place in the form of μ-chips. Consider Figure 5.19b, if

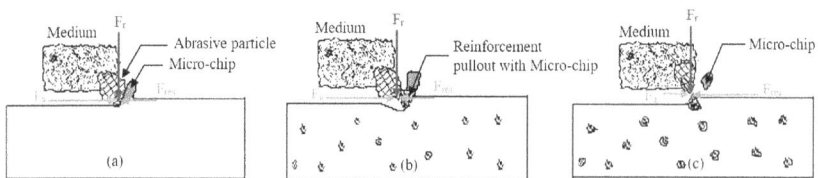

Figure 5.19 Material removal mechanism in AFM [32].

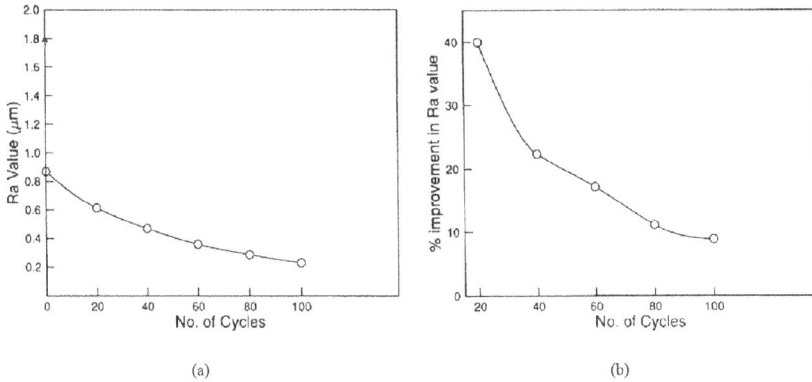

Figure 5.20 Variation of (a) surface roughness and (b) percentage improvement in surface roughness with the number of cycles during finishing [34].

$F'_{req} > F_a$, then there occurs no removal of material and only penetration of the abrasive takes place due to a normal force. The abrasive particle rotates in order to lower the depth of indentation until the condition given in Figure 5.19a is satisfied, thereafter removal of material takes place.

5.5.2 Process Parameters of AFM

The process performance of AFM is evaluated by the percentage change in surface roughness ($\%\Delta R_a$) during finishing. In AFM, one to-and-fro motion is referred to as a cycle. Thus, as the number of cycles increases there occurs a decrease in the surface roughness value or an improved finish is obtained Figure 5.20a. The $\%\Delta R_a$ at any instant is calculated by:

$$\%\Delta R_a = \frac{\text{Initial } R_a \text{ value} - \text{Final } R_a \text{ value}}{\text{Initial } R_a \text{ value}} \times 100 \tag{5.4}$$

As the number of cycles increases, hence the finishing time, so the $\%\Delta R_a$ value decreases. This implies that, initially, the improvement is greater but that later it decreased with the number of cycles, as shown in Figure 5.20b. The size of the abrasive also determines the process performance; bigger sized abrasive particles show more improvement in the roughness value as compared to smaller sized particles (Figure 5.21a). As the concentration of the abrasive particles increases there occurs an increase in the change of the roughness value for a similar set of parameters; see Figure 5.21b.

5.5.3 Developments and Application of AFM

AFM offers several advantages over other finishing processes but it has some limitations such as a low finishing rate. The finishing capabilities

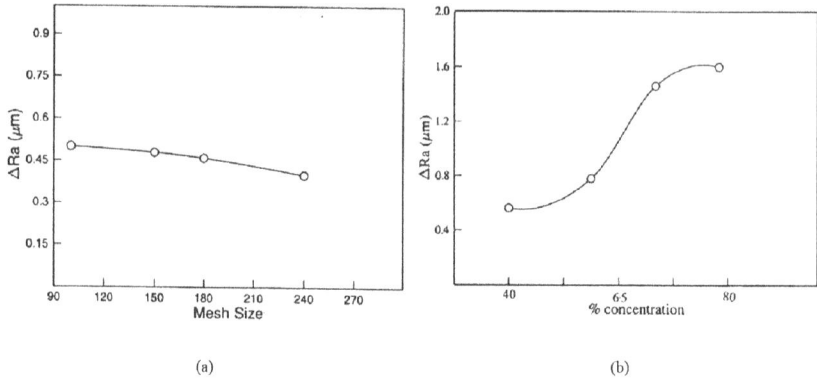

(a)   (b)

Figure 5.21 Change of surface roughness with (a) mesh size and (b) percentage concentration of abrasive particles during finishing [34].

with a medium decays with the number of finishing cycles, which further decreases the finishing rate. Thus, many attempts have been made by researchers to improve the rate of finishing. The development of variants like electrochemical-aided AFM [35] utilize polymeric electrolytes such as water gels or a gelated polymer carrier medium. Rotational AFM process [36] have been developed in which the workpiece fixture is rotated by an external means. The rotation of the workpiece increases the contact of the abrasive with the workpiece surface and a helical path is followed by the abrasive particles. Magneto abrasive flow finishing (MAFM) [30] employs a magnetic carrier and applies a magnetic field by an electromagnet placed around the workpiece. The AFM process is applied to the operation for the deburring, polishing and removal of a recast layer [34]. Xu et al. [37] applied AFM to finish the helical gear shown in Figure 5.22.

The photograph in Figure 5.23 illustrates a medical implant finished by AFM; an observation of complex part features provides a notion of process capabilities and versatility.

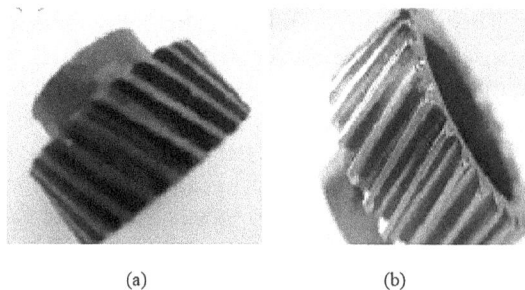

(a)   (b)

Figure 5.22 Helical gear finishing by AFM, (a) before and (b) after [38].

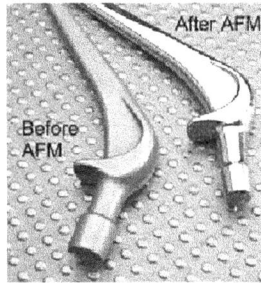

Figure 5.23 Photograph of a medical implant (hip joint) finished by AFM [38].

5.5.4 Future Scope of AFM

In AFM, tool design is a major problem for the adoption of new applications as it requires that the tooling be redesigned. Generally, inferior tool design leads to the problem of leakage, loss of finishing time, as well as resources. Efforts are being made to develop a reconfigurable fixturing arrangement that can be applied to varying surface configurations instead of a customized fixture. Effort is also needed to produce a cost-effective abrasive medium, as the existing one is quite expensive.

5.6 CONCLUSION

This chapter has illustrated mechanically based non-conventional machining processes, namely AJM, AWJM, MAF and AFM. These processes use mechanical means to remove the material by utilizing abrasive particles as the main cutting tool. The following conclusions may be made.

- AJM as well as AWJM can be applied to cut, drill and machine metal and non-metals. The force produced by AWJM is very high compared to AJM. AWJM can also be applied to cut very hard material, like stones.
- The geometrical accuracy of the machined surface obtained by AJM and AWJM is a concern, so care should be taken while selecting the process parameters. The multi-pass operation can be a solution to make deeper slots.
- MAF and AFM are nano-finishing techniques that use flexible abrasive particles that can conform to a complex-shaped work surface.
- During MAF, the finishing forces are mainly regulated by varying the current or supply voltage followed by a working gap. This can enable the finishing of magnetic as well as non-magnetic materials.

- In AFM, a pressurized flexible abrasive medium between the pistons is used as a flexible abrasive tool. A complex surface like gears can be finished in a single set-up of tooling using AFM. In AFM, a special fixture is required for every component for finishing; also the design of the fixture for varying components is a major challenge.

Though there are many advantages to the presented processes, there is still further possibility for research work on mechanically based non-conventional machining processes.

REFERENCES

[1] Ramachandran N, Ramakrishnan N. A review of abrasive jet machining. *J Mater Process Tech* 1993; 39: 21–31.
[2] Jain NK, Jain VK. Modeling of material removal in mechanical type advanced machining processes: A state-of-art review. *Int J Mach Tools Manuf* 2001; 41: 1573–1635.
[3] Melentiev R, Fang F. Recent advances and challenges of abrasive jet machining. *CIRP J Manuf Sci Technol* 2018; 22: 1–20.
[4] Marshall DB, Lawn BR, Evans AG. Elastic/plastic indentation damage in ceramics: The lateral crack system. *J Am Ceram Soc* 1982; 65: 561–566.
[5] Buijs M. Erosion of glass as modeled by indentation theory. *J Am Ceram Soc* 1994; 77: 1676–1678.
[6] Finnie I. Erosion of surfaces by solid particles. *Wear* 1960; 3: 87–103.
[7] Verma AP, Lal GK. An experimental study of abrasive jet machining. *Int J Mach Tool Des Res* 1984; 24: 19–29.
[8] Haldar B, Ghara T, Ansari R, et al. Abrasive jet system and its various applications in abrasive jet machining, erosion testing, shot-peening, and fast cleaning. *Mater Today Proc* 2018; 5: 13061–13068.
[9] Belloy E, Thurre S, Walckiers E, et al. The introduction of powder blasting for sensor and microsystem applications. *Sensors Actuators A Phys* 2000; 84: 330–337.
[10] Costa HL, Hutchings IM. Some innovative surface texturing techniques for tribological purposes. *Proc Inst Mech Eng Part J J Eng Tribol* 2015; 229: 429–448.
[11] Slikkerveer PJ, Bouten PCP, De Haas FCM. High quality mechanical etching of brittle materials by powder blasting. *Sensors Actuators, A Phys* 2000; 85: 296–303.
[12] Belloy E, Walckiers E, Sayah A, et al. Introduction of powder blasting for sensor and microsystem applications. *Sensors Actuators, A Phys* 2000; 84: 330–337.
[13] Natarajan Y, Murugesan PK, Mohan M, et al. Abrasive water jet machining process: A state of art of review. *J Manuf Process* 2020; 49: 271–322.
[14] Hashish M. A modeling study of metal cutting with abrasive waterjets. *J Eng Mater Technol Trans ASME* 1984; 106: 88–100.
[15] Zagórski I, Kłonica M, Kulisz M, et al. Effect of the AWJM method on the machined surface layer of AZ91D magnesium alloy and simulation of roughness parameters using neural networks. *Materials (Basel)* 2018; 11: 2111.

[16] Wang J, Guo DM. The cutting performance in multipass abrasive water-jet machining of industrial ceramics. *J Mater Process Technol* 2003; 133: 371–377.

[17] Frotscher M, Kahleyss F, Simon T, et al. Achieving small structures in thin NiTi sheets for medical applications with water jet and micro machining: A comparison. *J Mater Eng Perform* 2011; 20: 776–782.

[18] Folkes J. Waterjet-an innovative tool for manufacturing. *J Mater Process Technol* 2009; 209: 6181–6189.

[19] Tan KL, Yeo SH, Ong CH. Nontraditional finishing processes for internal surfaces and passages: A review. *Proc Inst Mech Eng Part B J Eng Manuf* 2016; 231: 2302–2316.

[20] Chang G-W, Yan B-H, Hsu R-T. Study on cylindrical magnetic abrasive finishing using unbonded magnetic abrasives. *Int J Mach Tools Manuf* 2002; 42: 575–583.

[21] Jain VK. Magnetic field assisted abrasive based micro-/nano-finishing. *J Mater Process Technol* 2009; 209: 6022–6038.

[22] Misra A, Pandey PM, Dixit USS, et al. Modeling of material removal in ultrasonic assisted magnetic abrasive finishing process. *Int J Mech Sci* 2017; 131–132: 853–867.

[23] Misra A, Pandey PM, Dixit US, et al. Modeling of finishing force and torque in ultrasonic-assisted magnetic abrasive finishing process. *Proc Inst Mech Eng Part B J Eng Manuf* 2019; 233: 411–425.

[24] Mulik RS, Pandey PM. Experimental investigations and optimization of ultrasonic assisted magnetic abrasive finishing process. *Proc Inst Mech Eng Part B J Eng Manuf* 2011; 225: 1347–1362.

[25] Singh DK, Jain VK, Raghuram V. Parametric study of magnetic abrasive finishing process. *J Mater Process Technol* 2004; 149: 22–29.

[26] Yin S, Shinmura T. Vertical vibration-assisted magnetic abrasive finishing and deburring for magnesium alloy. *Int J Mach Tools Manuf* 2004; 44: 1297–1303.

[27] Misra A, Pandey PM, Dixit US, et al. Multi-objective optimization of ultrasonic-assisted magnetic abrasive finishing process. *Int J Adv Manuf Technol* 2019; 101: 1661–1670.

[28] Xing B, Zou Y. Investigation of finishing aluminum alloy A5052 using the magnetic abrasive finishing combined with electrolytic process. *Machines* 2020; 8: 1–14.

[29] Yamaguchi H, Shinmura T. Internal finishing process for alumina ceramic components by a magnetic field assisted finishing process. *Precis Eng* 2004; 28: 135–142.

[30] Yan BH, Lin YC, Huang FY. Development of magneto abrasive flow machining process. *Int J Mach Tools Manuf* 2002; 42: 953–959.

[31] Goyal A, Singh H, Goyal R, et al. Recent advancements in abrasive flow machining and abrasive materials: A review. *Mater Today Proc*. Epub ahead of print December 2021. doi:10.1016/j.matpr.2021.12.109.

[32] Basha SM, Basha MM, Venkaiah N, et al. A review on abrasive flow finishing of metal matrix composites. *Mater Today Proc* 2021; 44: 579–586.

[33] Jain VK, Sidpara A, Sankar MR, et al. Nano-finishing techniques: A review. *Proc Inst Mech Eng Part C J Mech Eng Sci* 2012; 226: 327–346.

[34] Jain VK, Adsul SG. Experimental investigations into abrasive flow machining (AFM). *Int J Mach Tools Manuf* 2000; 40: 1003–1021.

108 Additive and Subtractive Manufacturing Processes

[35] Dabrowski L, Marciniak M, Szewczyk T. Analysis of abrasive flow machining with an electrochemical process aid. *Proc Inst Mech Eng Part B J Eng Manuf* 2006; 220: 397–403.

[36] Sankar MR, Jain VK, Ramkumar J. Experimental investigations into rotating workpiece abrasive flow finishing. *Wear* 2009; 267: 43–51.

[37] Xu Y, Zhang K, Lu S, et al. Experimental investigations into abrasive flow machining of helical gear. *Key Eng Mater* 2013; 546: 65–69.

[38] Kumari C, Chak SK. Study on influential parameters of hybrid AFM processes: A review. *Manuf Rev* 2019; 6: 23.

Chapter 6

Thermal-Energy-Based Advanced Manufacturing Processes

Hardik Beravala
Birla Vishvakarma Mahavidhyalaya, Vallabhvidyanagar, India

Nishant K. Singh
Hindustan College of Science and Technology, Mathura, India

CONTENTS

6.1 INTRODUCTION

The development of advanced materials requires the evolution of techno-logical enhancements within the existing machining and so offers the production of quality products at the least cost. These materials are mainly high-temperature alloys, ceramics and composites having superior strength, hardness and wear-resistance properties. They are extensively used in the manufacturing of parts for the tooling, aerospace, automobile and medical industries. However, the machining of such materials through conventional machining processes is the biggest challenge. Thermal energy-based advanced machining processes have the potential to shape tough-to-cut materials. In these processes, the source of heat is focused on a localized region of the workpiece which raises the temperature. This causes the heating, melting and evaporation of the workpiece in that region. The fine control of the process parameters achieves the geometrical and dimensional requirements and a high surface finish in the machined parts. These processes remove material independent of the hardness and strength of the workpiece. The classification of thermal energy-based advanced manufacturing processes, according to the source of energy, is shown in Table 6.1.

DOI: 10.1201/9781003327394-6

Table 6.1 Thermal Energy-Based Advanced Manufacturing Processes

Name of Process	Source of Thermal Energy
Plasma arc machining	Plasma
Laser beam machining	Laser beam
Electron beam machining	Electron beam
Electrical discharge machining	Electric spark
Ion beam machining	Ion beam

Despite all the advantages, these processes remove the materials at a very slow rate compared to the conventional way of machining [1]. The extreme localized heating harms surface integrity and creates a heat affected zone (HAZ) and micro-cracks. It is very important to identify suitable process variables to achieve a higher material removal rate (MRR) and surface finish and low HAZ, micro-cracks, and so on. The MRR is the ratio of the weight of material loss from a workpiece to the total machining time. In order to enhance performance and eliminate adverse effects, over the last two decades, the advancement of processes has been made by the hybridization of these processes. These integrated processes are classified into combined and assisted machining processes [2].

Electrical discharge machining was first introduced by Boris Romanovich Lazarenko in 1944. The advancement of the pulse generator and computer numerical control in electrical discharge machining (EDM) tremendously improved the efficiency of machining operations [3]. However, several drawbacks of this process—in terms of low machining rate, defects in machined geometry due to electrode wear, the development of micro-cracks and HAZ on the machined surface, and the environmental issues in the disposal of a dielectric—reduce its scope in the manufacturing sectors [4]. To enhance the process efficacy, several modifications have been made in the EDM. The air/gas assisted EDM and magnetic field assisted EDM have been developed as the EDM assisted processes over the last decade. The present chapter discusses the effects of air/gas assistance and a magnetic field on surface integrity, the electrode wear rate (EWR) and MRR. The present chapter also discusses the effect of the combined assistance of air/gas and a magnetic field on MRR and surface integrity.

6.2 AIR/GAS ASSISTED EDM

The thermal properties of a dielectric performs a vital role in the melting and cooling of a machined surface during EDM [5]. This affects the machined surface and sub-surface quality [6]. Table 6.2 shows the properties of

Table 6.2 Properties of Various Dielectric Fluids [5]

	Liquid Dielectric Fluid		Gaseous Dielectric Fluid			
Property	EDM Oil	De-ionized Water	Air	N_2	O_2	Ar
Heat capacity (J/gK)	2.16	4.19	1.04	1.04	0.92	0.52
Thermal conductivity (W/mK)	0.149	0.606	0.026	0.025	0.026	0.016
Dielectric strength (MV/m)	14–22	13.00	3.00	1.00	0.92	0.18

different liquid and gaseous dielectric fluids. Liquid dielectric fluid (EDM oil) has a higher heat capacity and thermal conductivity than gaseous fluids. More discharge energy is required for a liquid dielectric due to its high breakdown strength compared to a gaseous dielectric. The tailoring of dielectric properties helps to achieve machining goals, inclusive of a high MRR and integrity, and low roughness. This may be executed with the aid of a liquid-gaseous mixed dielectric fluid as an alternative to liquid fluid, a process which is called air/gas assisted EDM. Figure 6.1 shows the setup of the air/gas assisted EDM process.

In the air/gas assisted EDM process, the liquid-gaseous dielectric is supplied to the machining gap with the help of a tubular or multi-hole electrode. Kunieda et al. [7] studied the performance of the liquid-gaseous fluid

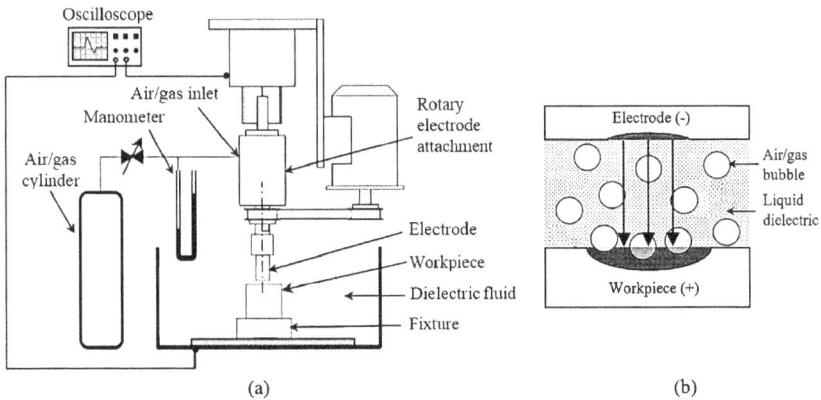

Figure 6.1 (a) Air/gas assisted EDM process setup; (b) presence of air/gas bubbles and liquid dielectric fluid in the machining gap.

in EDM. A mixture of oxygen and water was supplied to the machining gap through a tubular graphite electrode to machine S45C steel work material. It was reported that the exothermic reactions between the oxygen and the work material produced heat that was 5.6% of the energy supplied by an electric pulse. This resulted in a higher MRR using a liquid-gas mixed dielectric compared to conventional EDM. Kunieda et al. [8] machined 3D shaped specimens using dry EDM. Air and oxygen was supplied through a thin wall pipe electrode during machining. A high MRR was found using oxygen gas compared to EDM using an air and liquid dielectric. The electrode wear ratio was observed to be near zero while using air, and a very high electrode wear ratio was observed while using a liquid dielectric. Zhang et al. [9] examined the effect of ultrasonic vibration on dry EDM. Air was delivered through the tubular electrode in the machining area. It was found that the ultrasonic vibration and air improved the MRR as compared to a liquid dielectric. The high MRR was obtained at the higher amplitude of vibration, pulse-on time, current and gap voltage. In addition, a drastic reduction in the reattachment of debris to the surface was noticed. The influence of a mixed-phase dielectric fluid in electrical discharge drilling was examined by Kao et al. [10]. Experimentation was conducted using air, water–air mixture (mist) and water, which were used as gaseous, liquid-gaseous and liquid phase dielectric fluids respectively. The water–air mixed dielectric improved the MRR and reduced debris deposition on the workpiece as compared to air. The water–air mixed dielectric also provided a higher MRR in EDM compared to water. However, high electrode wear in the electrical discharge drilling was observed while using a water–air dielectric over a water dielectric.

Tao et al. [5] examined the performance of electrical discharge milling under dry and near-dry dielectric conditions using gas and gas–liquid mixed dielectric fluids respectively. Air, oxygen, helium and nitrogen were mixed with water in EDM to produce the near-dry condition of a dielectric. A high surface finish was noticed when helium and nitrogen gases were used in place of air and oxygen gases. Air and oxygen gas increased the MRR in near-dry conditions. However, heavy debris adhered to the electrode and workpiece while using oxygen gas. Fujiki et al. [11] addressed the problem of debris attachment to the surface machined in a dry EDM. To overcome this issue, a mist of air–kerosene as a dielectric was used in EDM. This reduced debris deposition and arcing in the gap significantly. Moreover, the MRR increased and the electrode wear ratio decreased with an increase in the dielectric flow rate. Puthumana and Joshi [12] observed clogging in machined grooves by the debris particles in dry EDM. The slots were cut at the periphery of the electrode to facilitate debris ejection from the gap. This caused a reduction in debris deposition and improved MRR and reduced EWR.

Teimouri and Baseri found that debris expulsion from the gap in dry EDM improved by the multi-hole electrode [13]. An increase in the number of holes improved the energy density of the spark which increased MRR and EWR [14]. Furthermore, the pulse current, electrode rpm and pulse duration were found to be the most effective factors on MRR, EWR and surface roughness. In dry EDM, Liqing and Yingjie [15] investigated the effect of different gases and cryogenic cooling on the workpiece. Air, nitrogen and argon gases were added to the oxygen and supplied by the tubular electrode during machining. It was reported that the MRR improved by more than 200% when nitrogen and argon gases were mixed with oxygen. In addition, the MRR increased by 234–281% when the oxygen gas was mixed with the air. The cryogenic cooling also improved MRR and surface roughness in contrast to the workpiece without cooling.

The effect of different dielectric gases on process performance and spark stability in dry EDM was studied by Roth et al. [16]. Air, carbon dioxide, oxygen and nitrogen gases were supplied at a pressure of 20 bar during EDM. The oxygen gas and air formed electrically non-conductive oxides during machining and suppressed the arcing phenomenon effectively. Furthermore, the MRR was found to be highest while using oxygen gas and to be lowest while using nitrogen gas in dry EDM. Shen et al. used air and liquid dielectrics in electrical discharge milling on Inconel 718 [6]. Fewer micro-cracks and large-size shallow craters were observed on the surface when air was used as a dielectric. This was due to the reduction of the cooling rate of the metal in the discharge crater by the air. In addition, the air formed oxides, mainly Fe_2O_3 and Cr_2O_3, in the recast layer. Bozdana and Ulutas [17] used a multi-tube electrode in dry EDM. This effectively conveyed most debris far away and raised the MRR and surface finish. Singh et al. [18] examined the effectiveness of a liquid-air dielectric in EDM during the machining of an HCHCr die steel material. The process was termed air assisted EDM (AEDM). Low pressure air was supplied to the inter-electrode gap through a perforated electrode during machining. A high MRR, low electrode wear ratio and a reduced number of surface cracks were reported while using a liquid-air mixed dielectric in place of a liquid dielectric. Bai et al. [19] experimented using a three-phase dielectric medium which was a mixture of air, liquid and powder. It was concluded that the powder concentration in the dielectric influenced the MRR. Furthermore, the high powder concentration caused an abnormal discharge of the spark which reduced the MRR. The MRR, surface roughness and tool wear rate (TWR) were improved by using a three-phase dielectric with respect to a conventional dielectric in EDM.

6.3 MAGNETIC FIELD ASSISTED EDM

In the last decade, the magnetic field was introduced to EDM to enhance its efficacy. The process was termed magnetic field assisted EDM (see to Figure 6.2). The introduction of a cross-magnetic field into the machining

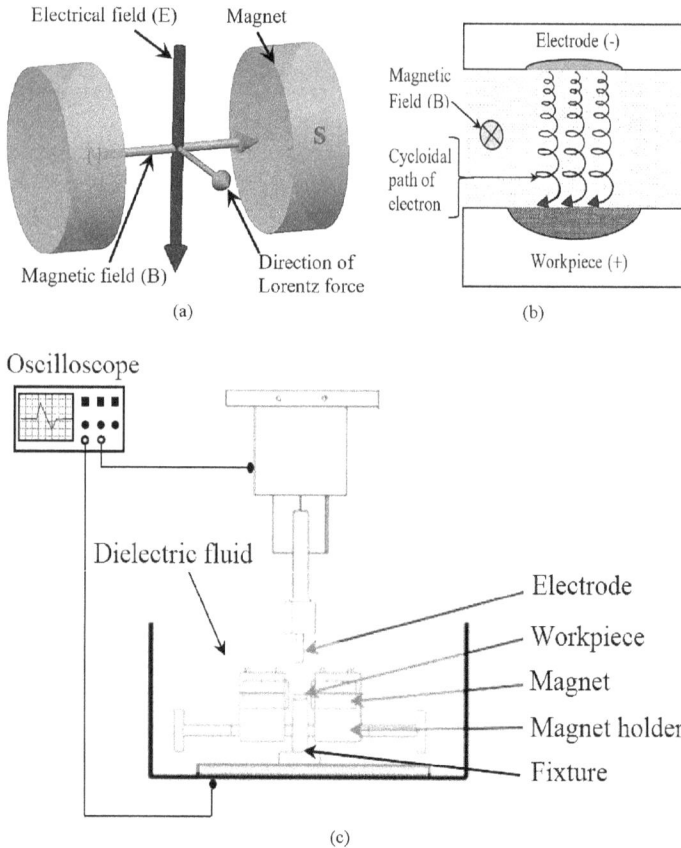

Figure 6.2 Schematic diagram showing (a) the development of Lorentz force, (b) the motion of the electrons in the cycloidal path and (c) the setup of magnetic field assisted EDM.

gap increases ionizing collisions in EDM. This is because of the alteration of the electron's path from a straight line to a cycloidal one under the effect of the cross-magnetic field [20], which resulted in an increase in the temperature of the spark. A similar observation was made by Heinz et al. [21] when the magnetic field was applied to micro-EDM. The rise in the spark temperature increased material removal by 50%. Furthermore, the magnetic field developed Lorentz force which improved the debris expulsion around the discharge crater and, hence, improved the surface morphology. Jafferson et al. [22] noticed that the Lorentz force produced by the cross-magnetic field efficiently cleaned debris from the gap.

During deep hole drilling in the magnetic-field-assisted μ-EDM, the effective ejection of debris was noticed by Yeo et al. [23]. Moreover, the hole depth increased by 26% and high EWR was found in the process compared to the drilling of a hole in the absence of a magnetic field. Chattopadhyay et al. [24] reported a rise in the spark temperature while machining EN8 steel, assisted by a magnetic field. It also effectively flushed away debris, reduced short-circuits and improved MRR. The application of the rotating magnetic field improved MRR by about three times and reduced the surface cracks and the thickness of the recast layer [25]. Lin et al. [26] noticed that the magnetic field improved the MRR and the finishing in EDM. This is because of a substantial reduction in debris attached to the surface by the magnetic field.

Joshi et al. [27] performed dry EDM on a machined stainless steel material. The magnetic field was applied in the form of a pulse during machining. An improvement in the MRR by 130% and a reduction in work material deposition to the electrode by 20% was recorded when the process was assisted by a magnetic field. Teimouri and Baseri [28] used a magnetic field in rotary EDM to examine its consequence on the EWR and overcut. It was reported that the magnetic field and the electrode's rotational speed influenced the EWR and overcut significantly. The electrode's rotation and magnetic field jointly enhanced debris expulsion compared to using a stationary electrode. The same researchers noticed in another experiment that debris flushing enhanced the MRR and surface finish using a multi-hole electrode in dry EDM [13]. It was found that the flushing of the debris improved when more holes were present in the electrode.

Bhattacharya et al. [29] evaluated the influence on the material transfer mechanism due to the magnetic field during the EDM process. It was observed that the assistance of a magnetic field formed large-size deep craters. Furthermore, the magnetic field improved the expulsion of work material and restricted its deposition to the electrode. Bhatt et al. [30] introduced a magnetic field in a powder-mixed EDM to enhance process performance. The results showed that the MRR, overcut and micro-hardness were improved by the magnetic field. It also effectively removed the debris before it redeposited on a surface. Bains et al. [31] introduced a magnetic field in EDM to cut an SiCp/A359 metal matrix composite. The magnetic-field-improved spark energy resulted in an improvement in MRR by 12.5%. In addition, it also reduced debris attachment to the work surface which in turn reduced the thickness of the resolidifying layer. Zhang et al. [32] cut a Ti6Al4V material using wire-cut EDM. During cutting, they provided the assistance of ultrasonic vibration and a magnetic field. It was reported that the magnetic field developed Lorentz force which flushed the non-magnetic debris and improved the discharge condition, though the magnetic field and ultrasonic vibration together did not show any effect on the performance of wire-cut electrical discharge machining (WEDM).

Govindan et al. [33] derived a mathematical model to estimate the crater size formed by mono-spark discharge in magnetic-field-assisted dry EDM. The model included plasma confinement and reduction in the mean electron free path due to the magnetic field. The model could predict crater size within a 5% error. Gupta and Joshi [34] developed an analytical model to calculate the crater size in dry EDM when the machining was performed under a pulsating magnetic field. The magnetic flux density of 0.1–0.3 T was produced by the electromagnets during experimentation. The model was successfully validated and found to have an error of 9–10% in the predicted crater size with respect to experimental values.

6.4 MAGNETIC FIELD AND AIR/GAS ASSISTED EDM

The previous discussion illustrates that the use of air/gas improves the efficacy of EDM. Furthermore, a magnetic field enhances MRR and surface integrity. Therefore, to obtain the advantages of both processes, a hybrid assisted process has recently been developed, which is termed the magnetic field and air/gas assisted EDM process [35]. In this process, the assistance of a magnetic field and a liquid-gaseous dual-phase dielectric fluid is provided during machining. There are two variants of this process that have been developed. In the first variant, the liquid-air mixture is used as a dielectric in the magnetic field assisted process. The process is termed magnetic field air assisted EDM (MAEDM). In the second variant, the mixture of liquid and argon gas is used as a dielectric. The process is termed magnetic field gas assisted EDM (MGEDM).

To evaluate the performance of the MAEDM and MGEDM processes, a rotating multi-hole electrode was used in the experiment and compared to AEDM and liquid dielectric based EDM processes [35]. AISI 304 was selected as the work material. The air/argon gas pressure and magnetic flux density were changed during the experiment. It was noticed that air in the AEDM process enhanced the MRR by 50%. This was because of exothermic reactions between the air and molten metal, which developed surplus heat that contributed to melting and vaporization during the process. The MRR was further improved by 34% due to the assistance of the magnetic field in MAEDM. An improvement in ionization by the magnetic field raised the temperature of the plasma. The MRR was found to be low in MGEDM compared to MAEDM. The magnetic field attracted debris so that it was evacuated from the gap before being deposited on the finished surface. This mechanism enhanced the surface finish by 18% in MAEDM and 32% in MGEDM [36]. Argon gas and air provided cooling to the electrode and hence reduced EWR. As compared to AEDM, the magnetic field raised the temperature of the electrode and increased EWR by 39% in MAEDM.

The magnetic field improved the surface morphological characteristics in terms of suppressing the formation of micro-pores. However, the magnetic flux density increased the width of micro-cracks and produced a thicker HAZ and redeposit layer. The dielectric of mixed liquid-air and liquid-argon gas reduced crack width by 55%, and 72% in the case of a liquid dielectric. Air and argon gas with a liquid dielectric increased the recast layer thickness by 83% and 43% respectively, due to the slow heat dissipation rate, compared to a liquid dielectric [36]. The thickness of the recast layer and HAZ increased with an increase in air/gas pressure and formed thin micro-cracks in MAEDM and MGEDM. The oxides were detected in an XRD analysis of specimens machined by AEDM and MGEDM processes. This resulted from exothermic reactions by air to the metal that formed the oxides.

The debris forms in MAEDM and MGEDM were collected and examined [37]. It was found that the magnetic saturation in debris was found to be lower in MAEDM than in MGEDM. This was because of the formation of oxides, mainly FeO, Fe_2O_3, Fe_3O_4 and CrO also found in debris. Though the work material was paramagnetic, the presence of ferromagnetic phases caused debris to be attracted by magnets. Therefore a reduction in frequent arcing facilitated the uniform and stable discharging of sparks in both MAEDM and MGEDM.

A mathematic model has been developed to determine the MRR in the MAEDM process [38]. The physical phenomena—mainly the reduction in the energy density of the spark due to expansion, the assistance of a liquid-air dielectric and the reduction in the electron mean free path due to the magnetic field—have been incorporated in the model:

$$MRR_{MAEDM} = M\left[K_1 \frac{V_g I^{K_2} t_{on}^{K_3} \rho_w}{\pi l_a} \left(\frac{1}{t_{on} + t_{off}} \right) + K_4 A_1 \sqrt{\frac{2P\rho_a}{\left(\dfrac{A_1}{A_2}\right)^2 - 1}} \right] \quad (6.1)$$

where $K_1 = 7.6 \times 10^{-6}$, $K_2 = 2.8066$, $K_3 = 0.8821$, $K_4 = 0.1020$, M is the magnetic flux density (T), V_g is voltage in the inter-electrode gap (V), t_{on} is the pulse-on time (µs), t_{off} is the pulse-off time (µs), A_1 is the cross-sectional area of the hole in the electrode inlet, A_2 is the cross-sectional area of the hole in the electrode outlet, P is air pressure (mm of Hg) and ρ_a is the density of air (kg/m³). This model is also applicable to conventional EDM and AEDM processes.

6.5 CONCLUSIONS

Thermal-energy-based advanced manufacturing processes have the potential to remove material from difficult-to-machine advanced materials in a controlled manner. Hybridization has always improved the performance of advanced manufacturing processes. MAEDM and MGEDM are the most recent advancement in the EDM. The magnetic field and liquid-gaseous mixed dielectric fluid together improved the performance of EDM. The magnetic field enhances MRR and the surface finish. The assistance of argon gas reduces EWR. Controlling the magnetic flux density and air/gas pressure in a liquid dielectric improves MRR, surface morphology and reduces surface roughness and EWR. The developed model also helps to predict the MRR when the assistance of air and a magnetic field is applied to EDM with reasonable accuracy.

REFERENCES

[1] N. Mohd Abbas, D. G. Solomon, and M. Fuad Bahari, "A review on current research trends in electrical discharge machining (EDM)," *Int. J. Mach. Tools Manuf.*, vol. 47, no. 7–8, pp. 1214–1228, 2007.

[2] B. Lauwers, F. Klocke, A. Klink, A. E. Tekkaya, R. Neugebauer, and D. McIntosh, "Hybrid processes in manufacturing," *CIRP Ann. - Manuf. Technol.*, vol. 63, no. 2, pp. 561–583, 2014.

[3] H. A. El-Hofy, *Fundamentals of machining processes: conventional and non-convetional processes*, 2nd ed. CRC Press, 2014.

[4] V. K. Jain, *Advanced machining processes*. Allied Publishers, 2007.

[5] J. Tao, A. J. Shih, and J. Ni, "Experimental study of the dry and near-dry electrical discharge milling processes," *J. Manuf. Sci. Eng.*, vol. 130, no. 1, pp. 011002-1–011002-9, 2008.

[6] Y. Shen, Y. Liu, H. Dong et al. Surface integrity of Inconel 718 in high-speed electrical discharge machining milling using air dielectric. *Int. J. Adv. Manuf. Technol.*, 90, pp. 691–698, 2017. https://doi.org/10.1007/s00170-016-9332-7

[7] M. Kunieda, S. Furuoya, and N. Taniguchi, "Improvement of EDM efficiency by supplying oxygen gas into gap," *CIRP Ann. - Manuf. Technol.*, vol. 40, no. 1, pp. 215–218, 1991.

[8] M. Kunieda, M. Yoshida, and N. Taniguchi, "Electrical discharge machining in gas," *CIRP Ann. - Manuf. Technol.*, vol. 46, no. 1, pp. 143–146, 1997.

[9] Q. H. Zhang, J. H. Zhang, S. F. Ren, J. X. Deng, and X. Ai, "Study on technology of ultrasonic vibration aided electrical discharge machining in gas," *J. Mater. Process. Technol.*, vol. 149, no. 1–3, pp. 640–644, 2004.

[10] C. C. Kao, J. Tao, and A. J. Shih, "Near dry electrical discharge machining," *Int. J. Mach. Tools Manuf.*, vol. 47, no. 15, pp. 2273–2281, 2007.

[11] M. Fujiki, J. Ni, and A. J. Shih, "Investigation of the effects of electrode orientation and fluid flow rate in near-dry EDM milling," *Int. J. Mach. Tools Manuf.*, vol. 49, no. 10, pp. 749–758, 2009.

[12] G. Puthumana and S. S. Joshi, "Investigations into performance of dry EDM using slotted electrodes," *Int. J. Precis. Eng. Manuf.*, vol. 12, no. 6, pp. 957–963, 2011.

[13] R. Teimouri and H. Baseri, "Experimental study of rotary magnetic field-assisted dry EDM with ultrasonic vibration of workpiece," *Int. J. Adv. Manuf. Technol.*, vol. 67, no. 5–8, pp. 1371–1384, 2013.

[14] E. Aliakbari and H. Baseri, "Optimization of machining parameters in rotary EDM process by using the Taguchi method," *Int. J. Adv. Manuf. Technol.*, vol. 62, no. 9–12, pp. 1041–1053, 2012.

[15] L. Liqing and S. Yingjie, "Study of dry EDM with oxygen-mixed and cryogenic cooling approaches," *Procedia CIRP*, vol. 6, pp. 344–350, 2013.

[16] R. Roth, F. Kuster, and K. Wegener, "Influence of oxidizing gas on the stability of dry electrical discharge machining process," *Procedia CIRP*, vol. 6, pp. 338–343, 2013.

[17] A. T. Bozdana and T. Ulutas, "The effectiveness of multichannel electrodes on drilling blind holes on Inconel 718 by EDM process," *Mater. Manuf. Process.*, vol. 31, no. 4, pp. 504–513, 2016.

[18] N. K. Singh, P. M. Pandey, and K. K. Singh, "EDM with air assisted multi-hole rotating tool," *Mater. Manuf. Process.*, vol. 31, no. 14, pp. 1872–1878, 2016.

[19] X. Bai, T. Yang, and Q. Zhang, "Experimental study on the electrical discharge machining with three-phase flow dielectric medium," *Int. J. Adv. Manuf. Technol.*, 96, 2003–2011, 2018.

[20] F. F. Chen, *Introduction to plasma physics and controlled fusion*, vol. 1, Plasma. Plenum press, 1984.

[21] K. Heinz, S. G. Kapoor, R. E. DeVor, and V. Surla, "An investigation of magnetic-field-assisted material removal in micro-EDM for nonmagnetic materials," *J. Manuf. Sci. Eng.*, vol. 133, no. April 2011, pp. 021002:1–9, 2011.

[22] J. M. Jafferson, P. Hariharan, and J. R. Kumar, "Effects of ultrasonic vibration and magnetic field in micro-EDM milling of nonmagnetic material," *Mater. Manuf. Process.*, vol. 29, pp. 357–363, 2014.

[23] S. H. Yeo, M. Murali, and H. T. Cheah, "Magnetic field assisted micro electro-discharge machining," *J. Micromechanics Microengineering*, vol. 14, pp. 1526–1529, 2004.

[24] K. D. Chattopadhyay, P. S. Satsangi, S. Verma, and P. C. Sharma, "Analysis of rotary electrical discharge machining characteristics in reversal magnetic field for copper-en8 steel system," *Int. J. Adv. Manuf. Technol.*, vol. 38, no. 9–10, pp. 925–937, 2008.

[25] Y. C. Lin and H. S. Lee, "Machining characteristics of magnetic force-assisted EDM," *Int. J. Mach. Tools Manuf.*, vol. 48, no. 11, pp. 1179–1186, 2008.

[26] Y. C. Lin, Y. F. Chen, D. A. Wang, and H. S. Lee, "Optimization of machining parameters in magnetic force assisted EDM based on Taguchi method," *J. Mater. Process. Technol.*, vol. 209, no. 7, pp. 3374–3383, 2009.

[27] S. Joshi, P. Govindan, A. Malshe, and K. Rajurkar, "Experimental characterization of dry EDM performed in a pulsating magnetic field," *CIRP Ann. - Manuf. Technol.*, vol. 60, no. 1, pp. 239–242, 2011.

[28] R. Teimouri and H. Baseri, "Study of tool wear and overcut in EDM process with rotary tool and magnetic field," *Adv. Tribol.*, 2012.

[29] A. Bhattacharya, A. Batish, and G. Bhatt, "Material transfer mechanism during magnetic field-assisted electric discharge machining of AISI D2, D3 and H13 die steel," *Proc. Inst. Mech. Eng. Part B J. Eng. Manuf.*, vol. 229, no. 1, pp. 62–74, 2015.

[30] G. Bhatt, A. Batish, and A. Bhattacharya, "Experimental investigation of magnetic field assisted powder mixed electric discharge machining," *Part. Sci. Technol.*, vol. 33, no. 3, pp. 246–256, 2015.

[31] P.S. Bains, S.S. Sidhu, and H.S. Payal Magnetic field assisted EDM: New horizons for improved surface properties. *Silicon*, 10, 1275–1282 2018. https://doi.org/10.1007/s12633-017-9600-7

[32] Z. Zhang, H. Huang, W. Ming, Z. Xu, Y. Huang, and G. Zhang, "Study on machining characteristics of WEDM with ultrasonic vibration and magnetic field assisted techniques," *J. Mater. Process. Technol.*, vol. 234, pp. 342–352, 2016.

[33] P. Govindan, A. Gupta, S. S. Joshi, A. Malshe, and K. P. Rajurkar, "Single-spark analysis of removal phenomenon in magnetic field assisted dry EDM," *J. Mater. Process. Tech.*, vol. 213, no. 7, pp. 1048–1058, 2013.

[34] A. Gupta and S. S. Joshi, "Modelling effect of magnetic field on material removal in dry electrical discharge machining," *Plasma Sci. Technol.*, vol. 19, no. 2, pp. 1–10, 2017.

[35] H. Beravala and P. M. Pandey, "Experimental investigations to evaluate the effect of magnetic field on the performance of air and argon gas assisted EDM processes," *J. Manuf. Process.*, vol. 34, pp. 356–373, 2018.

[36] H. Beravala and P. M. Pandey, "Experimental investigations to evaluate the surface integrity in the magnetic field and air/gas-assisted EDM," *J. Brazilian Soc. Mech. Sci. Eng.*, vol. 43, no. 4, 2021.

[37] H. Beravala and P. M. Pandey, "Characterization of debris formed in magnetic field-assisted EDM using two-phase dielectric fluid," *J. Adv. Manuf. Syst.*, vol. 19, no. 9, pp. 629–640, 2020.

[38] H. Beravala and P. M. Pandey, "Modelling of material removal rate in the magnetic field and air-assisted electrical discharge machining," *Proc. Inst. Mech. Eng. Part C J. Mech. Eng. Sci.*, vol. 234, no. 7, pp. 1286–1297, 2020.

Chapter 7

Polymer-Based Additive Manufacturing

Narinder Singh
University of Salerno, Fisciano, Italy

Buta Singh
University of Miskolc, Hungary, UK

CONTENTS

7.1 INTRODUCTION

AM based on a polymeric material is primarily utilized in constructing prototypes and the practical evaluation of installations and equipment to achieve a degree of reliable control before engaging in AM methods [1]. It is often utilized to render limited modules such as braces, cable harnesses and ducts for non-load-bearing application ranges. AM based on polymeric materials, nonetheless, has been utilized in various treatments such as the fabrication of surgical instruments, components (especially optical), architectural components, athletic gear, smart textiles and soft robotics with the drastic development of specialized polymer products [2–5]. Advancements such as tailored instruments for the aviation sector often include AM-manufactured components. However, as the conversion of features during AM processing is yet to be thoroughly established, there are still issues regarding accuracy and reproducibility in AM's features [6]. It has been suggested that we ignore polymeric materials that are tailored to conventional process approaches to broaden the polymer content database for AM and to establish a priority polymer design that contributes to the kinetics and the dynamic multiphysics associated with the pathway of AM [7–10]. This approach to the development of molecules using AM will lead to success in understanding the diverse area of procedures based on AM. Thermoplastic equipment is used to manufacture a large share of industrial and commercial goods. Beyond prototyping for intermediate instruments and end-use materials, the aviation, pharmaceutical and industrial sectors have been early adopters of these new developments [11]. In addition to their convenience of manufacturing across mainstream conventional networks, such initial adopter companies also use thermoplastics for their structural influence and bio-inert qualities. In addition, similar industries also need their items to work in the presence of fires and solid solvents and heavy functional pressures at extreme temperatures. For these early adopters, HT engineering polymeric materials are also well-suited [12,13]. Within the scope of this chapter, we concentrate on HT materials, suggesting certain polymer materials with operating temperature requirements above 250°C. The processing of polymers for HT engineering has major complications, notably for processing at high temperatures. The complexities of HT processing may be machine-based or material-based [14].

7.2 VARIOUS TECHNIQUES USED IN AM

After broad research, various approaches were established for AM. Every AM technique shares a similar deposition method while a 2D structure is printed on the single plane (XY) to prepare a single layer, accompanied by the deposition of successive layers along the height (Z-axis) to create a 3D framework [15]. Since this is a standardized technique, a specific file (.STL) is widely used to produce a "layered" printing structure. There are different applications for drawing the desired 3D structure, like CAD and SketchUp. The machine software analyzes the structure and cuts it into several separate layers to translate it to .STL format [16]. The subsequent STL file is then transformed to the file form that various 3D printers can adapt. After loading the .STL file into the 3D system, the structure is prepared based on that .STL file.

7.2.1 Fused Deposition Modeling

One often utilized methodology in which the product feed is in a filament shape is FDM. In the context that the polymer is delivered into the cylindrical extruder in which it melts and is extruded, it is identical to a standard extruder [17,18]. The filament is heated moderately by the FDM to melt it, it is then extruded at about 0.25 mm in diameter to produce a thin fiber, and is then deposited on a framework in a 2D structure in which it loses temperature and immediately solidifies. The same process continues for the required number of layers. The resolution of this technique is low since it is dependent on the fiber thickness being extruded. However, basic manufacturing (printing) and setup mechanisms provide it with excellent performance. The drawback of FDM is that melting and extrusion are required for the printing process, so only polymers (thermoplastic) can be processed [19,20]. The filament, though, is soft for flexible thermoplastics that, even during feeding, could not actually survive the compression stress. Filament bending and feeding failure are often caused by this. Polymers of high performance typically have an elevated melting point and a high molecular weight. These are renowned for their exceptional mechanical strength and thermal stability. Finding it challenging to melt the filament, the extrusion needs a high compaction force and shear force owing to the high viscosity of the molten material. Numerous manufacturing companies have effectively printed high-performance polymers with FDM, including Stratasys and Arevo Labs, for example PPSU, PEEK and PAEK [21].

7.2.2 Stereolithography (SLA)

The SLA technique is one of the mostly utilized AM methods. It is centered on the polymerization print mechanism [22–25]. In this process, liquid resin mixed with a photo initiator, a photopolymer as well as other compounds is positioned in a container in which a framework monitors resin movement. Two print styles, top-down and bottom-up, are applied in various SLA printing techniques as seen in Figure 7.1.

In the top-down process, the base frame is raised to the resin bath surface such that the resin flows to the base and a thin resin sheet (a thickness equivalent to the concrete thickness) is formed on top of the foundation. During the whole printing process, a precise laser beam from the top end of

Top Down   Bottom Up

Continuous exposure   Sequential exposure
to light source   to light source

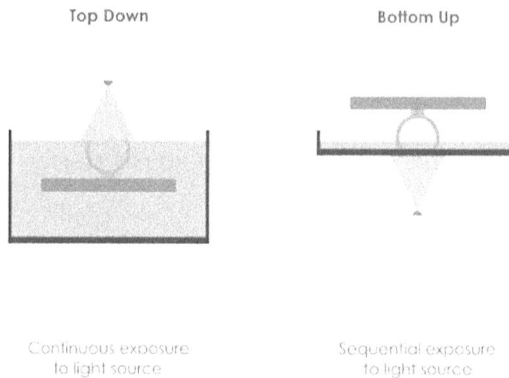

Figure 7.1 Top-down and bottom-up SLA technique.

the exposed resin glows on the surface to initiate the process of polymerization [26]. The platform is adjusted after printing each sheet and another
layer of resin is applied to the solidified substance for the curing of the following layers. In the bottom-up approach the building base is pushed to a
location near the bottom where a slight space between reservoir and platform is left to be a mount for another layer of thin resin under that space. A
beam of laser flashes below the tank cleans the uncovered resin. The platform is lifted after printing each sheet to expose an additional thin layer. The
gap between layers is automatically filled under the action of the gravity, so
the bottom-up SLA printer doesn't need a roller. With this kind of printing
technology very fine prints can be obtained with a high finish; it also
decreases harmful laser exposure, as the beam is confined inside the printer.
Printers equipped with the SLA technique have one of the best resolutions
of all AM processes, where the volume of photons applied is controlled [27].
Normally, one photon is utilized to trigger polymerization, whereas two
photons gain increasing focus due to better spatial power. Much higher efficiency (100 nm) is achievable with unique initiation. In addition, Old World
Laboratory introduced a 100 nm SLA printer that can be utilized to create
complex structures. Unfortunately, the material that SLA can print should
be photocurable, limiting the use of HPP printing, as most HPPs are not
photocurable [28].

7.2.3 Laminated Object Manufacturing (LOM)

Subtractive and additive manufacturing processes are examples of LOM. To
create a 2D pattern, the technique involves first cutting and then laminat-
ing the surface of a bulk material. The problem is the production of waste,
which restricts LOM's efficiency of material use. In order to form a 3D
framework, the additive portion approach involves adhering one level of

a bulk material layer to another layer [29]. Typically, before adding to the previous step, adhesives materials are spread on every consecutive sheet; the bonding force is relatively small, leading to poor material characteristics all along the z-axis. Its efficient method of printing makes LOM very cost-effective. Resolution can be lowered by controlling the z-direction since it depends on the depth of the layers. Furthermore, to remove waste materials and further improve printing efficiency, post-processing is required [30].

7.2.4 Selective Laser Sintering (SLS)

SLS is yet another technique which is based on powdered materials. Much like 3DP, materials are delivered to the platform in a powder phase [31]. SLS beams a high intensity laser instead of using a fluid binder to systematically sinter the particles, allowing the material fusion to form a 2D structure. Slightly under the melting point of building powders, the foundation is also heated so that heat dissipation and degradation can be minimized [32]. The materials being printed do not need a support structure, as in SLA or FDM, which have supportive elements for overhanging frameworks, though complex formations can be conveniently printed. Since the procedure used during printing is sintering, it is only possible to print inorganic substances such as ceramics and metal with an organic binder addition and thermoplastic [33,34]. An electron beam is often used for materials with a higher melting temperature to provide a temperature limit. Although better than 3DP, the surface finish of SLS is still not as decent in SLA, and printing costs are higher because of the expensive laser [35].

7.2.5 Laser Engineered Net Shaping (LENS)

Yet another technique based on a powdered material, though a nozzle is employed to disperse the powder on a base rather than placing the powder on the surface, and a highly intensified laser beam is used for melting the scattered powder simultaneously. To construct a 3D structure, the subsequent melted layer will settle on the former layer [36]. With this technology complex parts can be made, and repair work can also be carried out since there is no need for building foundations. HPPs are very difficult to print with this technology, so there is little literature available on it. The extremely high mechanical strength of these polymers requires a large amount of energy for powder formation.

7.2.6 Polyjet

Another printing method for the polymerization process is polyjet. The resin in polyjet, unlike SLA, is supplied to form a 2D shape by injection from a print head. In order to process and solidify the accumulated resin, UV light is centered right afterward [37]. At the same time, several printing heads can be used to spread various resins simultaneously, hence this method can be utilized to print multiple-resin materials together in one piece. Developed pieces are exposed to a water jet for further finishing. The resolution of polyjet is very high but it comes at the cost of lower mechanical properties than SLA or SLS [38].

7.3 HT THERMOPLASTICS IN ADDITIVE MANUFACTURING: STRUCTURE

New materials are continuously produced and tested for their printability and AM functionality [39]. They are classified as specialized polymer materials, plastics engineering and standard plastics due to the operational conditions and physical nature, as seen in Figure 7.2. Many traditional polymers are currently accessible and can be employed as feedstock for AM off-the-shelf, but AM's production of advanced plastics is still in progress [40].

Although, as we move up the pyramid, efficiency improves, the complexity of processing and related costs escalate rapidly.

AM, MatEx and PBF are common HT thermoplastics. The aromatic framework of such a polymer matrix is important to notice [41]. This feature increases the stiffness of the molecular level and introduces supplementary interactions such as stacking π-π, thus increasing the molecular motion's energy level, that is the temperature required for flow. Different forms of material and properties of thermoplastics widely used in MatEx and PBF, as defined in Table 7.1, are presented in this section. The data provided in Table 7.1 also lays the foundation for a comparison of the material characteristics of HT thermoplastics produced using traditional manufacturing methods with the characteristics of PBF and MatEx technological modes.

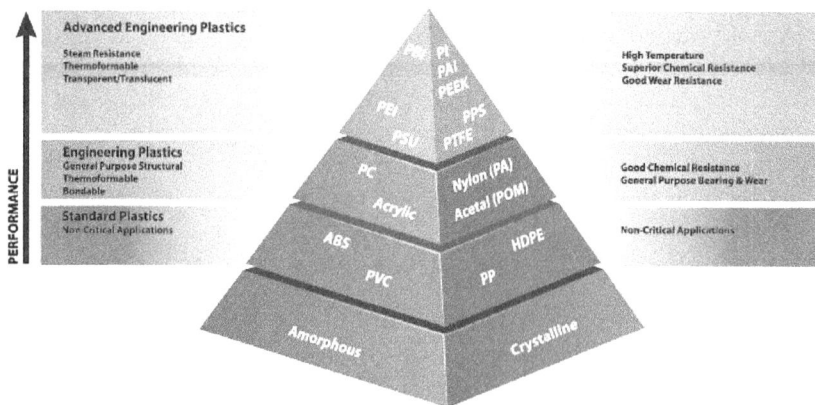

Figure 7.2 Overview of polymers as a performance feature [40].

Table 7.1 Properties of Thermoplastics Widely Used [40]

Polymer	Service Temperature	T_M (C)	Tg (°C)	Mechanical Properties	General Properties
PEK	260	373	162	Tensile modulus:4.0 GPa; tensile strength:105 MPa; ultimate strain: 15%	Superior mechanical characteristics; better thermal stability; resistance to chemicals.
PEKK	125–204	390	156	Tensile modulus:4.20 ± 0.81 GPa; Tensilestrength: −102.0 ± 3.9 MPa; Ultimate strain: 62.1 ± 2.5%	At extreme temperatures, this material retains its strength; it also absorbs little moisture. Industries associated with aerospace; surgical implants.
PEEK	204–232	343	143	Tensile modulus:1.28–4.10 GPa; Tensile strength: −92.3 MPa; Ultimate strain: −40%	Resistant to mechanical wear, creep deform and chemical deterioration. Sectors include automobile, aviation, textiles and biomedical devices.
PPS	204–232	285	85	Tensile modulus: −3.697 GPa; Ultimate strain: −3%; Tensile strength: 64–82 MPa	Chemically inert; electrically conducting; resistant to wear.
PEI	177–204	330–380 (amorphous)	217	Tensile modulus:2.934 GPa; Ultimate strain:18%; Tensile strength: −102.5 MPa	Excellent strength and stiffness; reusable; resistant to fire and creep.

7.4 HT ENGINEERING THERMOPLASTICS IN PBF

A summary of polymeric materials printed by PBF is provided in this section. Printing HT materials is challenging, as has been reported in the literature; however PBF has emerged as a top method for HT materials.

Table 7.2 outlines the printing variables and physical properties [40]. The quantity of the total reported PBF polymer analysis is less than the polymeric materials of MatEx; thus, the subset of previous literature is reasonably applicable to HT thermoplastics scales. PEK and PEEK are the subject of most of the published literature (PEEK). In addition to Toray's industrial "PBF-grade" PPS powder [41], some few reported scientific studies investigate the printing of PPS.

In Table 7.2 the tensile strength of such HT thermoplastics produced using PBF is demonstrated. There is still a scarcity in the existing literature on modification techniques or chemical synthesis or of HT polymeric materials that will particularly allow printable PBF. Similarly, any literature dealing with additives in HT polymers for PBF is solely associated with the modification of the material properties of the final element and is not a production assistant [42]. The current section will focus on the set of process variables and the subsequent physical characteristics of HT polymeric materials printed with PBF. A machine-specific analysis will then follow. A large gap in the current literature appears to focus on the consensus of disseminating the values of significant parameters to allow for an essential analysis and integration of the research results published. The authors assume that every variable (i.e., laser power, beam intensity, spot size and

Table 7.2 Tensile Strength of HT Thermoplastics (PBF Production) [40]

Polymer/Bed Temperature	Ultimate Tensile Strain (%)	Ultimate Tensile Strength (MPa)
PEKK		50
PEK (HP3)/368°C	3–4	90
PEK (HP3)/365°C	2.8	90
PEK (HP3)		88 ± 7
PEEK/200°C		79 ± 3
PEEK Victrex 150PF/338°C		41
PEEK Victrex 450 PF/332°C		31–37
PEEK + Carbon fiber (CF)		109 ± 1
EOS PEEK HP3	4.2 ± 0.2	88.7 ± 1.5
PEEK + Carbon Black (CB)/330°C		30–50
PPS/200°C		64.5
PPS/230°C	3.27 ± 0.22	61.8 ± 4
PEI/Room temperature	3.0–5.7	15–37

hatch spacing) expressed in the energy density equation and bed temperature for fair comparison and comprehension should be stated at least in each article. The EMR equation formed by a researcher requires the powder characteristics and other essential materials to be used and is discussed in the next section [43,44].

It is also possible to calculate these polymer properties at manufacturing temperature, not at room temperature. A quantitative study of proper relationships, including actual heat, often differs significantly from the glassy state and from the temperature of the glass transformation to the heating of the material melting. It is important to consider and report performance features to predict an accurate value of the processing parameters.

7.5 HIGH PERFORMANCE POLYMERS (HPPS)

HPPs are classified as polymeric materials that can maintain their favorable characteristics when subjected to extreme conditions including an aggressive environment, high pressure and high temperature [45–48]. HPPs used for 3D printing are of immense importance owing to the need for designs of excellent technical, chemical and physical durability at high strain and higher temperatures. To produce an HPP, consider the root of its power. Bond intensity is the most significant aspect [49]. Bond dissociation energy will quantify this. In this case the bond dissociation force is directly proportional to the splitting force of the polymer chain, and the polymer's strength and tolerance to a harsh climate [50]. Additionally, the tolerance and versatility of HPPs can be enhanced by including aromatic constituents all along the backbone via resonance stabilization. The incorporation of resonance-stabilized units is expected to add 40–70 kcal mol^{-1} in bond power [51]. Polymers with high aromatic composition when subjected to high temperatures do not yield volatile fragments, so breaking resonance-stabilized bonds is tougher. At elevated temperatures, when polymer degradation starts, radical substances are formed. After the reaction of the polymer with those radical substances they sever the polymeric chain [52–54]. If these substances are created from resonance-stabilized units, they are less reactive because through the π mechanism, radicals delocalize. The major drawback for the higher aromatic volume is that the polymer becomes even more rigid and it is very challenging to treat the polymer [55–57]. Heteroatoms including oxygen, nitrogen and sulfur are added to aromatic groups to address this issue. This allows the polymer to increase processability; however, due to the decrease in bond power, chemical resistance and thermal stability are lowered. The polymers are designed to provide supplementary forces including hydrogen bonding, polar contact and van der Waals forces to accommodate the reduced bond strength, all of which will greatly enhance HPPs' thermal stability due to the intermolecular interaction between polymer links [58,59]. The molecular mass of the polymer and crosslinking density, distribution, crystallinity, additives and reinforcements are other factors regarded as improving thermal resistance. HPPs meet these requirements:

1. After 10,000 hours of exposure at 177°C, the material should maintain its mechanical, chemical and electrical behavior.
2. At 450°C, there must only be around a 5% weight loss.
3. At high temperatures, there should be a limited weight loss rate.
4. The material should have a high Tg and excellent mechanical characteristics provided by a higher aromatic volume and fairly rigid sections.
5. The temperature level at which 10% deformation occurs under a load of 1.52 MPa should be not less than 177°C.

HPPs can be classified into semi-crystalline, crystalline, amorphous and liquid.

Amorphous polymers include PEI and PSU.

Semi-crystalline polymers include PPS, PEEK and numerous liquid crystalline polymers (Table 7.3).

7.5.1 Amorphous HPPs

There is no short and long-range order with polymers of randomly aligned molecules and called amorphous polymers. The amorphous product polymeric substances are PS, PVC, PC, PMMA, and PPC; PEI, PSU, PESU and PPSU comprise amorphous HPPs [60–63].

7.5.2 Polysulfone (PSU)

PSU is a category of HPPs; its structure consisting of aryl-SO2-aryl (sulfone). PSU has a T_g of 185°C, making it one of the best HPP temperatures for operation. At very low temperatures (–100 to –150°C), it also retains its mechanical and physical properties. Non-polar organic solvents such as dichloromethane, chloroform, acetone and aromatic hydrocarbons have low tolerance [64]. In terms of mechanical qualities, PSU exhibits exceptional resistance and a significant flexural modulus due to the polar groups that run along its links. It possesses an excellent hydrolytic stability, even when exposed to heated air steam or boiling water. PSU is utilized in medical steam sterilization and autoclave procedures [65].

7.5.3 Polyetherimide (PEI)

By melting bisphenol A dianhydride with diamine, polyetherimide (PEI) can be synthesized (usually m-phenylenediamine). A highly reactive, melt-processing material benefits from its efficiency. High thermal responsiveness, high strength and stiffness, dimensional stability, and impact resistance are the outstanding mechanical and chemical qualities of PEI. PEI is used to replace metal in some building materials [66–68]. It can survive attacks by chemical hydrocarbons, alcohol and halogenated solvents. It is often considered for flame retardation and very low smoke output. But it also suffers from several disadvantages, including the fact that it is notch-sensitive, requires elevated processing temperatures, and has very high material and operating costs, even though it has several advantageous characteristics [69,70].

7.5.4 Poly(phenylene sulfide) (PPS) and Semi-crystalline HPPs

Amorphous and crystalline regions comprise semi-crystalline polymeric materials. PE and PET are semi-crystalline commodity polymers, though cases of semi-crystalline HPPs include PEEK, PPS, PPA, PAEK, PEKK, nylon HT (HTN) and TPI [71].

PPS is a semi-crystalline polymer composed of recurrent para-substituted benzene rings joined by sulfur atoms (arylene sulfides). Phillips Petroleum commercialized PPS (under the name Ryton®) in 1972 by combining it with sodium sulfide in a polar aprotic solvent. Exceptionally immune to corrosion, PPS has been shown to endure steam, chemical solvents, bleach and solid bases [72]. The outstanding dimensional stability; thermal strength; tolerance to mildew, aging and sunlight; electrical conductivity; low friction coefficient; and abrasion resistance are other advantages of using PPS. A need for rather higher operating temperatures (melt temperatures), the fairly high cost, fragility and susceptibility to warping, however, limit the use of PPS. PPS brittleness with fibers and fillers will effectively be over-reinforced [73–75].

7.5.5 Polyether Ether Ketone (PEEK)

An example of a semi-crystalline aromatic substance that mixes ether and ketone is PEEK. It is usually formulated in a polar aprotic solution by reacting alkali-metal carbonates with 1.4-benzenediol with 4.4'-difluorobenzenophenone (e.g., diphenyl sulfone) [76]. PEEK is by far the most widely used member of the polyaryletherketone family and the first to achieve widespread adoption [77]. The response should be in an inert atmosphere and at a temperature greater than the polymer's melting point, often greater than 300°C. It can handle chemical attacks and explosions. It has powerful dielectric properties, a strong tolerance to wearing and hydrolytic stability. It has been used in a variety of industries, including shipping, aerospace and textiles. One downside to using PEEK, however, is that it is very expensive, including elevated extrusion and molding temperature.

Table 7.3 Different High-Performance
          Polymers

Polymer	Name
Oxygen	LCP
	PEEK
	PEK
	PEKK
Sulfur	PPS
	PSU
	PES
	PPSU
	PI
Nitrogen	PAI
	PEI

7.5.6 Liquid Crystalline Polymers (LCPs)

LCPs are partially crystalline-aromatic polyesters, with both melt and solid-state configurations that are tightly formed; they are based on p-hydroxybenzoic acid and associated monomers. The lyotropic (liquid crystallinity occurs after the polymer has been solved into a solvent) and thermotropic can be divided into two classes. The widely viable Kevlar is a standard case of a Lyotropic LCP. Kevlar possesses exceptional mechanical properties and chemical resistance as a result of the intermolecular hydrogen connection between carbonyl groups and NH centers and aromatic packing between adjacent polymeric chains. However, Vectran is the most widely used thermotropic LCP. Vectran is an aromatic polyester formed by the polycondensation of 4-hydroxybenzoic acid and 6-hydroxynaphthalene-2-carboxylic acid. HT mechanical flexibility is normally provided by LCPs. They have excellent resistance to chemical hazards, toxins and burning [78–80].

7.5.7 Nano-Based Materials/Innovative Polymers

Lately, lignin and cellulose have emerged as important topics in the manufacturing of polymer-based materials with features highly desired by plastic products (e.g., PET, PP, PS and PE), renewable and artificial; and bio-based and organic polymer-degradable materials such as polybutylene succinate and PHA. While these materials could not completely replace hydrocarbon-based plastics, more could be done to decrease the carbon emissions of AM products and the process lifespans. In order to enable them to succeed, some main features have to be creative, focusing on the specific market for the polymer-based material. For instance, the biomedical industry requires innovative printable and, most significantly, biocompatible polymeric materials, with sufficient kinetics for degradation and degradation by-products [81]. For better utility, more emphasis must be imposed on the automotive industry: on the machinability and tailored mechanical properties of the in-use polymer. We will now discuss the above-mentioned advanced polymers, following these requirements, and review the developments in the use of ME3DP polymer-based architectures.

7.5.8 Poly Butylene Succinate (PBS)

PBS is one of the most vital compostable aliphatic polymers of Bionolle. It is a biopolymer generated from the polycondensation of succinic acid and butanediol, providing a significant building block for biocompatible formulations and polymers to plastic producers. It has identical features to those widely used for injection, extrusion, compression or blow molding processes [82] for polypropylene, polyethylene, and so on. Substantially, the manufacturing ability of Bionolle is equivalent to conventional resins, such as bulk plastic. For some applications, Bionolle has been used, for example for shopping bags and farming items. There is a possible substitute for polyethylene terephthalate, PP, polyolefin, and polystyrene in some materials; it is therefore essential to examine PLA and starch which have already been developed in other varieties of this material [83] and used for the production of environmentally sustainable polymer composites with tailored properties. Chemically, Bionolle is harmless under average conditions, but in the presence of microorganisms found in composts, moist soil, freshwater, seawater and activated sludge, it can become biodegradable [84]. It can fully disintegrate into water and carbon dioxide, helping it to become eco-sustainable. Due to its lower crystallinity level, PBSA, a PBS copolymer, shows more degradability [85]. With a significant lack of research and promising features, PBS is a flexible biopolymer that could significantly affect the development of high-performance, eco-sustainable material products and systems. Their best features comprise [86]:

- For warm containers, packages and cookware, reasonably good operation conditions.
- High potential for heat-seal; provide at reduced temperature about the same sealing strength as conventional petroleum material.
- Low cost for the atmosphere relative to other artificial polymeric materials.
- Traditional compatibility with biopolymers.
- Enhanced printability despite significant pre-treatment.
- Excellent processing capabilities.

7.5.9 Polyhydroxyalkanoates (PHAs)

PHAs are indeed a group of sustainable polyesters developed using the bacterial fermentation of carbon-based raw materials: plastics that are simple and conveniently biodegradable [87]. Other PHAs are organic and sustainable with physical and mechanical characteristics similar to polypropylene, making it a better alternative for PLA and PBS in the production of the biopolymer method. PHA is broadly known as:

- Unsolvable and reasonably hydrolytic-resistant.
- Ultraviolet resistant, but acid or base resistance is reduced.
- Biocompatible, non-toxic and suitable for biomedical requirements and applications for food packing.

7.5.10 Lignin

Lignin is a strongly aromatic biomaterial, found primarily throughout the dealignment process in the fiber section of plants as a by-product of the wood pulping industry [88]. A neutral carbon footprint of neutral CO_2 makes lignin an attractive alternative for the materials of the next generation because of its antibacterial, lightweight, eco-sustainable, biodegradable and antioxidant properties. However, the current problems with widely viable lignin materials appear to be low-purity, heterogeneity, odor and pigment issues.

7.6 CHALLENGES IN PRINTING WITH HT ENGINEERING THERMOPLASTICS

The issues involving the PBF region of the matrix material can be separated into system and polymer concerns, but these challenges are related. The unit's issues are often faced by all polymer processing at extreme temperatures. These issues include heating certain system areas, while insulating critical components, such as electrical separation and the viewing position of the operator. In PBF, it is important to isolate or place the feed material at least at a cold temperature until it is time to spread over the building piston. This may be more difficult than molding techniques where the process time is shorter, that is the injection molding scenario, or where all charged polymer material is utilized, which is the rotational molding circumstance. It should be mentioned that in PBF, for some of these HT polymers, it is harder to reclaim and reuse unfused material than with polymeric materials like nylon 12. The key features of the efficient performance of PBF depend on the span of time a material is well over its T_g [89]. With each chemically specific polymer, specific powder lifespan tests are required. By reducing T_m and the required processing temperature, the work preceding the PEKK co-polymer marketing, influences powder lifespan. Therefore, identical effects can be obtained by more efficiently and selectively coalescing particles using photo-absorbing inhibitors or viscosity additives without changing the chemical structure of the particle in the feed or accompanying layer [90]. This could

also be done by innovative ideas on how to systematically isolate feed powder during development and design better chemically resistant polymers to
meet long-term PBF processing specifications. The knowledge of where to
use the heating systems and the light source to provide the power needed
to melt the polymer powder is yet another machine-related topic for future
work. The priority for future studies is not exclusive to high-temperature
thermoplastics, but it is potentially more important for their treatment. By
keeping the powder at a lower temperature during production, the ability
to provide more energy optically than thermally appears to raise the life of
unfused powders. This is due to the increased selectivity of a laser beam's
energy source compared to heaters [91]. Building on the laser beam would
also allow PBF to print HT polymers on less expensive machines than were
originally constructed to print polyamides. A systematic experimental design
would discover the interaction between each energy form in order to provide
the same volume to each polymer at high temperatures by means of different optical and thermal energy component formulations and to determine
the viability of optical energy production. In contrast, further research will
confirm that elevated chamber temperatures are required to ensure that both
the heated melt pool and the colder surrounding powder are created geo-
metrically correctly and minimize tensile stress. The HSS and MJF technologies enhance heater selectivity through the jetting of photothermal ink onto
the powder bed. In order to determine if these inks can escalate temperatures locally to enable complete coalescence, future research is required. The
researchers found that the main material obstacles and issues of HT polymer
printing by PBF are (i) powder production and (ii) accuracy in publication
reports. Powdered polymer matrix processing is also a problem for sectors
such as rotational and powder coating [40]. In these sectors, however, the
primary assets are high. During recuperation, PBF presents the challenge of
solid state powder movement. This not only includes scale requirements,
but also powders that are spherically shaped. One benefit of HT polymers
is that an elevated T_g allows polymers to be processed by grinding. Study in
this field [92] is led by Berretta's thesis and co-authors. They analyzed indus-
trialized grades of PEK powder, HP3 and two PEEKs, Victrex 150 PF and
PF 450, which were not classified as PBF grade, and reported that non-PBF
grades were not optimized for machine fluidity, but that high printing output
was positive for future refining. Finally, starting to develop an understanding
of the PBF content development literature is essential for this field. When
individuals know about the work of Vasquez and others in their melt energy
ratio [93] and document all associated process parameter values regularly, a
synthesis of the structure–property–process–property relationship guidance
would boost opportunities. These are the fundamental parts of polymer processing research that will lead to the expansion of the catalog of PBF material. In particular, the broader range of products would offer more prospects
for enhanced industrial applications as traditional production methods,
including injection molding, arrive at the material procurement stage.

7.7 CONCLUSIONS

In this chapter emphasis has been given to elaborate and various AM technologies that use polymers as feed stock. Along with that various other polymers such as high-performance polymers and high temperature polymers have been discussed. It has been seen that FDM is not a very good contender for high end applications because of low printing resolution and low accuracy. On the other hand it has been seen that SLA and PBF have turned out to be some of the highly accurate technologies with very high resolution. Still a gap has been seen in the printing of high-performance polymers including high temperature polymers, the reason being that high performance polymers require fairly high processing temperatures and high-end machines. Further testing and analysis require a high skill set along with the required testing equipment. For example, PEEK and PAEK materials have comparatively high mechanical properties but processing temperatures are well above 350°C. Hence it becomes difficult to process these materials in powder form. Some other classes of materials such as amorphous, crystalline and semi-crystalline materials along with the nano- and innovative polymers have also been discussed in this chapter for better understanding of the application domain for which polymers have to be used. At the end some of the challenges posed and faced by 3D printing technologies were discussed and elaborated. It has been seen that dependency of the polymer on the application is a crucial part of the process and it has to be confirmed from the existing literature or preliminary testing which polymer fits best for the required application.

ABBREVIATIONS

3DP	3 dimensional printing
AM	Additive manufacturing
CAD	Computer added design
EMR	Energy melt ratio
FDM	Fused deposition modelling
HPPs	High performance polymers
HT	High temperature
KPBSA	Poly butylene succinate
LCP	Liquid crystal polymer
LCPs	Liquid crystalline polymers
LENS	Laser engineered net shaping
LOM	Laminated object manufacturing
LPBF	Laser powder bed fusion
MatEx	Material extrusion
MJF	Multi Jet fusion
PAEK	Polyaryletherketone
PAI	Polyamide-imide

PBF	Powder bed fusion
PBS	Poly butylene succinate
PC	Polycarbonate
PEEK	Polyether ether ketone
PEI	Polyetherimide
PEK	Polyether ketone
PESU	Polyethersulfone
PHA	Poly-hydroxyalkcanoates
PI	Polyimide
PMMA	Poly(methyl methacrylate)
PPC	Polypropylene carbonate
PPS	Poly(phenylene sulfide)
PPSU	Poly phenyl sulfone
PSU	Polysulfone
PVC	Poly(vinyl chloride)
SLA	Stereolithography
SLS	Selective laser sintering
T_g	Glass transition temperature
T_m	Melting temperature
PS	Polystyrene
MJF	Multi-jet fusion
TPI	Thermoplastic polyimide

REFERENCES

[1] Kishore, R.A. and Priya, S., 2018. A review on low-grade thermal energy harvesting: materials, methods and devices. *Materials*, 11(8), p. 1433.

[2] Ligon, S.C., Liska, R., Stampfl, J., Gurr, M. and Mülhaupt, R., 2017. Polymers for 3D printing and customized additive manufacturing. *Chemical Reviews*, 117(15), pp. 10212–10290.

[3] Zhang, Z., Demir, K.G. and Gu, G.X., 2019. Developments in 4D-printing: a review on current smart materials, technologies, and applications. *International Journal of Smart and Nano Materials*, 10(3), pp.205–224.

[4] Idumah, C.I., Nwuzor, I. and Odera, S.R., 2020. Recent advancements in self-healing polymeric hydrogels, shape memory, and stretchable materials. *International Journal of Polymeric Materials and Polymeric Biomaterials*, 70(13) pp.1–26.

[5] Kamarudin, S.F., Mustapha, M. and Kim, J.K., 2021. Green strategies to printed sensors for healthcare applications. *Polymer Reviews*, 61(1), pp.116–156.

[6] Kumar, M.B. and Sathiya, P., 2020. Methods and materials for additive manufacturing: A critical review on advancements and challenges. *Thin-Walled Structures*, 159, p. 107228.

[7] Ngo, T.D., Kashani, A., Imbalzano, G., Nguyen, K.T. and Hui, D., 2018. Additive manufacturing (3D printing): A review of materials, methods, applications and challenges. *Composites Part B: Engineering*, 143, pp.172–196.

[8] Chatham, C.A., Long, T.E. and Williams, C.B., 2019. A review of the process physics and material screening methods for polymer powder bed fusion additive manufacturing. *Progress in Polymer Science*, 93, pp.68–95.

[9] Bandyopadhyay, A. and Heer, B., 2018. Additive manufacturing of multi-material structures. *Materials Science and Engineering: R: Reports*, 129, pp.1–16.

[10] Yuan, S., Shen, F., Chua, C.K. and Zhou, K., 2019. Polymeric composites for powder-based additive manufacturing: Materials and applications. *Progress in Polymer Science*, 91, pp. 141–168.

[11] Rosato, D.V., 2011. *Plastics end use applications*. Springer Science & Business Media.

[12] Mohanty, A.K., Misra, M.A. and Hinrichsen, G.I., 2000. Biofibres, biodegradable polymers and biocomposites: An overview. *Macromolecular Materials and Engineering*, 276(1), pp.1–24.

[13] Tsao, C.W. and DeVoe, D.L., 2009. Bonding of thermoplastic polymer microfluidics. *Microfluidics and Nanofluidics*, 6(1), pp.1–16.

[14] Goebel, R., Glaser, T. and Skiborowski, M., 2020. Machine-based learning of predictive models in organic solvent nanofiltration: Solute rejection in pure and mixed solvents. *Separation and Purification Technology*, 248, p. 117046.

[15] Singh, N., Hui, D., Singh, R., Ahuja, I.P.S., Feo, L. and Fraternali, F., 2017. Recycling of plastic solid waste: A state of art review and future applications. *Composites Part B: Engineering*, 115, pp. 409–422.

[16] Singh, R., Singh, N., Fabbrocino, F., Fraternali, F. and Ahuja, I.P.S., 2016. Waste management by recycling of polymers with reinforcement of metal powder. *Composites Part B: Engineering*, 105, pp. 23–29.

[17] Singh, R., Singh, N., Amendola, A. and Fraternali, F., 2017. On the wear properties of Nylon6-SiC-Al2O3 based fused deposition modelling feed stock filament. *Composites Part B: Engineering*, 119, pp. 125–131.

[18] Singh, S., Singh, N., Gupta, M., Prakash, C. and Singh, R., 2019. Mechanical feasibility of ABS/HIPS-based multi-material structures primed by low-cost polymer printer. *Rapid Prototyping Journal*, 25(1), pp. 152–161.

[19] Singh, N., Singh, R. and Ahuja, I.P.S., 2018. Recycling of polymer waste with SiC/Al2O3 reinforcement for rapid tooling applications. *Materials Today Communications*, 15, pp. 124–127.

[20] Singh, N., Singh, R., Ahuja, I.P.S., Farina, I. and Fraternali, F., 2019. Metal matrix composite from recycled materials by using additive manufacturing assisted investment casting. *Composite Structures*, 207, pp. 129–135.

[21] de Leon, A.C., Chen, Q., Palaganas, N.B., Palaganas, J.O., Manapat, J. and Advincula, R.C., 2016. High performance polymer nanocomposites for additive manufacturing applications. *Reactive and Functional Polymers*, 103, pp. 141–155.

[22] González-Henríquez, C.M., Sarabia-Vallejos, M.A. and Rodriguez-Hernandez, J., 2019. Polymers for additive manufacturing and 4D-printing: Materials, methodologies, and biomedical applications. *Progress in Polymer Science*, 94, pp. 57–116.

[23] Manapat, J.Z., Chen, Q., Ye, P. and Advincula, R.C., 2017. 3D printing of polymer nanocomposites via stereolithography. *Macromolecular Materials and Engineering*, 302(9), p. 1600553.

[24] Wang, X., Jiang, M., Zhou, Z., Gou, J. and Hui, D., 2017. 3D printing of polymer matrix composites: A review and prospective. *Composites Part B: Engineering*, 110, pp. 442–458.

[25] Saroia, J., Wang, Y., Wei, Q., Lei, M., Li, X., Guo, Y. and Zhang, K., 2020. A review on 3D printed matrix polymer composites: its potential and future challenges. *The International Journal of Advanced Manufacturing Technology*, 106(5), pp. 1695–1721.

[26] Lee, K.S., Kim, R.H., Yang, D.Y. and Park, S.H., 2008. Advances in 3D nano/microfabrication using two-photon initiated polymerization. *Progress in Polymer Science*, 33(6), pp. 631–681.

[27] Jafari, R., Cloutier, C., Allahdini, A. and Momen, G., 2019. Recent progress and challenges with 3D printing of patterned hydrophobic and superhydrophobic surfaces. *The International Journal of Advanced Manufacturing Technology*, 103(1), pp. 1225–1238.

[28] Quan, H., Zhang, T., Xu, H., Luo, S., Nie, J. and Zhu, X., 2020. Photocuring 3D printing technique and its challenges. *Bioactive Materials*, 5(1), pp. 110–115.

[29] Singh, N., Singh, R. and Ahuja, I.P.S., 2018. On development of functionally graded material through fused deposition modelling assisted investment casting from Al 2 O 3/SiC reinforced waste low density polyethylene. *Transactions of the Indian Institute of Metals*, 71(10), pp. 2479–2485.

[30] Singh, N., Singh, R. and Ahuja, I.P.S., 2019. Thermomechanical investigations of SiC and Al2O3–reinforced HDPE. *Journal of Thermoplastic Composite Materials*, 32(10), pp. 1347–1360.

[31] Sun, J., Peng, Z., Zhou, W., Fuh, J.Y., Hong, G.S. and Chiu, A., 2015. A review on 3D printing for customized food fabrication. *Procedia Manufacturing*, 1, pp. 308–319.

[32] Memon, S.A., Lo, T.Y., Shi, X., Barbhuiya, S. and Cui, H., 2013. Preparation, characterization and thermal properties of Lauryl alcohol/Kaolin as novel form-stable composite phase change material for thermal energy storage in buildings. *Applied Thermal Engineering*, 59(1–2), pp. 336–347.

[33] Chen, Z., Li, Z., Li, J., Liu, C., Lao, C., Fu, Y., Liu, C., Li, Y., Wang, P. and He, Y., 2019. 3D printing of ceramics: A review. *Journal of the European Ceramic Society*, 39(4), pp. 661–687.

[34] Valino, A.D., Dizon, J.R.C., Espera Jr, A.H., Chen, Q., Messman, J. and Advincula, R.C., 2019. Advances in 3D printing of thermoplastic polymer composites and nanocomposites. *Progress in Polymer Science*, 98, p. 101162.

[35] Choudhari, C.M. and Patil, V.D., 2016, September. Product development and its comparative analysis by SLA, SLS and FDM rapid prototyping processes. In *IOP Conference Series: Materials Science and Engineering* (Vol. 149, No. 1, p. 012009). IOP Publishing.

[36] Atwood, Clint, Michelle Griffith, Lane Harwell, Eric Schlienger, Mark Ensz, John Smugeresky, Tony Romero, Don Greene, and Daryl Reckaway, 1998. "Laser engineered net shaping (LENS™): A tool for direct fabrication of metal parts." In *International Congress on Applications of Lasers & Electro-Optics* (Vol. 1998, no. 1, pp. E1–E7). Laser Institute of America.

[37] Ibrahim, D., Broilo, T.L., Heitz, C., de Oliveira, M.G., de Oliveira, H.W., Nobre, S.M.W., dos Santos Filho, J.H.G. and Silva, D.N., 2009. Dimensional error of selective laser sintering, three-dimensional printing and PolyJet™ models in the reproduction of mandibular anatomy. *Journal of Cranio-Maxillofacial Surgery*, 37(3), pp. 167–173.

[38] Barclift, M.W. and Williams, C.B., 2012, August. Examining variability in the mechanical properties of parts manufactured via polyjet direct 3D printing. In

International Solid Freeform Fabrication Symposium (pp. 6–8). University of Texas at Austin Austin, Texas.

[39] Gabrion, X., Placet, V., Trivaudey, F. and Boubakar, L., 2016. About the thermomechanical behaviour of a carbon fibre reinforced high-temperature thermoplastic composite. *Composites Part B: Engineering*, 95, pp.386–394.

[40] Das, A., Chatham, C.A., Fallon, J.J., Zawaski, C.E., Gilmer, E.L., Williams, C.B. and Bortner, M.J., 2020. Current understanding and challenges in high temperature additive manufacturing of engineering thermoplastic polymers. *Additive Manufacturing*, 34, p. 101218.

[41] Koerner, H., Price, G., Pearce, N.A., Alexander, M. and Vaia, R.A., 2004. Remotely actuated polymer nanocomposites—stress-recovery of carbon-nanotube-filled thermoplastic elastomers. *Nature Materials*, 3(2), pp. 115–120.

[42] Mousa, A.A., 2016. Experimental investigations of curling phenomenon in selective laser sintering process. *Rapid Prototyping Journal*, 22(2), pp. 405–415.

[43] Vasquez, M., Haworth, B. and Hopkinson, N., 2011. Optimum sintering region for laser sintered nylon-12. *Proceedings of the Institution of Mechanical Engineers, Part B: Journal of Engineering Manufacture*, 225(12), pp. 2240–2248.

[44] Chatham, C.A., Long, T.E. and Williams, C.B., 2019. Powder bed fusion of poly (phenylene sulfide) at bed temperatures significantly below melting. *Additive Manufacturing*, 28, pp. 506–516.

[45] Fink, J.K., 2014. *High performance polymers*. William Andrew.

[46] Seymour, R.B. ed., 1986. *High performance polymers: their origin and development*. Elsevier.

[47] Swager, T.M., 2008. Iptycenes in the design of high performance polymers. *Accounts of Chemical Research*, 41(9), pp. 1181–1189.

[48] Robeson, L.M., Burgoyne, W.F., Langsam, M., Savoca, A.C. and Tien, C.F., 1994. High performance polymers for membrane separation. *Polymer*, 35(23), pp. 4970–4978.

[49] Braga, R.R., Ballester, R.Y. and Ferracane, J.L., 2005. Factors involved in the development of polymerization shrinkage stress in resin-composites: a systematic review. *Dental Materials*, 21(10), pp. 962–970.

[50] Masuelli, M.A., 2013. Introduction of fibre-reinforced polymers– polymers and composites: Concepts, properties and processes. In *Fiber reinforced polymers-the technology applied for concrete repair*. IntechOpen.

[51] Beer, L., Brusso, J.L., Cordes, A.W., Haddon, R.C., Itkis, M.E., Kirschbaum, K., MacGregor, D.S., Oakley, R.T., Pinkerton, A.A. and Reed, R.W., 2002. Resonance-stabilized 1, 2, 3-dithiazolo-1, 2, 3-dithiazolyls as neutral π-radical conductors. *Journal of the American Chemical Society*, 124(32), pp. 9498–9509.

[52] Gewert, B., Plassmann, M.M. and MacLeod, M., 2015. Pathways for degradation of plastic polymers floating in the marine environment. *Environmental Science: Processes & Impacts*, 17(9), pp. 1513–1521.

[53] McNeill, I.C., Memetea, L. and Cole, W.J., 1995. A study of the products of PVC thermal degradation. *Polymer Degradation and Stability*, 49(1), pp. 181–191.

[54] Grassie, N. and Scott, G., 1988. *Polymer degradation and stabilisation*. CUP Archive.

[55] Bose, S., Kuila, T., Nguyen, T.X.H., Kim, N.H., Lau, K.T. and Lee, J.H., 2011. Polymer membranes for high temperature proton exchange membrane fuel cell: recent advances and challenges. *Progress in Polymer Science*, 36(6), pp. 813–843.

[56] Gregory, G.L., Sulley, G.S., Carrodeguas, L.P., Chen, T.T., Santmarti, A., Terrill, N.J., Lee, K.Y. and Williams, C.K., 2020. Triblock polyester thermoplastic elastomers with semi-aromatic polymer end blocks by ring-opening copolymerization. *Chemical Science*, 11(25), pp. 6567–6581.

[57] Upton, B.M. and Kasko, A.M., 2016. Strategies for the conversion of lignin to high-value polymeric materials: review and perspective. *Chemical Reviews*, 116(4), pp. 2275–2306.

[58] Hergenrother, P.M., 2003. The use, design, synthesis, and properties of high performance/high temperature polymers: an overview. *High Performance Polymers*, 15(1), pp. 3–45.

[59] Azwa, Z.N., Yousif, B.F., Manalo, A.C. and Karunasena, W., 2013. A review on the degradability of polymeric composites based on natural fibres. *Materials & Design*, 47, pp. 424–442.

[60] Abenojar, J., Torregrosa-Coque, R., Martínez, M.A. and Martín-Martínez, J.M., 2009. Surface modifications of polycarbonate (PC) and acrylonitrile butadiene styrene (ABS) copolymer by treatment with atmospheric plasma. *Surface and Coatings Technology*, 203(16), pp. 2173–2180.

[61] Jia, Z., Wang, Z., Xu, C., Liang, J., Wei, B., Wu, D. and Zhu, S., 1999. Study on poly (methyl methacrylate)/carbon nanotube composites. *Materials Science and Engineering: A*, 271(1–2), pp. 395–400.

[62] Feng, Y., Ashok, B., Madhukar, K., Zhang, J., Zhang, J., Reddy, K.O. and Rajulu, A.V., 2014. Preparation and characterization of polypropylene carbonate bio-filler (eggshell powder) composite films. *International Journal of Polymer Analysis and Characterization*, 19(7), pp. 637–647.

[63] Hourston, D.J. and Lane, J.M., 1992. The toughening of epoxy resins with thermoplastics: 1. Trifunctional epoxy resin-polyetherimide blends. *Polymer*, 33(7), pp. 1379–1383.

[64] Liu, J., Jeong, H., Liu, J., Lee, K., Park, J.Y., Ahn, Y. and Lee, S., 2010. Reduction of functionalized graphite oxides by trioctylphosphine in non-polar organic solvents. *Carbon*, 48(8), pp. 2282–2289.

[65] McKeen, L.W., 2014. Plastics used in medical devices. In *Handbook of polymer applications in medicine and medical devices* (pp. 21–53). William Andrew Publishing.

[66] Rattan, R. and Bijwe, J., 2007. Influence of impingement angle on solid particle erosion of carbon fabric reinforced polyetherimide composite. *Wear*, 262(5–6), pp. 568–574.

[67] Bijwe, J., Rattan, R. and Fahim, M., 2007. Abrasive wear performance of carbon fabric reinforced polyetherimide composites: influence of content and orientation of fabric. *Tribology International*, 40(5), pp. 844–854.

[68] Lyu, M.Y. and Choi, T.G., 2015. Research trends in polymer materials for use in lightweight vehicles. *International Journal of Precision Engineering and Manufacturing*, 16(1), pp. 213–220.

[69] Wang, P., Zou, B., Xiao, H., Ding, S. and Huang, C., 2019. Effects of printing parameters of fused deposition modeling on mechanical properties,

surface quality, and microstructure of PEEK. *Journal of Materials Processing Technology*, 271, pp. 62–74.

[70] Stokes-Griffin, C.M. and Compston, P., 2015. The effect of processing temperature and placement rate on the short beam strength of carbon fibre–PEEK manufactured using a laser tape placement process. *Composites Part A: Applied Science and Manufacturing*, 78, pp. 274–283.

[71] Touny, A.H., Joseph, L.G., Jones, A.D. and Bhaduri, S.B., 2010. Effect of electrospinning parameters on the characterization of PLA/HNT nanocomposite fibers. *Journal of Materials Research*, 25(5), pp. 857–865.

[72] Sudarsan, V., 2017. Materials for Hostile Chemical Environments. In *Materials Under Extreme Conditions* (pp. 129–158). Elsevier.

[73] Motavalli, M., Czaderski, C., Schumacher, A. and Gsell, D., 2010. Fibre reinforced polymer composite materials for building and construction. In *Textiles, polymers and composites for buildings* (pp. 69–128). Woodhead Publishing.

[74] Hasnat, A., 2014. *Experimental investigation on flexural capacity of reinforced concrete beams strengthened with carbon fiber reinforced polymer strips*. Thesis presented at the Department of Civil Engineering, Bangladesh University of Engineering and Technology, Dhaka., pp. 1–233.

[75] Harris, B., 1999. *Engineering composite materials*. Institute of Materials, London, England.

[76] Fortney, A. and Fossum, E., 2012. Soluble, semi-crystalline PEEK analogs based on 3, 5-difluorobenzophenone: Synthesis and characterization. *Polymer*, 53(12), pp.2327–2333.

[77] Abdulhamid, M.A., Park, S.H., Vovusha, H., Akhtar, F.H., Ng, K.C., Schwingenschlögl, U. and Szekely, G., 2020. Molecular engineering of high-performance nanofiltration membranes from intrinsically microporous poly (ether-ether-ketone). *Journal of Materials Chemistry A*, 8(46), pp. 24445–24454.

[78] Donald, A.M., Windle, A.H. and Hanna, S., 2006. *Liquid crystalline polymers*. Cambridge University Press.

[79] Wang, X.J. and Zhou, Q.F., 2004. *Liquid crystalline polymers*. World Scientific.

[80] Finkelmann, H., 1987. Liquid crystalline polymers. *Angewandte Chemie International Edition in English*, 26(9), pp. 816–824.

[81] Daminabo, S.C., Goel, S., Grammatikos, S.A., Nezhad, H.Y. and Thakur, V.K., 2020. Fused deposition modeling-based additive manufacturing (3D printing): Techniques for polymer material systems. *Materials Today Chemistry*, 16, p. 100248.

[82] Liu, L., Yu, J., Cheng, L. and Qu, W., 2009. Mechanical properties of poly (butylene succinate)(PBS) biocomposites reinforced with surface modified jute fibre. *Composites Part A: Applied Science and Manufacturing*, 40(5), pp. 669–674.

[83] Tawakkal, I.S., Cran, M.J., Miltz, J. and Bigger, S.W., 2014. A review of poly (lactic acid)-based materials for antimicrobial packaging. *Journal of Food Science*, 79(8), pp. R1477–R1490.

[84] Lichocik, M., Owczarek, M., Miros, P., Guzińska, K., Gutowska, A., Ciechańska, D., Krucińska, I. and Siwek, P., 2012. Impact of PBSA (Bionolle) biodegradation products on the soil microbiological structure. *Fibres & Textiles in Eastern Europe*, 6B(96), pp. 179–185.

[85] Puchalski, M., Szparaga, G., Biela, T., Gutowska, A., Sztajnowski, S. and Krucińska, I., 2018. Molecular and supramolecular changes in polybutylene succinate (PBS) and polybutylene succinate adipate (PBSA) copolymer during degradation in various environmental conditions. *Polymers*, 10(3), p. 251.

[86] Reddy, C.S.K., Ghai, R. and Kalia, V., 2003. Polyhydroxyalkanoates: an overview. *Bioresource Technology*, 87(2), pp. 137–146.

[87] Argyropoulos, D.S. and Menachem, S.B., 1998. Lignin. In *Biopolymers from renewable resources* (pp. 292–322). Springer, Berlin, Heidelberg.

[88] Vanholme, R., Morreel, K., Ralph, J. and Boerjan, W., 2008. Lignin engineering. *Current Opinion in Plant Biology*, 11(3), pp. 278–285.

[89] Cárdenas, B. and León, N., 2013. High temperature latent heat thermal energy storage: Phase change materials, design considerations and performance enhancement techniques. *Renewable and Sustainable Energy Reviews*, 27, pp. 724–737.

[90] Dilara, P.A. and Briassoulis, D., 2000. Degradation and stabilization of low-density polyethylene films used as greenhouse covering materials. *Journal of Agricultural Engineering Research*, 76(4), pp. 309–321.

[91] Gokuldoss, P.K., Kolla, S. and Eckert, J., 2017. Additive manufacturing processes: Selective laser melting, electron beam melting and binder jetting—Selection guidelines. *Materials*, 10(6), p. 672.

[92] Bagherifard, S., Beretta, N., Monti, S., Riccio, M., Bandini, M., and Guagliano, M., 2018. On the fatigue strength enhancement of additive manufactured AlSi10Mg parts by mechanical and thermal post-processing. *Materials & Design*, 5(145), pp. 28–41.

[93] Tonelli, L., Fortunato, A. and Ceschini, L., 2020. CoCr alloy processed by Selective Laser Melting (SLM): Effect of Laser Energy Density on microstructure, surface morphology, and hardness. *Journal of Manufacturing Processes*, 52, pp. 106–119.

Chapter 8

Recent Research Progress and Future Prospects in the Additive Manufacturing of Biomedical Magnesium and Titanium Implants

Haytham Elgazzar and Khalid Abdelghany
Institute (CMRDI), Cairo, Egypt

CONTENTS

8.1 INTRODUCTION

Healthcare remains one of the most important subjects for every country in the world. Every year, thousands of people are injured due to numerous types of accidents, such as work and traffic accidents. Traffic accidents are considered the major type of accident, where more than a million people are killed around the world and millions more are injured every year. This causes a global health, economic and social crisis. The treatment of injured people requires sometimes metallic implants made from various materials, for example cobalt-based alloys, magnesium (Mg) alloys, stainless steel alloys and titanium (Ti) alloys (Putra et al. 2020, Sezer et.al. 2020). Among various metallic materials, Ti and its alloys, particularly Ti6Al4V alloy, are increasingly used due to their interesting characteristics such as the high strength to weight ratio, biocompatibility and an acceptable resistance to corrosion (Bartolomeu et al. 2021, Putra et al. 2020). However, the modulus of elasticity of such alloys ($\approx$110 GPa) is greater than that of human cortical bone tissue (0.5 to 30 GPa) (Bartolomeu et al. 2021, Kabir et al. 2021, Putra et al. 2020). Moreover, the corrosion process produces potentially harmful products to human body tissues. Metal implants usually need to be removed after completion of the tissue healing process. This is usually done through a new surgical procedure which may increase patient morbidity and the over-all cost of healing (Kabir et al. 2021, Putra et al. 2020, Sezer et.al. 2020). Alternatively, biodegradable metal (BM) implants such as Mg and zinc (Zn), as third generation biomaterials, are becoming a subject of interest due to their ability to dissolve gradually and be absorbed into the human body after implantation (Kabir et al. 2021, Putra et al. 2020, Sezer et.al. 2020, Wang et al. 2020). The use of BM implants provides a scientific solution to the problems of metallic implants such as irritation, the inability to adjust to growth and other continuous form changes in the human body. In addi-tion, it also avoids the associated problems of releasing harmful ions and/or particles resulting from the corrosion process (Kabir et al. 2021, Putra et al. 2020, Sezer et.al. 2020, Wang et al. 2020, Qin et al. 2019). These kinds of materials are currently considered to be essential tools in the treatment for common bone ailments. Moreover, the use of BM implants is favorable to patients due to the acceptable cost and other aesthetic and convenience rea-sons (Kabir et al. 2021, Putra et al. 2020, Sezer et.al. 2020, Wang et al. 2020, Qin et al. 2019). Therefore, the development of BM implants is considered the most important research area in medical science.

Mg and its alloys have recently received significant interest as BM implants because of their biocompatibility, biodegradability and excellent mechanical characteristics, that is, having high strength and an elastic mod-ulus close to that of bone (Kabir et al. 2021, Putra et al. 2020, Sezer et.al. 2020, Tao et al. 2020, Wang et al. 2020, Qin et al. 2019). Moreover, Mg and its alloys are excellent biomaterials compared to permanent metallic implants that cause numerous problems such as the release of toxic ele-ments, physical irritation and behavior mismatches between the implant and the human body (Kabir et al. 2021, Putra et al. 2020, Sezer et.al. 2020, Tao et al. 2020, Wang et al. 2020, Qin et al. 2019). Mg and its alloys, as a BM, do not remain as permanent metallic implants in the human body and, hence, there is no need for a second surgical process after the tissue is healed. These intriguing characteristics make Mg the most potentially biodegrad-able implant for orthopedic applications and vascular stents (Kabir et al. 2021, Putra et al. 2020, Sezer et.al. 2020, Tao et al. 2020, Wang et al. 2020, Qin et al. 2019). These unique and interesting characteristics have inspired researchers to develop such a material and its alloys as a new class of bio-degradable material (Putra et al. 2020, Sezer et al. 2020, Tao et al. 2020, Wang et al. 2020, Qin et al. 2019).

However, the major drawback of Mg and its alloys, as well as Ti, is that they tend to corrode very quickly in the human body and lose their mechani-cal integrity before the end of the time required for complete healing (Kabir et al. 2021, Putra et al. 2020, Sezer et al. 2020, Tao et al. 2020, Wang et al. 2020, Qin et al. 2019). During the degradation process, the majority of alloying elements will dissociate into the human body where some harmful elements such as aluminum (Al) are released (Kabir et al. 2021, Manne et al. 2018, Ataee et al. 2018). It is known that Al has toxicological effects and is considered an important factor in Alzheimer's disease. The toxicological limit of Al has been determined according to its neurotoxicity, which should be at a lower limit than that associated with Alzheimer's disease (Kabir et al. 2021, Manne et al. 2018). Thus, the fast corrosion and side effects of the dissolved alloying elements of Mg and its alloys are considered to be the main restrictions on using such materials in medical applications (Kabir et al. 2021, Putra et al. 2020, Sezer et.al. 2020, Tao et al. 2020, Wang et al. 2020, Manne et al. 2018, Qin et al. 2019). Therefore, improving the corrosion properties is of greatest demand for these materials so they can be used as biomedical implants. Numerous solutions have been studied to overcome the high mismatch between the mechanical properties of Ti alloys and human bone, as well as the low corrosion resistance problems of Mg alloy and Ti alloy implants.

Recently, porous biomaterials have become a valuable choice to reduce the modulus of elasticity where the size of pore and porosity significantly affect the mechanical properties and the biological behavior of the implants and, thus, minimize or eliminate osteonecrosis and osteogenesis distortion around the metallic implants (Bartolomeu et al. 2021, Putra et al. 2020, Sezer et.al. 2020). Developing new alloys, through adding elements such as Ca, Zn, Zr or rare-earth elements such as Ce, Gd, Nd and/or incorporating bioactive ceramic materials such as hydroxyaptite, has recently been investigated as an interesting solution to the corrosion problem (Gao et al. 2020, Liu et al. 2020, Putra et al. 2020, Sezer et.al. 2020, Tao et al. 2020, Manne et al. 2018, Qin et al. 2019). However, the implementation of such solutions is still hindered by the difficulty of manufacturing complex 3D implants with superior properties using conventional manufacturing processes. Hence, additive manufacturing (AM) technology has become a breakthrough solution to producing high added-value biomedical metal implants.

8.2 ADDITIVE MANUFACTURING AND FABRICATION CHALLENGES OF BIOMEDICAL METAL IMPLANTS

AM is a mandatory tool for the fourth generation industrial revolution and plays an important role in developments in the medical industry concerning manufacturing high performance medical products. AM technology offers significant benefits for the medical industry, such as the manufacturing of designed custom implants according to patient need, design flexibility, accelerating design cycles, lower costs and less waste (Bandyopadhyay, Zhang, and Bose 2020, Cooke et al. 2020, Dev Singh, Mahender, and Reddy 2020). Various types of AM processes are currently available on the market and are classified, according to the feedstock material form and the suppling energy source, as direct energy deposition (DED), powder bed fusion (PBF) and sheet lamination processes (Lakhdar et al. 2021, Bandyopadhyay, Zhang, and Bose 2020, Cooke et al. 2020, Dev Singh, Mahender, and Reddy 2020,

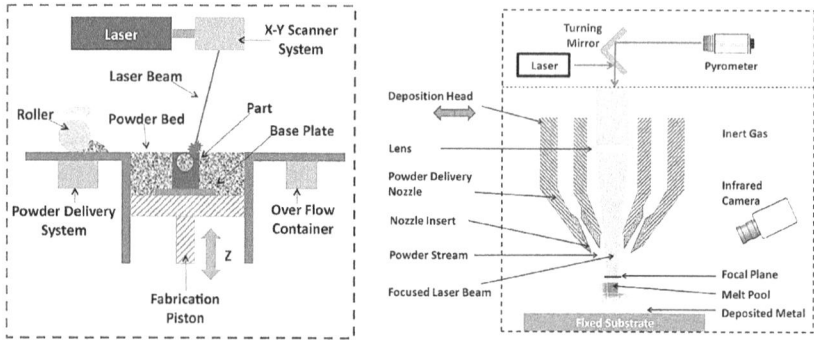

Figure 8.1 Schematic of (a) SLM and (b) LMD processes (Thompson et al. 2015).

Zhou et al. 2020). Among AM processes, laser-based AM processes, in practice selective laser melting (SLM) and laser metal deposition (LMD), as shown in Figure 8.1a, b (Thompson et al. 2015), have received much interest regarding manufacturing metal parts (Bandyopadhyay, Zhang, and Bose 2020, Cooke et al. 2020, Dev Singh, Mahender, and Reddy 2020, Zhang et al. 2020a). LMD and SLM are two techniques that can be used to manufacture complex 3D parts from metal powder (Bandyopadhyay, Zhang, and Bose 2020, Cooke et al. 2020, Dev Singh, Mahender, and Reddy 2020).

In the LMD process, also known as laser engineering net-shape (LENS), laser cladding or direct metal deposition (DMD), a focused laser beam is used to locally melt the original material that has the shape of powder or a wire to build up 3D shapes, while in the SLM process, a powder layer is spread onto a substrate plate which then selectively melts the powder using a focused laser beam to create 3D parts.

Both processes involve rapid heating and cooling cycles which result in products that have fine grains with a homogeneous structure which greatly improves the overall product's properties. The differences between these two techniques are layer thickness, accuracy and the powder feeding system (Bandyopadhyay, Zhang, and Bose 2020, Cooke et al. 2020). Table 8.1 (Langefeld 2013) shows a comparison between the two techniques. However, each technique has its own characteristics and merits.

However, the SLM process is more commonly used for manufacturing medical implants due to its high accuracy and ability to manufacture small dimensional complex implants compared to the LMD process. The fabrication process of SLM implants has several steps, as shown in Figure 8.2 (Gorsse et al. 2017). The first step is the preprocessing step which deals with powder preparation, and the characterization and design of the required products using CAD software. The second step deals with the fabrication step where the powder is loaded into the machine and the product design is transferred to the SLM machine. The third step is post-processing which deals with product finishing and quality inspections.

Table 8.1 Comparison Between the SLM and LMD Techniques (Langefeld 2013)

Criteria	SLM	LMD
Build speed	5–20 cm³/h (~40–160 g/h)	Up to 0.5 kg/h (~70 cm³/h)
Accuracy	±0.02–0.05 mm/25 mm	±0.125–0.25 mm/25 mm
Detail capability	0.04–0.2 mm	0.5–1.0 mm
Surface quality	Ra 4–10 μm	Ra 7–20 μm
Maximum part size	500 mm × 280 mm × 325 mm	2000 mm × 1500 mm × 750 mm
Average system price	EUR 450–600 K	EUR 500–800 K
Applications	Rapid prototyping, manufacturing on demand	Repair, modification

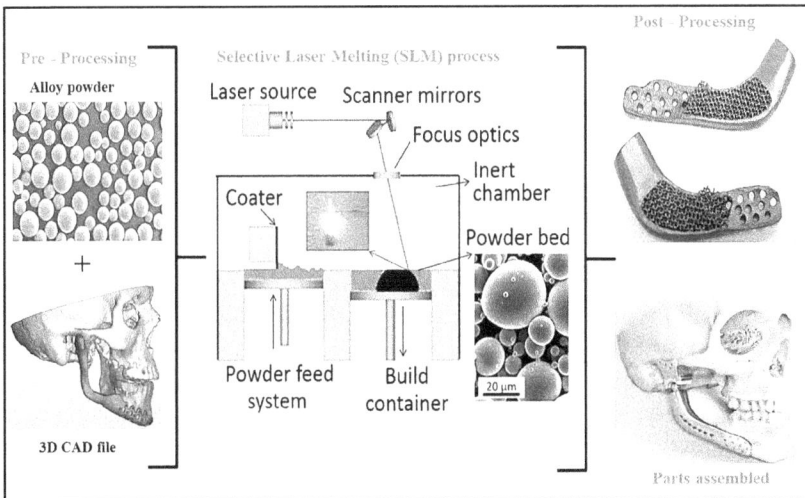

Figure 8.2 The SLM process steps of biomedical implants.

Numerous parameters should be carefully controlled for the laser based AM processes to achieve the desired properties. These include laser beam power, laser scanning speed, powder shape and particle size, characteristics of the interface, matrix grain size and phases in the matrix. Thus, it is important to compromise between all these parameters. In addition, the formation of undesirable intermetallic compounds should be minimized or avoided. To realize the influence of parameters on the quality of the products, it is important to consider the flow chart of the laser-based AM processes shown in Figure 8.3 (Sames et al. 2016). The main inputs of the process are machine hardware and software, the CAD file, laser power, scan strategy, feedstock material, protective gas type and pressure. Also included are the thermal

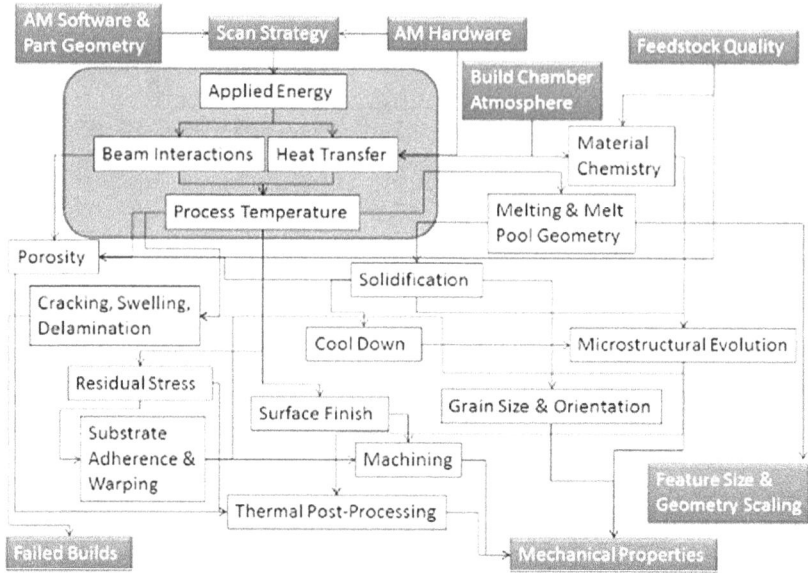

Figure 8.3 The relationship between the input parameters of laser based AM processes and the outcomes (Sames et al. 2016).

effects due to the changes of the process temperature as a result of the applied laser beam energy, the laser beam interactions with powder/substrates and heat transfer.

The important outputs of the process are the enhancement of the mechanical properties of the products, preventing/minimizing product defects and geometric conformity. However, AM products suffer from various defects such as porosity, lack of fusion, residual stresses and cracks (Zhang et al. 2020b, Ahmed 2019, He et al. 2019, Li et al. 2019). Figure 8.4 (Collins et al. 2016) shows the common defects that could exist in AM products. These defects originally come from the powder material and/or the improper

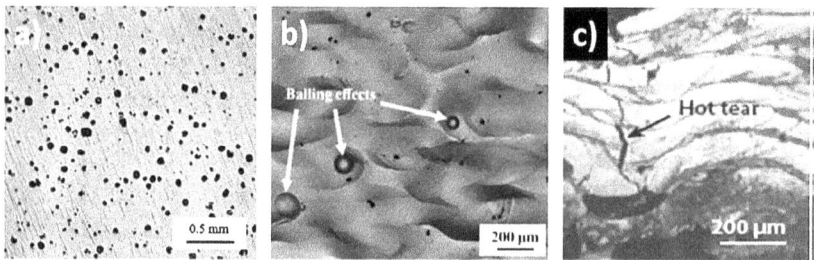

Figure 8.4 Examples of defects that can form in AM products: (a) porosity, (b) balling and (c) hot tears (Gorsse et al. 2017).

selection of the process parameters (Zhang et al. 2020b, Ahmed 2019, He et al. 2019, Li et al. 2019). Reviewing the previous scientific work shows three different approaches to overcome these defects. The first one focuses on the synthesis and production of metal powders with emphasis on powder composition, structure size, shape, density and flowability (Cordova et al. 2020, Zhang et al. 2020b, Li et al. 2019, Nagarajan et al. 2019, Qin et al. 2019). The second approach deals with the development of simulation models for different AM processes and powders in order to reduce or minimize these defects (Bartolomeu et al. 2020, Cooke et al. 2020, Maconachie et al. 2019, Yuan, Ding, and Wen 2019).

A process simulation model can predict the temperature history, which can be coupled with a phase transformation simulation to foresee undesired process results. In the third approach, the process parameters are optimized to enhance the product quality, precision and reliability of the manufacturing process (Bartolomeu et al. 2021, Bandyopadhyay, Zhang, and Bose 2020, Bartolomeu et al. 2020, Esmaily et al. 2020, Li et al. 2020, Putra et al. 2020, Sezer et.al. 2020, Shimizu et al. 2020, Zhang et al. 2020b, Qin et al. 2019).

The fabrication of metallic implants using the SLM process should meet the design requirements set by physicists where the high elastic modulus of the metallic materials must be lowered by raising their porosity (Bartolomeu et al. 2021, Putra et al. 2020). Increasing the porosity of SLM metallic products also improves the bone adhesion process and facilitates the healing process.

8.3 THE FABRICATION OF Ti6Al4V IMPLANTS USING THE SLM PROCESS

The fabrication of porous Ti6Al4V implants by the SLM process has been investigated by numerous research groups. Bartolomeu et al. (2021) fabricated porous Ti6Al4V implants that had an elastic modulus near that of bone tissue, which can be applied in orthopedic applications. Chen et al. (2020) investigated the noticeable changes in the mechanical and biological properties of Ti6Al4V porous scaffolds depending on the influence of the dimensions of pores and porosity. The results showed that a Ti6Al4V scaffold that has a 500 μm pore size and 60% porosity is the most preferred design for cell proliferation, osteogenic differentiation and bone ingrowth. These results agree with Taniguchi et al. (2016) who reported that a 600 μm pore size is the best porous structure for SLM-Ti6Al4V orthopedic implants.

Chen et al. (2019) studied the impact of the additive angle on the surface properties and biocompatibility of the porous Ti6Al4V cage. The results showed a smooth surface roughness obtained with a decreased additive angle, while changing the additive angle had no effect on cell adhesion. Sarker et al. (2018) investigated the influence of various additive angles on surface topography, surface chemistry and cell attachment. The results agree with Chen et al.'s (2019) findings regarding smooth surface roughness obtained with a decreased additive angle. However, Sarker et al. showed that cell attachment and corrosion resistance are improved by increasing the additive angle. Zhao et al. (2017) studied the effect of the fabrication process on the corrosion resistance of Ti6Al4V. The authors showed that the SLM process provided Ti6Al4V implants with higher corrosion resistance compared to that obtained by an electron beam process. Despite the numerous benefits of Ti6Al4V for biomedical implants, it is still restricted to specific medical applications because it is not biodegradable in the human body and brings long-term complications (Bartolomeu et al. 2021, Shimizu et al. 2020).

8.4 BIOMEDICAL TI6AL4V IMPLANTS: CASE STUDIES

A porous structure of Ti6Al4V implants that has 60% porosity was fabricated by an SLM system aiming to decrease the implant–bone modulus mismatch and provide acceptable osteointegration. A medical-grade powder of Ti6Al4V has an average grain size of ~30 μm and was selected as the basic material. The patient's preoperative radiographic records consisted of computed tomography (CT) which was conducted by a cone beam CT scanner. The patients' CTs were transferred to surgical planning software (Mimics 10.0: Materialise, Belgium) to create a 3D model that exported using a specially designed convert tool that allows the creation of standard tessellation language (STL) files. The STL data was then edited using editor software (Magics, Materialise, Belgium) to check errors in the STL files, such as bad edges, contours and surfaces. The Magics software slices the 3D model into thin layers after any data record repair. The slices are created and saved as a (*.cls) file that is transferred to the SLM machine (M3Linear – Concept Laser, Germany). The fiber laser was used to melt the Ti6Al4V powder to create the desired 2D layer. Subsequently, the new layer was spread, then the process was repeated until the implant was constructed. Figure 8.5 shows the fabricated implant for a zygomatic-orbito-maxillary defect reconstructed using a Ti6Al4V implant. The continuous follow-ups revealed good implant integration with no complications.

The design and fabrication of patient specific prostheses from a Ti6Al4V alloy for the zygomatic bone and upper maxilla for a patient who had a tumor in the mandible is another case study that showed the benefits of SLM technology. In this case, a 3D CAD model was created for the zygomatic bone for the cancer tumor patient after being reconstructed from the patient CT scan data, as shown in Figure 8.6, while Figure 8.7 shows the created Ti6Al4V prostheses using the SLM process.

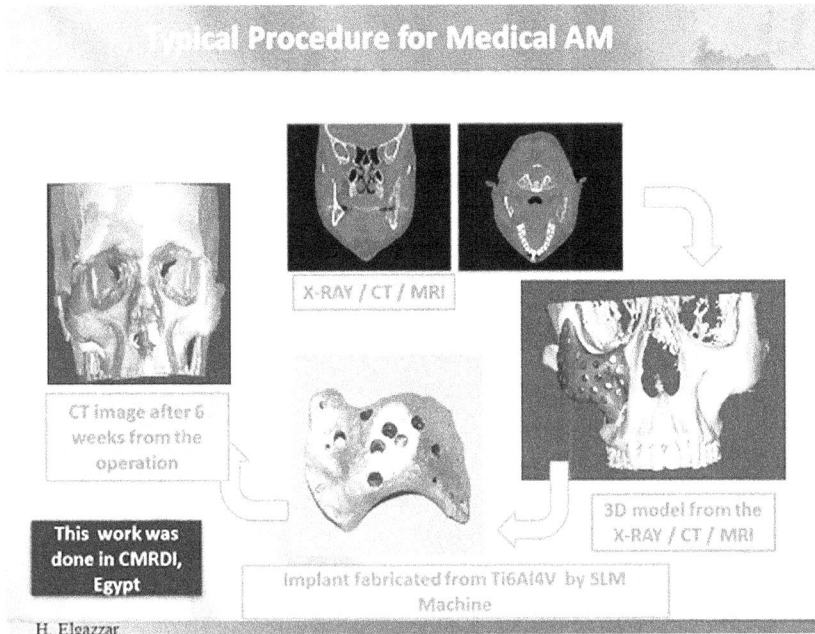

Figure 8.5 The SLM procedure for the fabrication of Ti6Al4V implants for a reconstructed zygomatic-orbito-maxillary defect.

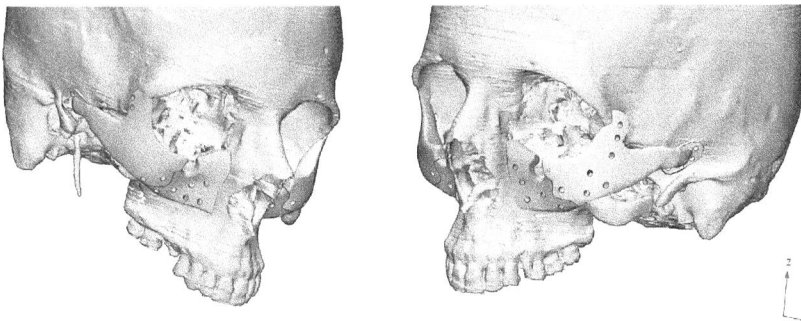

Figure 8.6 The 3D CAD model and the conceptual design for the prostheses.

8.5 THE FABRICATION OF MG IMPLANTS USING THE SLM PROCESS

The fabrication of Mg implants using SLM has been a more difficult process due to the flammability of Mg powder as well as its tendency to oxidize and the difficulty of preparing Mg powders (Gao et al. 2020, Liu et al. 2020, Putra et al. 2020, Sezer et.al. 2020, Wang et al. 2020, Qin et al. 2019, Manne et al. 2018, Tao et al. 2020). Numerous attempts have been made to

Figure 8.7 The Ti6Al4V prostheses fabricated by the SLM process at the Central Metallurgical Research and Development Institute (CMRDI).

overcome these challenges through controlling the Mg alloy powder composition during the powder processing step and optimizing the parameters of the SLM process, such as laser beam power, laser scan speed, thickness of layer and the hatch space (Gao et al. 2020, Putra et al. 2020, Sezer et.al. 2020, Tao et al. 2020, Wang et al. 2020, Qin et al. 2019, Manne et al. 2018). Extensive investigations indicated that both AZ series alloy and WE series alloy systems might satisfy the common requirements for biodegradable materials, including good mechanical properties, biocompatibilities and bio-degradation properties (Gao et al. 2020, Liu et al. 2020, Putra et al. 2020, Sezer et.al. 2020, Tao et al. 2020, Wang et al. 2020, Qin et al. 2019, Manne et al. 2018). Several Mg alloys have been developed by adding functional elements such as Ca, Mn and Zn to be used for biomedical applications. These alloying elements have been selected due to their significant biological roles in the human body. The Mg–Ca binary alloys, and the Mg–Zn–Mn–Ca, Mg–Zn–Y and Mg–Mn–Zn alloys, are several examples of the developed Mg alloys that have been investigated and evaluated extensively so as to be applicable to biomedical implants (Gao et al. 2020, Liu et al. 2020, Putra et al. 2020, Sezer et.al. 2020, Tao et al. 2020, Wang et al. 2020, Qin et al. 2019, Manne et al. 2018). Liu et al. (2017) fabricated porous Mg–Ca alloys with superior properties to that of as-cast pure Mg due to the grain refinement obtained as a result of the rapid solidification process. Wang et al. (2020) fabricated Mg scaffolds using newly developed Mg–Nd–Zn–Zr alloy powder. The obtained specimens showed fully interconnected structures with good mechanical and chemical properties. Moreover, the obtained specimens were subsequently coated with a di-calcium phosphate de-hydrate (DCPD) to enhance corrosion resistance and biocompatibility.

Tao et al. (2020) fabricated SLM-ZK30-xGO composites with 0.6 wt% graphene oxide (GO) as the reinforcement phase. The obtained specimens exhibited good cytocompatibility, mechanical properties and corrosion resistance due to grain refinement.

8.6 POST-PROCESSING OF SLM PRODUCTS

The surface treatment of the SLM implant products, as a post-processing step, is important to enhance the mechanical performance and satisfy the required biologic functions through reducing residual thermal stress and minimizing the microstructural defects. It is important also to overcome the corrosion problem of biomedical implants since weak corrosion resistance is a common defect in most biomedical metal implants especially in Mg alloys and Ti alloys. The surface properties of the biomedical implants are important, where the interaction between them and the biological environment usually occurs on the surface of these materials. Moreover, the response of the living tissues to these materials is greatly influenced by surface properties. Thus, improving the surface properties of the biomedical implants through various modification techniques is in great demand as the chemical, mechanical and biocompatibility properties can be extremely enhanced (Putra et al. 2020, Sezer et.al. 2020, Tao et al. 2020, Wang et al. 2020, Bartolomeu et al. 2019, Todea et al. 2019, Qin et al. 2019, Zhang and Chen 2019).

In this regard, Shimizu et al. (2020) investigated the influence of solution and heat treatment to improve Ti6Al4V implant adhesion by stimulating strontium (Sr) ions from Ti6Al4V implants fabricated by the SLM process. The author showed that Sr ions formed a uniform network in a nanostructure which formed bone-like apatite that improves implant adhesion. Todea et al. (2019) evaluated the bioactivity of the SLM-Ti6Al7Nb implants that were subjected to various surface treatments. The results showed that after the applied treatments, the texture of the surfaces and composition selectively influenced the apatite layer grown on the sample surface. Kelly et al. (2019) investigated the influence of designed porosity and surface treatment on the mechanical properties of SLM-Ti6Al4V.

The results showed an improvement in the fatigue strength as a result of reduction in the surface roughness of the Ti6Al4V after treatment for solid specimens. While porous specimens showed only a reduction in surface roughness, no improvement in fatigue strength was observed. Yan (2020) significantly raised the corrosion resistance of Ti6Al4V implants by transforming the α phase to the β phase using a vacuum annealing process. In addition, the authors successfully obtained almost fully dense products using hot isostatic pressing (HIP) that also enhanced the corrosion resistance of SLM Ti6Al4V implants.

A similar investigation was made by Gangireddy et al. (2019) who applied HIP treatment for SLM products made of a WE43 Mg alloy. Similar to Yan et al., these authors densified high porosity level specimens effectively, while the impact of the HIP treatment was negligible at a low porosity level because of the pores' enclosed nature.

The properties of Ti6Al4V and Mg implants are also considerably improved by a suitable coating of the metal surface. The coating layers must be uniform, well adhesive, free from defects, resistant to corrosion and possess favorable biocompatibility (Putra et al. 2020, Sezer et.al. 2020, Tao et al. 2020, Wang et al. 2020, Qin et al. 2019, Todea et al. 2019). Therefore, it would be accepted as a bone-like substitute by the living cells. It can be rapidly integrated and may fulfill its purpose for a long time without mechanical loosening. There are many techniques which coat the Ti6Al4V and Mg implants, such as electrochemical plating, chemical conversion coating, physical vapor deposition, LMD and anodic oxidation (Putra et al. 2020, Julmi et al. 2019, Qin et al. 2019, Todea et al. 2019, Zhang and Chen 2019), where each technique has its own merits and drawbacks.

Bioactive coating materials such as bioceramic coatings are inorganic compounds having ionic or covalent interatomic bonding that originated at high temperatures. A wide range of such materials is applied for modifying the surface of implanted devices that are used in hard tissue or skeletal repair. Bioceramics could be classified into the bioinert, bioactive, bioresorbable and porous coating. Examples of a bioinert coating are alumina and zirconia, while bioactive ones are materials such as bioactive glasses, hydroxyapatite and glass-ceramics. Tricalcium phosphate is a good example of bioresorbable materials, while porous coating materials such as hydroxyapatite coating and bioglass coating are used for tissue ingrowth (Zhang and Chen 2019). These types of coating materials have the ability to facilitate the healing process through stimulating bone regeneration and in the growth of bone at the implant–tissue interface without a need for a layer of intermediate fibrous tissue (Putra et al. 2020, Sezer et.al. 2020, Julmi et al. 2019, Todea et al. 2019, Yin et al. 2019, Zhang and Chen 2019). Bioactive ceramic coatings such as calcium–phosphate (Ca–P) compounds are of higher importance than other coating types for inducing modification to the surfaces of the metallic implants. Hence, there is great demand for the use of Ca–P-based surface coatings on Mg and Ti6Al4V implants for load-bearing applications such as hip and knee joint prostheses, and dental implants (Putra et al. 2020, Julmi et al. 2019, Qin et al. 2019, Todea et al. 2019, Yin et al. 2019, Zhang and Chen 2019, Sahasrabudhe and Bandyopadhyay 2018, Bandyopadhyay et al. 2016).

8.7 SUMMARY AND FUTURE WORK

AM is the ideal technology to produce customized medical implants depending on the anatomic data of the patient. Among various AM processes, the SLM process provides the manufacturing of complicated 3D geometries that have excellent precision construction. The recent progress and future prospects of the manufacturing of biomedical implants of Mg alloys and Ti alloys using AM technology has been represented in this chapter. In addition, two case studies have been briefly discussed, demonstrating the benefits of AM technology. The importance and benefits of the AM process for the fabrication of customized medical implants made of Ti6Al4V and Mg alloys, and the challenge for fabrication of such biomedical implants, have been briefly discussed. Recent studies show the developing of new biocompatible powder alloy compositions and/or bioactive coating layers that facilitate the healing process are at the top of current research activities. Recent trends and prospective research regarding biodegradable Mg based alloy implants have also been highlighted. Through the success of using SLM for the fabrication of Ti6Al4V and Mg alloy implants, more attention should be given to raw powder materials as they affect the product quality and integrity. In-site process monitoring and controlling, to avoid the evaporation of Mg and other volatile elements, are highly desirable. Furthermore, a few research articles focused on the simulation and modeling of the SLM process which should become an essential tool to guarantee optimal control of the whole process. In addition, combining the nanomaterial in the Mg alloy and Ti alloy structures and overcoming the difficulties of powder preparation and the SLM fabrication process are great challenges that have to be deeply investigated.

REFERENCES

Ahmed, Naveed. 2019. "Direct metal fabrication in rapid prototyping: A review." *Journal of Manufacturing Processes* 42:167–191. https://doi.org/10.1016/j.jmapro. 2019.05.001

Ataee, Arash, Yuncang Li, Milan Brandt, and Cuie Wen. 2018. "Ultrahigh-strength titanium gyroid scaffolds manufactured by selective laser melting (SLM) for bone implant applications." *Acta Materialia* 158:354–368. https://doi.org/10.1016/j. actamat.2018.08.005

Bandyopadhyay, Amit, Stanley Dittrick, Thomas Gualtieri, Jeffrey Wu, and Susmita Bose. 2016. "Calcium phosphate–titanium composites for articulating surfaces of load-bearing implants." *Journal of the Mechanical Behavior of Biomedical Materials* 57:280–288. https://doi.org/10.1016/j.jmbbm.2015.11.022

Bandyopadhyay, Amit, Yanning Zhang, and Susmita Bose. 2020. "Recent developments in metal additive manufacturing." *Current Opinion in Chemical Engineering* 28:96–104. https://doi.org/10.1016/j.coche.2020.03.001

Bartolomeu, F., C. S. Abreu, C. G. Moura, M. M. Costa, N. Alves, F. S. Silva, and G. Miranda. 2019. "Ti6Al4V-PEEK multi-material structures – design, fabrication and tribological characterization focused on orthopedic implants." *Tribology International* 131:672–678. https://doi.org/10.1016/j.triboint.2018.11.017

Bartolomeu, F., M. M. Costa, N. Alves, G. Miranda, and F. S. Silva. 2021. "Selective Laser Melting of Ti6Al4V sub-millimetric cellular structures: Prediction of dimensional deviations and mechanical performance." *Journal of the Mechanical Behavior of Biomedical Materials* 113:104123. https://doi.org/10.1016/j.jmbbm. 2020.104123

Bartolomeu, F., N. Dourado, F. Pereira, N. Alves, G. Miranda, and F. S. Silva. 2020. "Additive manufactured porous biomaterials targeting orthopedic implants: A suitable combination of mechanical, physical and topological properties."

Materials Science and Engineering: C 107:110342. https://doi.org/10.1016/j.msec.2019.110342

Chen, Cen, Ya Hao, Xue Bai, Junjie Ni, Sung-Min Chung, Fan Liu, and In-Seop Lee. 2019. "3D printed porous Ti6Al4V cage: Effects of additive angle on surface properties and biocompatibility; bone ingrowth in Beagle tibia model." *Materials & Design* 175:107824. https://doi.org/10.1016/j.matdes.2019.107824

Chen Ziyu, Xingchen Yan, Shuo Yin, Liangliang Liu, Xin Liu, Guorui Zhao, Wenyou Ma, Weizhong Qi, Zhongming Ren, Hanlin Liao, Min Liu, Daozhang Cai, and Hang Fang. 2020. "Influence of the pore size and porosity of selective laser melted Ti6Al4V ELI porous scaffold on cell proliferation, osteogenesis and bone ingrowth." *Materials Science and Engineering: C* 106:110289. https://doi.org/10.1016/j.msec.2019.110289

Collins, P. C., D. A. Brice, P. Samimi, I. Ghamarian, and H. L. Fraser. 2016. "Microstructural control of additively manufactured metallic materials." *Annual Review of Materials Research* 46 (1):63–91. https://doi.org/10.1146/annurev-matsci-070115-031816

Cooke, Shaun, Keivan Ahmadi, Stephanie Willerth, and Rodney Herring. 2020. "Metal additive manufacturing: Technology, metallurgy and modelling." *Journal of Manufacturing Processes* 57:978–1003. https://doi.org/10.1016/j.jmapro.2020.07.025

Cordova, Laura, Ton Bor, Marc de Smit, Simone Carmignato, Mónica Campos, and Tiedo Tinga. 2020. "Effects of powder reuse on the microstructure and mechanical behaviour of Al–Mg–Sc–Zr alloy processed by laser powder bed fusion (LPBF)." *Additive Manufacturing*:1–13. https://doi.org/10.1016/j.addma.2020.101625

Dev Singh, D., T. Mahender, and Avala Raji Reddy. 2020. "Powder bed fusion process: A brief review." *Materials Today: Proceedings*:1–6. https://doi.org/10.1016/j.matpr.2020.08.415

Esmaily, M., Z. Zeng, A. N. Mortazavi, A. Gullino, S. Choudhary, T. Derra, F. Benn, F. D'Elia, M. Müther, S. Thomas, A. Huang, A. Allanore, A. Kopp, and N. Birbilis. 2020. "A detailed microstructural and corrosion analysis of magnesium alloy WE43 manufactured by selective laser melting." *Additive Manufacturing* 35:101321. https://doi.org/10.1016/j.addma.2020.101321

Gangireddy, Sindhura, Bharat Gwalani, Kaimiao Liu, Eric J. Faierson, and Rajiv S. Mishra. 2019. "Microstructure and mechanical behavior of an additive manufactured (AM) WE43-Mg alloy." *Additive Manufacturing* 26:53–64. https://doi.org/10.1016/j.addma.2018.12.015

Gao, Chengde, Sheng Li, Long Liu, Shizhen Bin, Youwen Yang, Shuping Peng, and Cijun Shuai. 2020. "Dual alloying improves the corrosion resistance of biodegradable Mg alloys prepared by selective laser melting." *Journal of Magnesium and Alloys.* https://doi.org/10.1016/j.jma.2020.03.016

Gorsse, Stéphane, Christopher Hutchinson, Mohamed Gouné, and Rajarshi Banerjee. 2017. "Additive manufacturing of metals: a brief review of the characteristic microstructures and properties of steels, Ti-6Al-4V and high-entropy alloys." *Science and Technology of Advanced Materials* 18 (1):584–610. https://doi.org/10.1080/14686996.2017.1361305

He, Wei, Wenxiong Shi, Jiaqiang Li, and Huimin Xie. 2019. "In-situ monitoring and deformation characterization by optical techniques; part I: Laser-aided direct metal deposition for additive manufacturing." *Optics and Lasers in Engineering* 122:74–88. https://doi.org/10.1016/j.optlaseng.2019.05.020

Julmi, Stefan, Ann-Kathrin Krüger, Anja-Christina Waselau, Andrea Meyer-Lindenberg, Peter Wriggers, Christian Klose, and Hans Jürgen Maier. 2019. "Processing and coating of open-pored absorbable magnesium-based bone implants." *Materials Science and Engineering: C* 98:1073–1086. https://doi.org/10.1016/j.msec.2018.12.125

Kabir, Humayun, Khurram Munir, Cuie Wen, and Yuncang Li. 2021. "Recent research and progress of biodegradable zinc alloys and composites for biomedical applications: Biomechanical and biocorrosion perspectives." *Bioactive Materials* 6 (3):836–879. https://doi.org/10.1016/j.bioactmat.2020.09.013

Kelly, Cambre N., Nathan T. Evans, Cameron W. Irvin, Savita C. Chapman, Ken Gall, and David L. Safranski. 2019. "The effect of surface topography and porosity on the tensile fatigue of 3D printed Ti-6Al-4V fabricated by selective laser melting." *Materials Science and Engineering: C* 98:726–736. https://doi.org/10.1016/j.msec.2019.01.024

Lakhdar, Y., C. Tuck, J. Binner, A. Terry, and R. Goodridge. 2021. "Additive manufacturing of advanced ceramic materials." *Progress in Materials Science* 116:100736. https://doi.org/10.1016/j.pmatsci.2020.100736

Langefeld Bernhard 2013. *Additive Manufacturing-A Game Changer for the Manufacturing Industry*." Roland Berger Strategy Consultants GmbH, Munich, Germany, 1(5.1). http://www.rolandberger.com/media/pdf/Roland_Berger_Additive_Manufacturing_20131129.pdf.

Li, Neng, Shuai Huang, Guodong Zhang, Renyao Qin, Wei Liu, Huaping Xiong, Gongqi Shi, and Jon Blackburn. 2019. "Progress in additive manufacturing on new materials: A review." *Journal of Materials Science & Technology* 35 (2):242–269. https://doi.org/10.1016/j.jmst.2018.09.002

Li, Yageng, Holger Jahr, Jie Zhou, and Amir Abbas Zadpoor. 2020. "Additively manufactured biodegradable porous metals." *Acta Biomaterialia* 115:29–50. https://doi.org/10.1016/j.actbio.2020.08.018

Liu, Chang, Min Zhang, and Changjun Chen. 2017. "Effect of laser processing parameters on porosity, microstructure and mechanical properties of porous Mg-Ca alloys produced by laser additive manufacturing." *Materials Science and Engineering: A* 703:359–371. https://doi.org/10.1016/j.msea.2017.07.031

Liu, Long, Haotian Ma, Chengde Gao, Cijun Shuai, and Shuping Peng. 2020. "Island-to-acicular alteration of second phase enhances the degradation resistance of biomedical AZ61 alloy." *Journal of Alloys and Compounds* 835:155397. https://doi.org/10.1016/j.jallcom.2020.155397

Maconachie, Tobias, Martin Leary, Bill Lozanovski, Xuezhe Zhang, Ma Qian, Omar Faruque, and Milan Brandt. 2019. "SLM lattice structures: Properties, performance, applications and challenges." *Materials & Design* 183:108137. https://doi.org/10.1016/j.matdes.2019.108137

Manne, Bhaskar, Harish Thiruvayapati, Srikanth Bontha, Ramesh Motagondanahalli Rangarasaiah, Mitun Das, and Vamsi Krishna Balla. 2018. "Surface design of Mg-Zn alloy temporary orthopaedic implants: Tailoring wettability and biodegradability using laser surface melting." *Surface and Coatings Technology* 347:337–349. https://doi.org/10.1016/j.surfcoat.2018.05.017

Nagarajan, Balasubramanian, Zhiheng Hu, Xu Song, Wei Zhai, and Jun Wei. 2019. "Development of Micro Selective Laser Melting: The State of the Art and Future Perspectives." *Engineering* 5 (4):702–720. https://doi.org/10.1016/j.eng.2019.07.002

Putra, N. E., M. J. Mirzaali, I. Apachitei, J. Zhou, and A. A. Zadpoor. 2020. "Multi-material additive manufacturing technologies for Ti-, Mg-, and Fe-based biomaterials for bone substitution." *Acta Biomaterialia* 109:1–20. https://doi.org/10.1016/j.actbio.2020.03.037

Qin, Y., P. Wen, H. Guo, D. Xia, Y. Zheng, L. Jauer, R. Poprawe, M. Voshage, and J. H. Schleifenbaum. 2019. "Additive manufacturing of biodegradable metals: Current research status and future perspectives." *Acta Biomater* 98:3–22. https://doi.org/10.1016/j.actbio.2019.04.046

Sahasrabudhe, Himanshu, and Amit Bandyopadhyay. 2018. "In situ reactive multi-material Ti6Al4V-calcium phosphate-nitride coatings for bio-tribological applications." *Journal of the Mechanical Behavior of Biomedical Materials* 85:1–11. https://doi.org/10.1016/j.jmbbm.2018.05.020

Sames, W. J., F. A. List, S. Pannala, R. R. Dehoff, and S. S. Babu. 2016. "The metallurgy and processing science of metal additive manufacturing." *International Materials Reviews* 61 (5):315–360. https://doi.org/10.1080/09506608.2015.1116649

Sarker, Avik, Nhiem Tran, Aaqil Rifai, Joe Elambasseril, Milan Brandt, Richard Williams, Martin Leary, and Kate Fox. 2018. "Angle defines attachment: Switching the biological response to titanium interfaces by modifying the inclination angle during selective laser melting." *Materials & Design* 154:326–339. https://doi.org/10.1016/j.matdes.2018.05.043

Sezer, Nurettin, Zafer Evis, and Muammer Koç. 2020. "Additive manufacturing of biodegradable magnesium implants and scaffolds: Review of the recent advances and research trends." *Journal of Magnesium and Alloys*. https://doi.org/10.1016/j.jma.2020.09.014

Shimizu, Yu, Shunsuke Fujibayashi, Seiji Yamaguchi, Shigeo Mori, Hisashi Kitagaki, Takayoshi Shimizu, Yaichiro Okuzu, Kazutaka Masamoto, Koji Goto, Bungo Otsuki, Toshiyuki Kawai, Kazuaki Morizane, Tomotoshi Kawata, and Shuichi Matsuda. 2020. "Bioactive effects of strontium loading on micro/nano surface Ti6Al4V components fabricated by selective laser melting." *Materials Science and Engineering: C* 109:110519. https://doi.org/10.1016/j.msec.2019.110519

Taniguchi, Naoya, Shunsuke Fujibayashi, Mitsuru Takemoto, Kiyoyuki Sasaki, Bungo Otsuki, Takashi Nakamura, Tomiharu Matsushita, Tadashi Kokubo, and Shuichi Matsuda. 2016. "Effect of pore size on bone ingrowth into porous titanium implants fabricated by additive manufacturing: An in vivo experiment." *Materials Science and Engineering: C* 59:690–701. https://doi.org/10.1016/j.msec.2015.10.069

Tao, Jun-Xi, Ming-Chun Zhao, Ying-Chao Zhao, Deng-Feng Yin, Long Liu, Chengde Gao, Cijun Shuai, and Andrej Atrens. 2020. "Influence of graphene oxide (GO) on microstructure and biodegradation of ZK30-xGO composites prepared by selective laser melting." *Journal of Magnesium and Alloys* 8 (3):952–962. https://doi.org/10.1016/j.jma.2019.10.004

Thompson, Scott M., Linkan Bian, Nima Shamsaei, and Aref Yadollahi. 2015. "An overview of Direct Laser Deposition for additive manufacturing; Part I: Transport phenomena, modeling and diagnostics." *Additive Manufacturing* 8:36–62. https://doi.org/10.1016/j.addma.2015.07.001

Todea, M., A. Vulpoi, C. Popa, P. Berce, and S. Simon. 2019. "Effect of different surface treatments on bioactivity of porous titanium implants." *Journal of Materials Science & Technology* 35 (3):418–426. https://doi.org/10.1016/j.jmst.2018.10.004

Wang, Yinchuan, Penghuai Fu, Nanqing Wang, Liming Peng, Bin Kang, Hui Zeng, Guangyin Yuan, and Wenjiang Ding. 2020. "Challenges and solutions for the additive manufacturing of biodegradable magnesium implants." *Engineering* 6 (11):1267–1275. https://doi.org/10.1016/j.eng.2020.02.015

Yan, Xingchen, Chunbao Shi, Taikai Liu, Yun Ye, Cheng Chang, Wenyou Ma, Chunming Deng, Shuo Yin, Hanlin Liao, and Min Liu. 2020. "Effect of heat treatment on the corrosion resistance behavior of selective laser melted Ti6Al4V ELI." *Surface and Coatings Technology* 396:125955. https://doi.org/10.1016/j.surfcoat.2020.125955

Yin, Yong, Qianli Huang, Luxin Liang, Xiaobo Hu, Tang Liu, Yuanzhi Weng, Teng Long, Yong Liu, Qingxiang Li, Shaoqiang Zhou, and Hong Wu. 2019. "In vitro degradation behavior and cytocompatibility of ZK30/bioactive glass composites fabricated by selective laser melting for biomedical applications." *Journal of Alloys and Compounds* 785:38–45. https://doi.org/10.1016/j.jallcom.2019.01.165

Yuan, Li, Songlin Ding, and Cuie Wen. 2019. "Additive manufacturing technology for porous metal implant applications and triple minimal surface structures: A review." *Bioactive Materials* 4:56–70. https://doi.org/10.1016/j.bioactmat.2018.12.003

Zhang, Lai-Chang, and Liang-Yu Chen. 2019. "A review on biomedical titanium alloys: recent progress and prospect." 21 (4):1801215. https://doi.org/10.1002/adem.201801215

Zhang, Wan-neng, Lin-zhi Wang, Zhong-xue Feng, and Yu-ming Chen. 2020a. "Research progress on selective laser melting (SLM) of magnesium alloys: A review." *Optik* 207:163842. https://doi.org/10.1016/j.ijleo.2019.163842

Zhang, Wan-neng, Lin-zhi Wang, Zhong-xue Feng, and Yu-ming Chen. 2020b. "Research progress on selective laser melting (SLM) of magnesium alloys: A review." *Optik* 207:1–15. https://doi.org/10.1016/j.ijleo.2019.163842

Zhao, Bingjing, Hong Wang, Ning Qiao, Chao Wang, and Min Hu. 2017. "Corrosion resistance characteristics of a Ti-6Al-4V alloy scaffold that is fabricated by electron beam melting and selective laser melting for implantation in vivo." *Materials Science and Engineering: C* 70:832–841. https://doi.org/10.1016/j.msec.2016.07.045

Zhou, M. Y., L. B. Ren, L. L. Fan, Y. W. X. Zhang, T. H. Lu, G. F. Quan, and M. Gupta. 2020. "Progress in research on hybrid metal matrix composites." *Journal of Alloys and Compounds* 838:155274. https://doi.org/10.1016/j.jallcom.2020.155274

Chapter 9

Indirect Rapid Tooling Methods in Additive Manufacturing

Gurminder Singh
University College Dublin, Dublin, Ireland

Pawan Sharma
Sardar Vallabhbhai National Institute of Technology, Surat, India

Kedarnath Rane
University of Strathcylde, Glasgow, United Kingdom

Sunpreet Singh
National University of Singapore, Singapore

CONTENTS

DOI: 10.1201/9781003327394-9

9.1 INTRODUCTION

For the past ten years additive manufacturing (AM) has been the most popular among manufacturing technologies. The opportunity to validate a specification within hours of finishing the computer aided design (CAD) data is among the key advantages of AM [1]. This has reduced the product production period for suppliers. The AM technique is increasingly common and used to create prototypes with significantly greater dimensional precision, while using much less time. Stereo-lithography (SLA), selective laser sintering (SLS), laminated object manufacturing (LOM), fused deposition modeling (FDM), direct metal laser sintering (DMLS) and 3D printing (3DP) are some AM techniques that are becoming popular in today's industry [2–5]. AM methods are beneficial in different research fields such as the biomedical [6,7], machining [8,9] and smart manufacturing [10]. The fundamental theory employed in AM techniques is layer by layer inclusion of the substance in the end product. Due to constraints on AM systems, however, producers quite frequently discover that the prototype cannot be obtained in the material of the end product needed. Furthermore, the technical features of the prototype vary somehow from those of the finished product generated by the final manufacturing phase because of the concept of prototype build. The technology of quick tooling is basically the technology that adopts and applies AM technology to tools and dies [11].

The fabrication of tools or patterns using AM for other manufacturing processes like elxtroforming, dies, casting and sintering is becoming popular. It provides ease in developing tools at low cost, rapidly and with accurately desired shapes and sizes. The process using tooling or patterns fabricated by AM is known as rapid tooling (RT) [12]. It is increasingly common and presents a significant challenge to the development of traditional techniques. Increasingly, manufacturers look at RT not only as an option for AM but also for quick runs, which do not merit the expenditure needed for traditional tools. Several RT innovations in the industry are now popular. Some of them generate the tool from the AM method directly. However, in a secondary phase, the majority of RT technologies manufacture the instrument utilizing the model produced by the AM process, such as patterns.

Three realistic applications are feasible for RT: (i) utilizing prototypes (e.g., SLA, FDM, SLS) as a mold, often after such changes as a coating; (ii) copying an AM shape onto metal, for example through investment casting; or (iii) specifically manufacturing metallic shapes or molds with modified rapid prototyping systems. In RT, the difficulty is due to the fact that the demands are more rigid than for prototyping.

A significant number of parts may be produced, sometimes by injection molding with SLA molds, but the reality is that, due to inherent material properties, there will still be a cap. It should be feasible to manufacture molds with metal prototyping/RT for more than ten times the quantity of parts than can be created with SLA molds. However, the difficulty lies in balancing power, precision and surface quality for quick tooling [13]. Widespread implementation may arrive as it becomes feasible to reach the reproducible standards for the different properties.

RT is an advancement on the existing AM technique. It could be defined as the ability to fabricate tool prototypes directly or indirectly using the existing polymer or metal-based AM systems. The RT techniques were developed with a major aim of developing long-term tooling which is consistent with market needs, that is a tool which is capable of producing several thousand or even millions of final products before finally it wears out completely. Referring to tooling, initially, mainly polymer-based injection molds were considered owing to their frequent usage as forming tools [15]. However, with further developments, RT technology has been used for the development of tooling for processes such as electric discharge machining (EDM), casting dies, molds for sintering, cutting tools and heat transfer applications. Figure 9.1 shows the details related to the various types of RT technologies and also the integration of RT with existing systems.

RT techniques are classified mainly in three basic categories as patterns for casting, and indirect and direct tooling [16]. Direct RT involves the direct use of a rapid prototyping (RP) system to fabricate tools and tooling inserts directly. On the other hand, in indirect RT techniques, RP systems are used to develop a pattern, based on which the tools and tooling inserts are fabricated. Finally, patterns for casting, also referred to as rapid casting, involves the use of RP patterns to fabricate final metal parts. These techniques are discussed below.

9.2 INDIRECT RAPID TOOLING

Indirect RT techniques for tool production, as discussed, require a minimum of one intermediate replication step. An intermediate step might cause a loss of part accuracy and also result in the increase of the time required for building the part [17]. Therefore, to overcome this limitation of the indirect RT technique, an alternative RT technique has been developed wherein the injection molding dies and inserts are built directly using 3D CAD models. These methods are known as direct RT techniques.

9.3 DIRECT RAPID TOOLING

Direct RT techniques allow the manufacturing of tools and die inserts and that are capable of producing parts varying from a few dozen to tens of thousands [18]. As per their use, direct RT techniques could be divided into two major groups, based on cost, lead time and life cycle of tooling. The first group involves a comparatively less expensive technique which has shorter

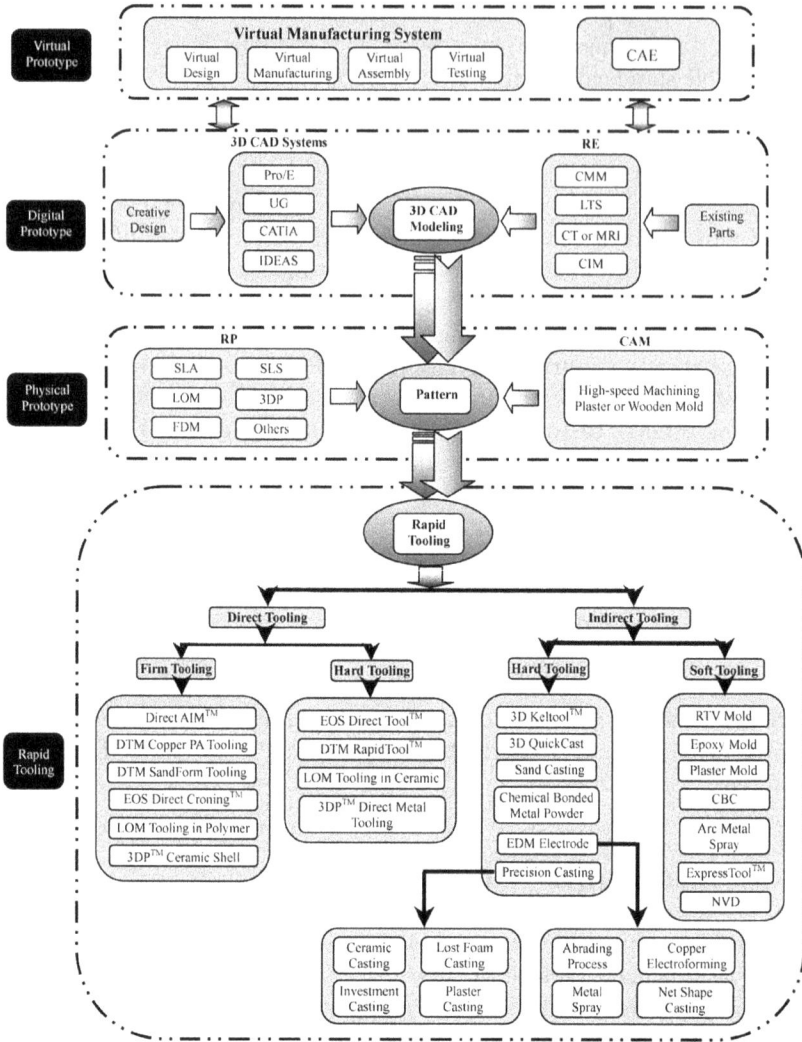

Figure 9.1 Details of the RT system [14].

lead-times. Direct RT techniques that belong to the first group are known as firm tooling, also commonly termed as bridge tooling [13,16,19]. Firm tooling based RT techniques generally bridge the gap that exists between soft tooling and hard tooling, thereby manufacturing tools having the capability of short prototype runs, varying approximately from fifty to a hundred parts. These parts are fabricated using a similar combination of material and production techniques as used for final parts.

The second group of direct RT techniques involves methods for the fabrication of inserts for pre-production as well as production tooling. This group of direct RT techniques is also referred to as hard tooling [13,16,19]. Hard tooling methods are generally based on the sintering of metal powders such as steel, iron and copper. The most commonly used direct RT techniques are Direct ACES™ Injection Moulds (AIM™), Laminate Tooling, LaserForm, SandForm™ tooling, EOS DirectTool™ process, direct metal tooling using 3D printing and Topographic Shape Formation (TSF) [19].

9.4 SOFT TOOLING

A reproduction of a positive pattern or master can be used to acquire soft instruments. The alternate concept is based on the rigidity and longevity of the tooling, which is known as polymeric tooling [20]. These two meanings are not identical, despite any overlap. We used the second concept, which allows soft tools to stand out from metallic or ceramic hard tools, as this topic was opened for scholarly debate.

9.5 PATTERN QUALITY BY AM PROCESS

Table 9.1 displays the consistency properties of untreated AM patterns. Many of the AM trends are well inside the appropriate tolerance range (±0.05 to ±0.254 mm) of most applications and surface quality (16–20 μm Ra). The investment casting (IC) explicitly interconnects the surface consistency of prototypes and castings. The relatively tough surfaces of untreated AM patterns are induced by layered design, construction methods (laser scanning, binder spray, etc.) and feed materials. The use of dyes (e.g., wax) to avoid slurry penetration during the processing of the shell includes patterns with pore surfaces (SLS, 3D-P, FDM). Polishing may be carried out to enhance surface consistency on most AM patterns. Screening or punching on a thin layer of methyl methacrylic (MMA), accompanied by light polishing, will intensify the surface of the acrylonitrile butadiene styrene (ABS) motifs. A surface ruggedness of 1–2 μm Ra was measured by us in MMA finished patterns. Similarly, epoxy-imprinted SLS sections accompanied by light polishing gave <1 μm Ra to the soil. Real patterns with a thin coat of fluid paraffin and polish may be completed to have an increased surface ruggedness of 0.8 μm Ra. Fine abrasive papers may further enhance the surface of other AM patterns (e.g., LOM paper patterns).

9.6 DIFFERENT RAPID TOOLING PROCESSES

Generally, RT techniques are based on existing AM systems such as SLA, FDM, LOM, SLS, binder jet 3D printing, inkjet printing, digital light processing (DLP printint) and solid ground curing (SGC). However, for the

Table 9.1 Pattern Features by Different AM Methods [21]

AM Method	Process	Material	Layer Thickness (mm)	Surface Roughness (μm)	Part Accuracy (mm)	Remaining Ash (%)
SLA	Photoploymerization	Epoxy, PMMA, ABS	0.1	12.5	±0.05	<0.01
Thermojet	Ink-Jet	Organic Polymer	0.04	5.090	NA	NA
SLS	Powder Sintering	Polyamide, Polycarbonate, Nylon	0.075	13	±0.25	<0.02
3DP	Ink-Jet	Starch	0.1	NA	±0.020	1–2
LOM	Paper Lamination	Paper	0.05	25	±0.25	NA
FFF	Melt Extrusion	ABS, Wax, PLA, PCL, PVA	0.05	12.5	±0.127	<1
SGC	Photocuring	Epoxy	0.06	25	0.1%	NA
DLP	Photocuring	PMMA, ABS, Epoxy	0.025	5.3	±0.05	<0.01
Binder Jetting 3DP	Binder Jetting	Sand, Polymamice, Metal	0.1	12	±0.127	<1

PMMA: poly(methyl methacrylate), ABS: acrylonitrile butadiene styrene, 3DP: 3D printing, FFF: fused filament fabrication, PLA: polylactic acid, PCL: polycaprolactone, PVA: polyvinyl alcohol, SGC: solid ground curing, DLP: digital light processing, SLS: selective laser sintering.

development of the RT process, conventional manufacturing techniques are used in conjunction with existing AM systems. There are many indirect RT methods that have been developed, such as silicon rubber tooling, epoxy resin tooling, cast metal tooling, spray metal tooling, 3D Keltool tooling, electroforming tooling and sintering tooling. The details of the conventional techniques used for RT technology development are discussed in the subsequent sections.

9.6.1 Electroforming

One of the conventional manufacturing methods used with the integration of RP systems is electroforming [22,23]. This technique can be applied to the fabrication of molds and precision tools. The electroforming process involves the electrodeposition of the metal onto the substrate. In RT using electroforming, a master pattern is printed using polymer-based RP systems followed by electrodeposition of the metal to fabricate an appropriate metal shell. The fabricated metal shell replicated using the master can be utilized for making molds and dies [23]. The process of rapid tooling using the electroforming technique (see Figure 9.2) is:

1. The master pattern having the desired shape, morphology, accuracy and surface finish is fabricated using polymer-based RP systems. The master pattern is the negative or complementary geometry of the required part to be used for some specific application. Furthermore, the master should possess enough rigidity to withstand the stress induced by the deposition process in electroforming.
2. Metallization of the polymer-based master pattern is carried out to achieve the electrically conductive master pattern. Generally, an electroless plating process is used for the metallization of the master pattern.
3. The metallized master is then immersed in a bath of electrolyte acting as the cathode. Thereafter, the deposition of a specified thickness of a metal layer, usually copper or nickel, on the metallized master is obtained.

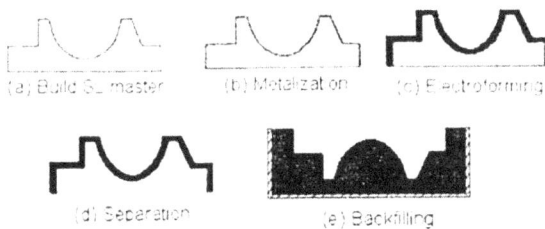

Figure 9.2 (a–e) Steps for RT using electroforming [23].

4. After deposition, the master is separated from the metal shell.
5. In a further step, the backing of the developed metal shell is achieved with some suitable materials to obtain a mold cavity or a tool such as an EDM electrode.

The emergence of polymer-based RP systems resulted in the development of new possibilities for the use of electroforming in RT technology. Since, using electroforming, geometries can be copied accurately and precisely, the integration of RP systems with electroforming may help in achieving complex shapes and cavities successfully.

9.6.2 Casting

The use of AM patterns in the casting of parts made up of metals is not always considered as RT. Rather it is referred to as rapid casting or rapid manufacturing. Nonetheless, it is a widely accepted application of AM in tooling developments. Conventional casting always requires a physical pattern having the morphological features of the required object. This pattern is used to produce molds into which metal is cast. Therefore, the use of AM fabricated patterns allows for the flexibility in the design of the product to be cast. Furthermore, complexity and customization in design become feasible using AM patterns in casting. It also supports the optimization of the casting design in terms of process as well as gating parameters. Hence, reduction in cost and time with more design flexibility is possible using AM with casting. The use of the AM system for different casting processes is discussed below.

9.6.3 Investment Casting

Investment casting is also known as sacrificial pattern casting wherein wax-based patterns are used for obtaining the desired product. Popular AM systems, such as FDM and SLS, are capable of fabricating patterns made of wax directly. Other AM systems such as SLA and Binder jet 3D printing can also be used for the fabrication of wax patterns for investment casting. Patterns made using SLA possess more accuracy and higher resolution compared to other AM systems, resulting in the manufacturing of highly accurate parts. However, the acrylic SLA patterns suffer from a limitation of expansion during the burn-out step thereby resulting in the cracking of the ceramic shell. To overcome this limitation castable resins have been developed for SLA machines. Another method to resolve this issue is the use of the QuickCast build style, wherein almost 95% of the part's internal mass, usually made of epoxy resin, is eliminated. During the burn-out step, the QuickCast pattern breaks down before the ceramic shell is overloaded,

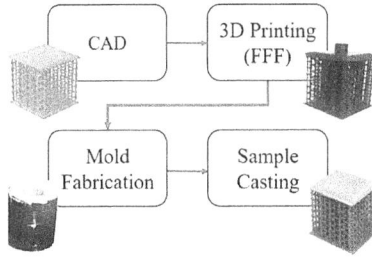

Figure 9.3 Work flow diagram for investment casting using the RT method [24].

producing only a small amount of ash. The work flow diagram for investment casting using the RT method is shown in Figure 9.3.

Pham and Dimov [25] used 3D printing methods like FDM and SLS to build an RT process for investment casting. However, issues with the subsequent removal of the sacrificial template and cracking of the shell mold have been recorded. Lee et al. [26] used an indirect investment casting model to examine the efficacy of the FDM-based RT approach (IC). The ABS-based sacrificial template was made using the FDM process and then used to prepare the mold for IC. The advantages of the new approach for the RT process were also documented, with significant cost and time reductions discovered. Also created were relatively precise items with an excellent surface polish. Bassoli et al. [27] investigated the effectiveness of 3D printed starch templates for the production of mold cavities for light alloy casting. It was found that the developed technique was effective in terms of the fabrication of prototypes for the casting of light alloys. Further, it was reported that the tooling phase was evaded which led to reduced fabrication time with dimensional tolerances within the limit of the required metal casting processes.

9.6.4 Sand Casting

A conventional sand casting process can also benefit from rapid prototyping and tooling technology. The fabrication of patterns and core boxes using AM systems helps in developing molds and cores, respectively. Furthermore, the addition of cores and internal cavities during the designing step may result in the reduction of processing time during casting. LOM is usually popular for this sand casting application, owing to the similarity of LOM fabricated patterns with that of wooden templates. Wang et al. [28] depicted the LOM based RT process to fabricate tooling for sand casting. It was reported that the geometry of the part is an important consideration in LOM based RT applications. Furthermore, it was found that the fabrication of geometrical features such as thin walls using the LOM based RT process is difficult. However, LOM-based RT was reported to be successful in achieving a reduction in the tooling cost and lead time.

9.7 SINTERING

Different methods of sintering, such as conventional sintering, pressureless sintering, microwave sintering and ultrasonic-assisted sintering, have been used successfully for the development of RT processes. These techniques are most commonly used for metal-based structures. Parts fabricated using a sintering based RT technique are used in applications such as biomedical implants, scaffolds and heat transfer. These techniques are discussed in detail in the subsequent sections.

9.7.1 Conventional Sintering

Conventional sintering can be used successfully with the integration of polymer-based 3D printing techniques for the development of RT techniques. Singh and Pandey [29] manufactured an open-cell-ordered interconnected porous structure made of metal using a combination of 3D printing and loose powder pressureless conventional sintering. They reported two different approaches for the fabrication of the metal structure. In the first approach, only metal particles having a spherical shape morphology were used. Metal particles and a solid metallic rod, on the other hand, were used in the second strategy. Staiger et al. [30] developed a six-step processing procedure. The initial stage in their research was to create a CAD model with the necessary architecture, which was then followed by the 3D printing of a positive polymer pattern based on the CAD model. Thereafter, NaCl paste was used which was further infiltrated using pressure inside the 3D printed template. After successful infiltration, the 3D printed pattern was removed, though burnt out, followed by sintering of NaCl to obtain the desired mold cavity. The negative NaCl mold was subsequently infiltrated using a liquid Mg (Timminco Ltd., commercially pure (99.98%)) at around 700 °C in the presence of argon (Grade 0). The infiltration pressure of 1.4–1.8 bar was used for pouring Mg into the NaCl mold. Finally, the desired Mg-based porous structure was manufactured through the dissolution of the NaCl mold in an NaOH solution. Singh and Pandey [31] provided a new method of RT using conventional sintering for bronze material. The fused filament fabrication (FFF) printed parts were used as a pattern for the mold preparation using investment powder. Later on a bronze powder was poured in to the mold and sintered in a furnace at high temperature. The method showed the ability to fabricate a solid bronze metal foam. The mechanical properties, density, dimensions and microstructure were evaluated.

9.7.2 Microwave Sintering

Microwave sintering (MS) has shown great promise in the production of engineered materials. The use of MS for metal processing, on the other hand, has been overlooked due to metal's inability to reflect microwaves.

To address this constraint, metal MS was obtained using silicon carbide or molybdenum disilicide susceptors [32]. After the microwave radiation is absorbed, the susceptors become heated. The heat created was then transmitted to the metal samples in order to achieve the necessary sintering activity. The MS of a copper and steel alloy was established by Anklekar et al. [33]. They compared the mechanical characteristics of MS alloys to those of commonly sintered copper steel alloys. It was found that, compared to traditional sintering, greater hardness, sintered density and flexural strength were attained. The presence of tiny, spherical-shaped and evenly distributed pores in MS samples is thought to be the cause. Saitou [34] investigated the shrinkage behavior of metal powders of iron, nickel, copper, cobalt and stainless steel in MS. When compared to traditional sintering, MS was found to induce greater shrinkage. In a comprehensive review of MS, Oghbaei and Mirzaee [35] discovered that sintering materials using microwave used less energy than traditional sintering. Furthermore, it was discovered that by employing microwaves, extremely high rates of heating may be achieved, resulting in a significant reduction in overall sintering time.

Sharma and Pandey [36–38] reported a novel processing route for the manufacturing of metal parts with complex and customized shapes using the systematic integration of polymer-based 3D printing and loose powder MS. The reported proceeding route comprises seven steps which begin with the development of the CAD model as per the desired application. Based on the CAD model, templates were printed using the stereolithography 3D printing technique. Thereafter, a mold using investment powders having a 3D printed template inside it was prepared. The desired mold cavity was then obtained through subsequent burn out of the polymer template. After the mold cavity was obtained, metal powders were placed inside it using ultrasonic vibrations followed by subsequent loose powder pressureless sintering in an MS furnace. Finally, the mold was removed and the metal part was taken out. The process flow diagram of the reported RT technique using MS is shown in Figure 9.4.

Recently, another RT process based on 3D printing and MS was developed by Mishra and Pandey [39]. In their work, a five-step RT methodology based on micro-extrusion 3D printing and MS was developed. The methodology begins with the creation of a 3D CAD model having the desired pore morphology followed by ink preparation. The ink consists of metal particles, necessary solvents, a polymer for 3D printing and a binding agent. After ink preparation, green parts were fabricated using micro-extrusion-based 3D printing. These parts were then sintered using a microwave furnace to fabricate the final metal part. Finally, post-processing was performed. Figure 9.5 depicts the stages involved in the reported methodology.

9.7.3 Ultrasonic Vibration Sintering

An ultrasonic vibration assisted sintering method was combined with SLA printing for the development of a unique RT method. The spherical

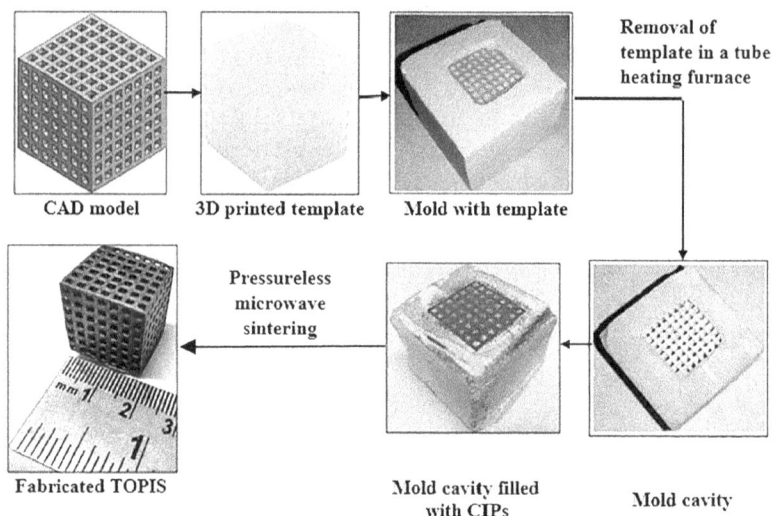

Figure 9.4 AM of iron foams using RT [36].

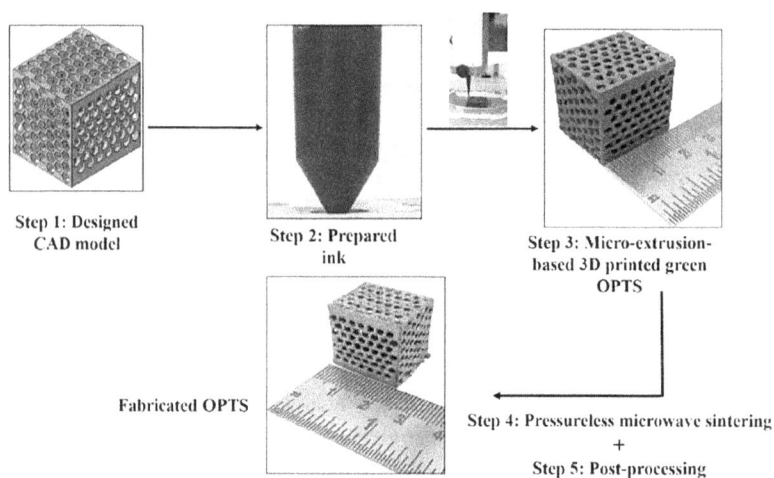

Figure 9.5 Additive manufacturing of iron foam using extrusion printing and sintering as an RT method [39].

morphology of particles gives sintering higher assurance and less dimensional shift in the product. These particles attach together to increase the density because of the decreasing surface energy. By utilizing ultrasonic resonance, the proportion density and compressive intensity of the material are very greatly improved. The changing of the densities may be attributed to many causes: ultrasonic waves of vibrations have been extended to particles within an ultrasonic horn by compression and rarefaction [40]. This helps in uniformly distributing the particles and raising a successful contact area of the particles in the sintering. This decrease in surface energy would contribute to far denser spheres. A price spike in newly moving property could be attributed to acoustic softening [41]. The exact cause of reducing the substance flow tension is the localized heating of particles around the dislocation with ultrasonic waves. The ultrasonic vibration is assumed to create local heat between the interaction of particles and the induced heat to decrease the tension from the material flow [42]. The incremental solidification of the foamed media culminated in the emergence of a thick system. To develop a proof of concept, a thermocouple was used and its temperature was tested [43]. It was found that the local temperature rose to an extreme 32°C with respect to room temperature with 100% ultrasonic control. As the sintering temperature rose, the ultrasonic vibration decreased the surface energy of the particles by losing the cumulative influence of the sintering temperature and ultrasonic vibrations. The mechanical properties of the samples significantly improved as the porosity reduced due to the ultrasonic vibration [44–46]—the fact that the filled deformation was localized on the necks or contacts between the particles. Singh and Pandey [47] used the ultrasonic vibration assisted sintering RT method for the fabrication of pure copper foams. The steps for the process are shown in Figure 9.6. The method results in the fabrication of pure copper foams with better surface roughness. It includes two different types of porosities: (i) a design porosity by CAD modeling and (ii) an internal micron porosity by pressureless sintering. The internal micron porosity can be varied by the sintering temperature for the different applications as per the requirement.

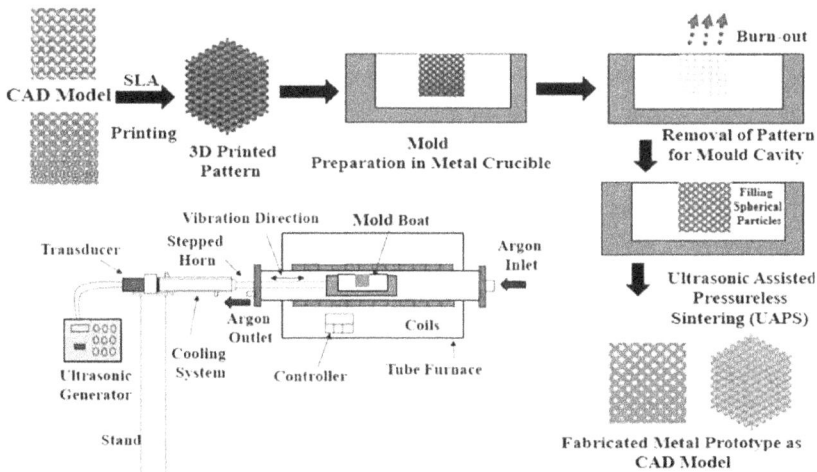

Figure 9.6 AM of pure copper foams using the RT method with ultrasonic sintering [47].

Table 9.2 Different Complex Shapes Fabricated by Indirect RT Methods

Material	Complex Shape Name	Application	Reference
Iron	Pyramid foam	Biomedical	[38]
Copper graphene composite	Simple cubic foam	Mechanical and thermal	[44]
Iron	Human skull scaffold	Biomedical	[48]
Bronze	Simple cubic foam	sound absorption and fluid flow control	[49]

9.8 APPLICATIONS OF INDIRECT RT METHODS

RT has various efficiencies that are used to manufacture the models in the current situation. RT is used to manufacture anything from toys and instruments to tools to patient-specific devices, like personalized implants. For example, you can effectively and efficiently create a solid implant while providing outstanding precision at an affordable cost by using RT, according to the five requirements outlined by the various applications. The five things are speed, expense, precision, ease of use, and content. RT gives freedom to fabricate complex shapes with different types of materials. Some of the examples of the complex shapes fabricated by indirect RT methods are given in Table 9.2.

9.8.1 Machining Tools

Indirect RT methods have a string capability to fabricate machining tools. The most popular application is for EDM electrode fabrication. EDM systems have switched to roughing, semiroughing and finishing because of their performance and precision in machining. The possibilities of AM electrodes in processing extremely complicated parts significantly boost them as the perfect solution for low cost EDM electrode production rather than the mechanical computer numerical control (CNC) milling that currently occurs.

Due to its high dimensional precision of epoxy materials and the excellent surface ruggedness on vertical and flat surfaces, the most successful AM technique, SLA, was first used by several researchers to manufacture RT electrodes with different variations. This contributed to the creation of scale model research equipment. For hard surface epoxy pieces, the electroless plating procedure is typically the preferred approach for changing the metal conductivity of plastic articles [50]. A flat paint covers better, especially on recesses and on details. Alternatively, electro-deposition is recommended for metal hardcoating. Copper is recommended over iron, nickel, zinc, aluminum, silver and other metals because of its strong electrical properties. Before surface metallization, it must be added to copper at room temperature to prevent the swelling of the copper shell owing to the epoxy providing a higher shrinkage factor. Due to the standardized deposition of copper during electroplating or electroforming, the current requires low average strain values, and the solution must have sufficient organic additives. Taking these into consideration, if a 0.6 mm thickness of copper is electroplated on a steel tool component, it will erode a 6 mm deep cavity in the tool part without being affected [51]. Nevertheless, the accuracy of electroplated copper electrodes is reduced owing to the irregular thickness of the copper shell. The electroplated shell thickness is influenced by the electric field and also the shape of the component. Besides these, a strictly regulated electrodeposition process should be used for an even shell thickness. Even if you change the standard tessellation language (STL) model to compensate for the thickness of the shell, a certain inaccuracy can occur because of the uneven shell thickness. This issue is more apparent on the corners of copper shell electrodes. This is attributed to the degradation and damage of RT electrodes. The sintering method combined with SLA as RT has been explored for the fabrication of complex shaped EDM electrodes with and without cooling channels [52,53].

9.8.2 Biomedical

In order to satisfy diverse applications ranging from hard to soft tissues, tissue interfaces and the construction of vascular networks, indirect scaffolds have been constructed based on a multiform style by an RT system and on a wide ranging of materials. By way of the integration of medical imaging and 3D processing techniques, tissue engineering implants may be produced for patient-specific and defect-specific purposes. RT has also been applied to create microfluidic and vascular networks. Two types of RT methods, that is internal indirect rapid prototyping (iiRP) and external indirect rapid prototyping (eiRP), are described in this section. In iiRP, the topological ordered implant can be fabricated by a casting method, in which a porous structure is fabricated by FFF methods and later a casted material is poured into the porous structure. The FFF printed structure is generally best fabricated by polyvinyl alcohol (PVA) which is water soluble. After cooling down the casting, the FFF printed part isused to print topologically ordered implants.

In the second method eiRP, two types of porosity are achieved. The shape of the 3D model is fabricated by the FFF method and the casted material mixed with space holder granules is poured into the FFF printed template. After cooling, both template and space holding granules are removed. This results in the fabrication of an implant with similar shape to the CAD model and also has internal porosity.

9.8.3 Other Applications

RT scientific breakthroughs are constantly revolutionizing casting. RT has been used for numerous investment casting products to date, varying from jewelry casting, sports equipment and surgical implants to high-performance components used in the automobile and aerospace industries, die-casting or injection molding industries.

9.9 BENEFITS OF RAPID TOOLING

While test models can now be created using different AM systems very easily, they are still not generated in the final product content and in the final phase of development. Such verification is now often requested by both designers and managers, before mass production starts. The development method, for instance the injection molding process, must have a prototype in the purest meaning of the term. This method of measurement and review is only realistic when AM is applied to instruments. Conventional injection molding tools need significant expenditure in resources and expenses. Today it is becoming more critical for producers to be the first on the market with their goods due to the globalization of customer markets and the consequent rise in the number of rivals confronting each individual producer. With RT, active case studies have shown that production time can be cut by at least half. For pre-series development, RT is ideally adapted. The development method is planned, albeit in limited numbers (about 500 pieces). Pre-series manufacturing typically includes checking development facilities and instruments and evaluating a product's consumer launch. The mechanical efficiency of a molded component depends on the specification, the properties of the material and the production process. As an example, certain output variables as position, fill pattern and comer radii, and thickness, decide the molecular orientation and the internal stresses of the plastic segment.

In the design of a plastic component, part geometry often plays an important role. Often wall parts may tend to be ideal for the shape to match requirements, but may not be moldable. The walls can be too thin to make a decent flux of plastics or the plastic can deform during the refreshing process in the case of thick parts. In the plastic pieces, sink marks can also occur. This happens if the exterior surface of the mold solidifies, though the internal material contraction allows the skin to depress below its expected profile. These forms of problems emphasize the significance of the geometry, components and molding processes of a component while prototyping. Notwithstanding major advancements in AM, the content available for prototype component manufacturing is still restricted to certain materials which can be used by different AM processes to create components. In certain instances, designers are involved in creating experiments in the components to be used in total manufacturing. This may be aluminum, glass or thermoplastics of different sorts. More research has been done to adapt RT to manufacturing the molds and instruments required for casting prototype pieces to resolve this issue.

9.10 FUTURE SCOPE AND SUMMARY

The tooling system selected is of little value from an industrial point of view. The final goal of RT creation is a highly efficient tool manufacturing process, which includes the manufacture of the CAD model, mold measurement, file fixing, smart and exact model sliding, transmission of data on RT equipment, pace and dependability of mold production, dimensional control and finishing equipment.

Several essential study avenues may be illustrated from the facts outlined in the previous section. At present, improvement is highly desired for the most successful techniques. SLS seems to have the greatest potential; the metal sintering method by RT is available at a developing stage, but more developments are expected in terms of mechanical properties. The best way of manufacturing extremely dense and solid elements should prevent as many post-treatments as possible (post-sintering, infiltration, polishing, etc.). This includes the sorting of a vast variety of products. A cure for high apparent density may be the continuous solidification of the front. Furthermore, there are other possible research directions for RT methods (see Figure 9.7). Pattern roughness improvement is still challenging even when working with SLA methods. Different surface finishing processes after pattern fabrication need to be explored. Also, to prevent any cracks during mold cavity formation, the use of low density materials are required for pattern 3D printing. Some other aspects such as *in situ* monitoring during processes like casting, sintering and electroforming could be explored for evaluating defects during the process. Furthermore, thermal treatments have not much been explored for the RT method using sintering to enhance or tune the properties of the fabricated materials. Industrial accreditation is most challenging for RT methods, which needs to be standardized for the growth of different RT methods at industry level.

Due to strong global competition, innovative technologies must be produced, assembled and sold rapidly and at a lower cost. While RT component production, with a view to aesthetics, ergonomics and fitment, has been helpful in attaining these goals for some time, newer approaches have added fresh perspectives to the mix.

RT, which was originally developed from AM, offers low-cost and short-time prototyping benefits to a particularly significant business market. Established techniques are various, and several of them have been effective in casting and injection molding. Researchers are devising different approaches to render tools directly on AM computers. The first manufacturing instruments were produced utilizing products that could be used for molds and dies, thereby expediting development. Famed corporations and

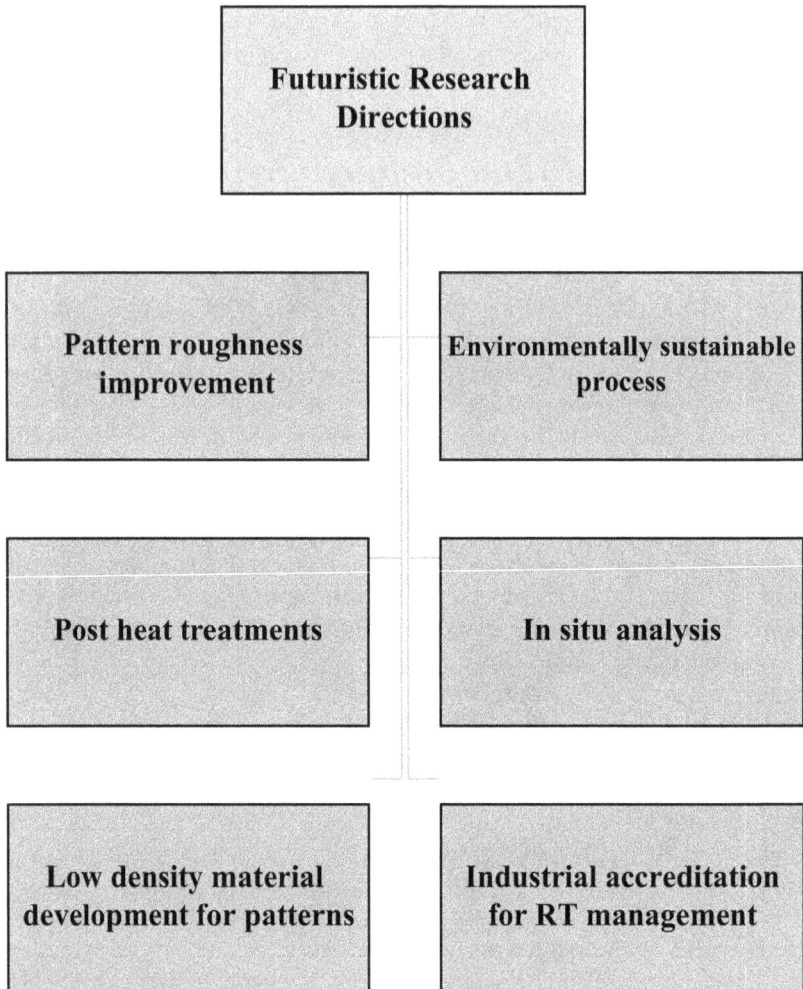

Figure 9.7 Possible research directions for RT methods.

colleges are highly engaged in the latest developments in RT. Many associated engineering practices have also been influenced by this latest development in production. Many things in RT are being explored for development at the commercial level.

REFERENCES

[1] K. Rane, M. Strano, A comprehensive review of extrusion-based additive manufacturing processes for rapid production of metallic and ceramic parts, *Adv. Manuf.* 7 (2019) 155–173. https://doi.org/10.1007/s40436-019-00253-6

[2] S. Singh, S. Ramakrishna, R. Singh, Material issues in additive manufacturing: A review, *J. Manuf. Process*. 25 (2017) 185–200. https://doi.org/10.1016/j.jmapro.2016.11.006

[3] S. Singh, G. Singh, C. Prakash, S. Ramakrishna, Current status and future directions of fused filament fabrication, *J. Manuf. Process*. 55 (2020) 288–306. https://doi.org/10.1016/j.jmapro.2020.04.049

[4] G. Singh, J.M. Missiaen, D. Bouvard, J.M. Chaix, Copper extrusion 3D printing using metal injection moulding feedstock: analysis of process parameters for green density and surface roughness optimization, *Addit. Manuf*. 38 (2021) 101778. https://doi.org/10.1016/j.addma.2020.101778

[5] G. Singh, J. Missiaen, D. Bouvard, J. Chaix, Copper additive manufacturing using MIM feedstock: adjustment of printing, debinding, and sintering parameters for processing dense and defectless parts, *Int. J. Adv. Manuf. Technol*. (2021). https://doi.org/10.1007/s00170-021-07188-y

[6] A. Pandey, G. Singh, S. Singh, K. Jha, C. Prakash, 3D printed biodegradable functional temperature-stimuli shape memory polymer for customized scaffoldings, *J. Mech. Behav. Biomed. Mater*. 108 (2020) 103781. https://doi.org/10.1016/j.jmbbm.2020.103781

[7] G. Singh, S. Singh, C. Prakash, R. Kumar, R. Kumar, S. Ramakrishna, Characterization of three-dimensional printed thermal- stimulus polylactic acid-hydroxyapatite-based shape memory scaffolds, *Polym. Compos*. (2020) 1–21. https://doi.org/10.1002/pc.25683

[8] K. Sandhu, G. Singh, S. Singh, R. Kumar, C. Prakash, S. Ramakrishna, G. Królczyk, C.I. Pruncu, Surface characteristics of machined polystyrene with 3D printed thermoplastic tool, *Materials (Basel)*. 13 (2020) 1–16. https://doi.org/10.3390/ma13122729

[9] S. Singh, G. Singh, K. Sandhu, C. Prakash, R. Singh, Materials Today: Proceedings Investigating the optimum parametric setting for MRR of expandable polystyrene machined with 3D printed end mill tool, *Mater. Today Proc*. (2020) 3–7. https://doi.org/10.1016/j.matpr.2020.03.465

[10] A. Kumar, G. Singh, R.P. Singh, P.M. Pandey, Role of Additive Manufacturing in Industry 4.0 for Maintenance Engineering, in: *Appl. Challenges Maint. Saf. Eng. Ind. 4.0*, (2020) 235–254.

[11] G. Singh, P.M. Pandey, Rapid manufacturing of copper-graphene composites using a novel rapid tooling technique, *Rapid Prototyp. J*. (2020). https://doi.org/10.1108/RPJ-10-2019-0258

[12] S. Ma, I. Gibson, G. Balaji, Q.J. Hu, Development of epoxy matrix composites for rapid tooling applications, *J. Mater. Process. Tech*. 193 (2007) 75–82. https://doi.org/10.1016/j.jmatprotec.2007.04.086

[13] C.K. Chua, K.H. Hong, S.L. Ho, Rapid Tooling Technology. Part 1. A Comparative Study, *Int. J. Adv. Manuf. Technol*. 15 (1999) 604–608.

[14] Y. Ding, H. Lan, J. Hong, D. Wu, An integrated manufacturing system for rapid tooling based on rapid prototyping, *Robot. Comput. Integr. Manuf*. 20 (2004) 281–288. https://doi.org/10.1016/j.rcim.2003.10.010

[15] G.N. Levy, R. Schindel, J.P. Kruth, Rapid manufacturing and rapid tooling with layer manufacturing (LM) technologies, state of the art and future perspectives, *CIRP Ann. - Manuf. Technol*. 52 (2003) 589–609. https://doi.org/10.1016/S0007-8506(07)60206-6

[16] A. Rosochowski, A. Matuszak, Rapid tooling: the state of the art, *J. Mater. Process.Technol.* 106 (2000) 191–198.https://doi.org/10.1016/S0924-0136(00) 00613-0

[17] A. Houben, J. Van Hoorick, J. Van Erps, H. Thienpont, S. Van Vlierberghe, P. Dubruel, Indirect Rapid Prototyping: Opening Up Unprecedented Opportunities in Scaffold Design and Applications, *Ann. Biomed. Eng.* 45 (2017) 58–83. https://doi.org/10.1007/s10439-016-1610-x

[18] L. Yi, C. Gläßner, J.C. Aurich, How to integrate additive manufacturing technologies into manufacturing systems successfully: A perspective from the commercial vehicle industry, *J. Manuf. Syst.* 53 (2019) 195–211. https://doi. org/10.1016/j.jmsy.2019.09.007

[19] D.T. Pham, S.S. Dimov, Rapid prototyping and rapid tooling — the key enablers, *Proc. Instn. Mech. Engrs.* 217, Part C (2003) 1–23.

[20] C.M.C.C.K. Chua, C.W.L.C. Feng, Rapid prototyping and tooling techniques: a review of applications for rapid, *J. Adv. Manuf. Technol.* (2005) 308–320. https://doi.org/10.1007/s00170-003-1840-6

[21] P.M. Dickens, R. Stangroom, M. Greul, B. Holmer, R. Hovtun, R. Neumann, D. Wimpenny, Conversion of RP models to investment castings, *Rapid Prototyp. J.* 1 (1995) 4–11.

[22] M. Monzón, A.N. Benítez, M.D. Marrero, N. Hernández, P. Hernández, J. Aisa, Validation of electrical discharge machining electrodes made with rapid tooling technologies, *J. Mater. Process. Technol.* 196 (2008) 109–114. https://doi. org/10.1016/j.jmatprotec.2007.05.025

[23] B. Yang, M.C. Leu, Integration of rapid prototyping and electroforming for tooling application, *CIRP Ann. - Manuf. Technol.* 48 (1999) 119–122. https:// doi.org/10.1016/S0007-8506(07)63145-X

[24] V.H. Carneiro, S.D. Rawson, H. Puga, J. Meireles, P.J. Withers, Additive manufacturing assisted investment casting: A low-cost method to fabricate periodic metallic cellular lattices, *Addit. Manuf.* 33 (2020) 101085. https://doi. org/10.1016/j.addma.2020.101085

[25] D.T. Pham, S.S. Dimov, Rapid prototyping and rapid tooling — the key enablers for rapid manufacturing, *Proc. Instn Mech. Engrs Part C J. Mech. Eng. Sci.* 217 (2002) 1–23.

[26] C.W. Lee, C.K. Chua, C.M. Cheah, L.H. Tan, C. Feng, Rapid investment casting: direct and indirect approaches via fused deposition modelling, *Int J Adv Manuf Technol.* 23 (2004) 93–101. https://doi.org/10.1007/ s00170-003-1694-y

[27] E. Bassoli, A. Gatto, L. Iuliano, M.G. Violante, 3D printing technique applied to rapid casting, *Rapid Prototyp. J.* 13 (2007) 148–155. https://doi. org/10.1108/13552540710750898

[28] W. Wang, J.G. Conley, H.W. Stoll, Rapid tooling for sand casting using laminated object manufacturing process, *Rapid Prototyp. J.* 5 (2007) 134–141.

[29] J.P. Singh, P.M. Pandey, Fabrication and characterization of open cell porous regular interconnected metallic structure with solid core, *Proc. Inst. Mech. Eng. Part B J. Eng. Manuf.* 232 (2018) 305–316. https://doi.org/10.1177/09544054 16641324

[30] M.P. Staiger, I. Kolbeinsson, N.T. Kirkland, T. Nguyen, G. Dias, T.B.F. Woodfield, Synthesis of topologically-ordered open-cell porous magnesium, *Mater. Lett.* 64 (2010) 2572–2574. https://doi.org/10.1016/j.matlet.2010.08.049

[31] J.P. Singh, P.M. Pandey, Fabrication and assessment of mechanical properties of open cell porous regular interconnected metallic structure through rapid manufacturing route Article information, *Rapid Prototyp. J.* 24 (2017) 138–149. https://doi.org/10.1108/RPJ-04-2015-0043

[32] R. Roy, D. Agrawal, J. Cheng, S. Gedevanishvili, Full sintering of powdered-metal bodies in a microwave field, *Nat. Publ. Gr.* 399 (1999) 668–670.

[33] R.M. Anklekar, D.K. Agrawal, R. Roy, Microwave sintering and mechanical properties of PM copper steel, *Powder Metallurgy.* 44 (2001) 355–362.

[34] K. Saitou, Microwave sintering of iron, cobalt, nickel, copper and stainless steel powders, *Scr. Mater.* 54 (2006) 875–879. https://doi.org/10.1016/j.scriptamat.2005.11.006

[35] M. Oghbaei, O. Mirzaee, Microwave versus conventional sintering: A review of fundamentals, advantages and applications, *J. Alloys Compd.* 494 (2010) 175–189. https://doi.org/10.1016/j.jallcom.2010.01.068

[36] P. Sharma, P.M. Pandey, A novel manufacturing route for the fabrication of topologically-ordered open-cell porous iron scaffold, *Mater. Lett.* 222 (2018) 160–163. https://doi.org/10.1016/j.matlet.2018.03.206

[37] P. Sharma, P.M. Pandey, Rapid manufacturing of biodegradable pure iron scaffold using amalgamation of three-dimensional printing and pressureless microwave sintering, *Proc. Inst. Mech. Eng. Part C J. Mech. Eng. Sci.* 0 (2018) 1–20. https://doi.org/10.1177/0954406218778304

[38] P. Sharma, P.M. Pandey, Morphological and mechanical characterization of topologically ordered open cell porous iron foam fabricated using 3D printing and pressureless microwave sintering, *Mater. Des.* 160 (2018) 442–454. https://doi.org/10.1016/j.matdes.2018.09.029

[39] D.K. Mishra, P.M. Pandey, Mechanical behaviour of 3D printed ordered pore topological iron scaffold, *Mater. Sci. Eng. A.* 783 (2020) 139293. https://doi.org/10.1016/j.msea.2020.139293

[40] G. Singh, S. Singh, J. Singh, P.M. Pandey, Parameters effect on electrical conductivity of copper fabricated by rapid manufacturing, *Mater. Manuf. Process.* 00 (2020) 1–12. https://doi.org/10.1080/10426914.2020.1784937

[41] G. Singh, P.M. Pandey, Design and Analysis of Long-Stepped Horn for Ultrasonic Assisted Sintering, in: *21st Int. Conf. Adv. Mater. Process. Technol.*, Dublin, Ireland, 2018.

[42] G. Singh, P.M. Pandey, Neck growth kinetics during ultrasonic-assisted sintering of copper powder, *Proc. Inst. Mech. Eng. Part C J. Mech. Eng. Sci.* 0 (2020) 1–11. https://doi.org/10.1177/0954406220904108

[43] G. Singh, P.M. Pandey, Rapid manufacturing of copper components using 3D printing and ultrasonic assisted pressureless sintering: experimental investigations and process optimization, *J. Manuf. Process.* 43 (2019) 253–269. https://doi.org/10.1016/j.jmapro.2019.05.010

[44] G. Singh, P.M. Pandey, Topological ordered copper graphene composite foam: Fabrication and compression properties study, *Mater. Lett.* 257 (2019) 1–5. https://doi.org/10.1016/j.matlet.2019.126712

[45] G. Singh, P.M. Pandey, Ultrasonic Assisted Pressureless Sintering for rapid manufacturing of complex copper components, *Mater. Lett.* 236 (2019) 276–280. https://doi.org/10.1016/j.matlet.2018.10.123

[46] G. Singh, P.M. Pandey, Experimental investigations into mechanical and thermal properties of rapid manufactured copper parts, *Proc. Inst. Mech. Eng.*

Part C J. Mech. Eng. Sci. 234 (2020) 82–95. https://doi.org/10.1177/0954
406219875483

[47] G. Singh, P.M. Pandey, Uniform and graded copper open cell ordered foams
fabricated by rapid manufacturing: surface morphology, mechanical properties
and energy absorption capacity, *Mater. Sci. Eng. A.* 761 (2019) 138035. https://
doi.org/10.1016/j.msea.2019.138035

[48] P. Sharma, P.M. Pandey, Rapid manufacturing of biodegradable pure iron scaffold using amalgamation of three-dimensional printing and pressureless microwave sintering, *Proc. Inst. Mech. Eng. Part C J. Mech. Eng. Sci.* 0 (2018) 1–20.
https://doi.org/10.1177/0954406218778304

[49] J.P. Singh, P.M. Pandey, Fabrication and assessment of mechanical properties
of open cell porous regular interconnected metallic structure through rapid
manufacturing route, *Rapid Prototyp. J.* 24 (2018) 138–149. https://doi.
org/10.1108/RPJ-04-2015-0043

[50] D.E. Dimla, N. Hopkinson, H. Rothe, Investigation of complex rapid EDM
electrodes for rapid tooling applications, *Int. J. Adv. Manuf. Technol.* 23 (2004)
249–255. https://doi.org/10.1007/s00170-003-1709-8

[51] S.J. Dover, A. Rennie, G.R. Bennett, Rapid prototyping using electrodeposition
of copper, 8 (1996) 191–198. http://eprints.lancs.ac.uk/50510/

[52] J. Singh, G. Singh, P.M. Pandey, Electric discharge machining using rapid
manufactured complex shape copper electrode with cryogenic cooling channel, *Proc. Inst. Mech. Eng. Part B J. Eng. Manuf.* (2020) 1–13. https://doi.
org/10.1177/0954405420949102

[53] J. Singh, G. Singh, P.M. Pandey, Electric discharge machining using rapid manufactured complex shape copper electrode: Parametric analysis and process
optimization for material removal rate, electrode wear rate and cavity dimensions, *Proc. Inst. Mech. Eng. Part C J. Mech. Eng. Sci.* 0 (2020) 1–15. https://
doi.org/10.1177/0954406220906445

Chapter 10

Laser Additive Manufacturing of Nickel Superalloys for Aerospace Applications

S. K. Nayak and A. N. Jinoop

Homi Bhabha National Institute, Mumbai, India

Raja Ramanna Centre for Advanced Technology, Indore, India

S. Shiva

Indian Institute of Technology Jammu, Jammu & Kashmir, India

C. P. Paul

Homi Bhabha National Institute, Mumbai, India

Raja Ramanna Centre for Advanced Technology, Indore, India

CONTENTS

DOI: 10.1201/9781003327394-10

10.1 INTRODUCTION

Traditionally, manufacturing in the industrial sectors has been dominated by subtractive and formative manufacturing techniques. Subtractive manufacturing techniques (like machining and polishing) use material removal for generating a 3D component, while formative manufacturing techniques (like injection moulding, die casting, pressing and stamping) involve forming the materials into the desired shape for fabricating a 3D component. Both subtractive and formative techniques face extreme challenges when manufacturing a component, which either has a complex shape or complex material distribution over the entire component volume. Industries generally use five-axis computerised numerical control (CNC) machining to build complex-shaped components. However, being a subtractive manufacturing technology it results in a lot of material wastage and a high buy-to-fly ratio. The answer to these challenges was provided by an innovative manufacturing technology known as additive manufacturing (AM). According to ISO/ASTM 52900, AM is defined as the "process of joining materials to make parts from 3d model data usually layer upon layer as opposed to subtractive manufacturing and formatting manufacturing methodologies" (*Additive Manufacturing-General Principles-Terminology*, 2018). The advent of AM proved to be a game-changer in the industries. With the deployment of AM, highly complex and customised components can be built in a layer-wise fashion with minimal lead time. Because of its highly customisable nature, different sectors in the industries started looking into AM for exploring new product designs to improve their output. Laser can provide sufficient energy to melt metals and good precision in the built components. Metal additive manufacturing (MAM) techniques using lasers are known as laser additive manufacturing (LAM). LAM techniques are increasingly being adopted commercially for engineering applications in different sectors. The aerospace sector is the one that has enthusiastically adopted LAM for meeting its requirements to build structures/components requiring complex profiles and light-weighting. Aerospace industries use LAM of different materials in powder or wire form for various applications. Some commercially popular alloys for LAM are Ti, Fe, Co and Ni based. Ni-based superalloys are the family of alloys having an Ni-matrix with a composition designed for very-high-temperature applications, that is the strength, corrosion and oxidation resistance of the Ni-superalloys are retained at elevated temperature conditions. This chapter describes different LAM processes, the LAM of an Ni-superalloy, commonly used Ni-superalloys in LAM, along with specific processing challenges and the latest research and applications in the aerospace sector.

10.2 THE LAM OF Ni-SUPERALLOYS

Ni-superalloys are one of the popular materials in the aerospace sector. These materials are often used to build components that are subjected to high-temperature conditions. These materials have excellent mechanical strength, creep resistance, and oxidation and corrosion resistance at elevated temperatures. The Ni-superalloys are primarily used for building aerospace engine components that need mechanical strength to be sustained at elevated temperature conditions over a long period. Since most of these components have complex geometries, LAM is an attractive manufacturing technology.

Generally, metallurgical studies of LAM-built Ni-superalloys show melt pool features throughout the entire volume of the built part, consisting of fine columnar growth that is generally observed due to the microstructural growth normal to the boundary of the melt pool. This is primarily due to rapid cooling and heat dissipation that typically takes place normal to the boundary of the melt pool. Due to this a strong texturing effect is generally seen in LAM-built components along the build direction, which can result in anisotropy in mechanical properties. In Ni-superalloys, preferential growth is generally seen along the (100) plane. The fine nature of the grains and higher dislocation density contribute to the higher mechanical strength of the LAM-built samples when compared with the wrought annealed samples and cast samples.

The defects in the LAM-built components can be primarily porosity, cracks or delamination. The porosity can be caused by improper consolidation of the melt pool during solidification or by the evolution of dissolved gases. Cracks are initiated in the LAM-built component when residual stresses in a particular region exceed the ultimate tensile stress (UTS). Cracking can also be due to elemental segregation, which leads to hot cracking. Delamination is another severe issue where the layers in the built component separate from each other when the inter-layer bond strength is overcome by the induced residual stress. These defects act as regions of stress concentration and often lead to a decrease in life during loading conditions. The defects can be reduced by adopting proper pre-processing, processing and post-processing techniques. Preheating the feedstock material or substrate are some of the pre-processing techniques that can reduce the residual stress generated during the process, which can inhibit the cracking and delamination to a greater extent. Processing methods for controlling defects include the selection of optimal parameters and the choice of scanning strategies to reduce defects and obtain dense components. Further improvement in the density or life of the component by removing the localised defects can be achieved by adopting post-processing techniques like hot isostatic pressing (HIPing). HIPing helps in closing the sub-surface defects, resulting in a nearly 100% dense component. As HIPing is generally carried out above the recrystallisation temperature, it can generate coarse recrystallised equiaxed grains that can improve the isotropy of the component. Heat treatments are the most common post-processing techniques for LAM-built components and the major aims include achieving a uniform microstructure, reducing residual stress and the generation of strengthening phases. The choice of proper heat treatment or HIPing conditions is critical as it governs the phases that would be redissolved in the matrix and the phases that would be precipitated/segregated, which might either affect the mechanical properties positively or negatively. Poor surface quality is another issue in LAM, which reduces the component life due to fatigue. Machining as a post-processing operation can be used for removing the rough surface and obtain a uniform surface. This also improves the aesthetics of the component.

10.3 LAM PROCESSES

A LAM process starts with a digital file as an input, which is further converted into machine-readable format (.stl, .obj, .amf, etc.) and sliced into multiple 2D layers as per the chosen layer thickness. Subsequently, the tool path is generated as per the 2D layers and it directs the laser movement which then consolidates and deposits material in a particular layer. This step is repeated for each and every layer until the entire component is fabricated. LAM can be mainly classified into two: (i) laser powder bed fusion (LPBF) and (ii) laser directed energy deposition (LDED) (Nayak, Mishra, Paul, et al., 2020b; Yadav et al., 2020).

LPBF involves laser movement over a preplaced bed of powder. A schematic of LPBF is shown in Figure 10.1a. A powder re-coater spreads powder over the build area. Subsequently, the laser movement is directed over the powder layer as per the 2D geometry for the particular layer using a Galvano-scanner. After the laser movement finishes consolidating the layer geometry, the build platform goes down and a new powder layer again spreads over it and the process repeats. In the process, a seamless binding mechanism ensures the adherence of the current layer with the previous layer (Paul et al., 2020). The parameters primarily governing the LPBF process are laser power (P), scan speed (v), powder layer thickness (t) and hatch spacing (h). The combined parameters commonly used to evaluate LPBF is laser energy density (LED), which is shown in Equation (10.1):

$$LED = \frac{P}{vht}.$$ (10.1)

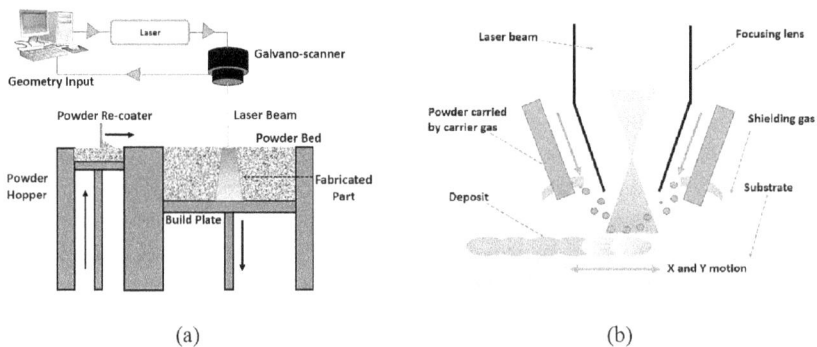

$$(a) \qquad\qquad (b)$$

Figure 10.1 Schematic diagram of (a) LPBF and (b) LDED.

The LDED process involves supplying material using a feedstock onto a melt-pool generated using a laser. Figure 10.1b show a schematic diagram of the LDED system. The laser head and the material feeding head move together for creating continuous material deposition. The material used for LDED can be either in the wire form or powder form. Accordingly, the process can be called wire-fed deposition or powder-fed deposition. The process parameters involved in the LDED process at a constant beam diameter are P, v, h and the powder feed rate f. The combined parameters commonly used to evaluate LDED are laser energy per unit length (LEL), powder feed per unit length (PFL) and laser energy per unit powder feed rate (LEPF) as shown in Equations (10.2)–(10.4):

$$LEL = \frac{P}{v} \tag{10.2}$$

$$PFL = \frac{f}{v} \tag{10.3}$$

$$LEPF = \frac{P}{f} \tag{10.4}$$

10.4 LAM-PROCESSED NI-SUPERALLOYS FOR AEROSPACE APPLICATIONS

10.4.1 Inconel 718 (IN718)

IN718 is an Ni–Cr alloy that is age hardenable and finds application in the temperature range –252 to 704°C. It possesses high strength and is highly corrosion-resistant. It contains a significant amount of Nb, Fe and Mo and small amounts of Ti and Al. The material possesses outstanding welding characteristics and has a high post-weld crack resistance; 50% of turbojet engine parts are made from this material and thus it is referred to as the workhorse alloy of the aerospace sector. Some examples of the parts built using IN718 are the discs, blades and casing of the high-pressure part of the compressors of turbines. The high toughness in cryogenic conditions protects rocket engine components from brittle tr ansformation, which attracts its use in the space sector, including applications in spacecraft and space shuttles. Other uses are in nuclear reactors, pumps, the seals of turbopumps and tooling (*INCONEL ® Alloy 718*, 2020). IN718's typical composition is given in Table 10.1.

The primary phase in IN718 is γ, an austenitic face-centred cubic Ni–Fe–Cr solid solution phase. Elements like Mn, Mo and C strengthen the γ matrix. The microstructure of IN718 can contain different phases and their formation depends upon the heat treatment cycle. The different

Table 10.1 Typical Composition of IN718 (wt.%)

Ni	Cr	Fe	Nb	Mo	Ti	Co	Al	Si
50–55	17–21	17	4.75–5.50	2.8–3.3	0.65–1.15	Max 1	0.2–0.8	Max 0.35

precipitates seen in IN718 are γ' (Ni(Ti,Al)), δ(Ni$_3$(Nb,Ti)), γ'' (Ni$_3$(Nb)), carbides (MC, M$_6$C, M$_{23}$C$_6$) and other secondary phases. γ'' is the major contributor towards the strength of the alloy and it is primarily formed during ageing treatment. The γ'' phase is metastable at room temperature and tends to transform from a body centred tetragonal phase to a δ phase that has an orthorhombic crystal structure on prolonged heating or exposure to sufficiently high temperatures (Donachie & Donachie, 2002; Reed, 2006).

10.4.2 LPBF of IN718

Parametric investigations indicate that the choice of scanning strategy affects the degree of distortion. It is observed that a constant scan pattern results in a 37.6% higher distortion (Dunbar et al., 2016). LPBF involves fast cooling rates and hence the γ' and γ'' phases are not observed in LPBF-built IN718 in most of the available literature. Studies have reported grain boundary and inter-dendritic precipitation of the Laves phase that makes them different from the microstructure of cast and wrought metals (see Figure 10.2). This necessitates the need for special heat treatment for LPBF-built IN718 that is different from the standard treatments used for conventional samples. The use of heat treatments can result in the formation of a strengthening of the γ' and γ'' phases (see Figure 10.3) (Deng et al., 2018).

Fayed et al. (2020) studied the LPBF-built IN718 in an as-built, homogenised and solution-treated condition over a wide range of soaking times. The homogenisation time directs the degree of recrystallisation and the volume and sizes of the different phases formed. A holding time of 1 h had no significant effect, whereas a holding time of 4 h resulted in complete recrystallisation and significant Laves phase dissolution. When the homogenisation time was increased to 7 h, grain growth was observed along with the coarsening of carbide precipitates. No detectable effect on the grain and the crystallographic structure was observed for the duration of the solution treatment time at 980°C, although the precipitation of the δ-phase was affected significantly. At 980°C, δ, γ' and γ'' are observed for a prolonged holding time. This resulted in increased hardness with solution treatment time up to a certain limit. The highest value of the hardness of the material was obtained after homogenisation for 1 h, whereas the prolonged heat treatments resulted in reduction in the hardness (Fayed et al., 2020). The literature also shows studies explaining the effect of HIPing and variations

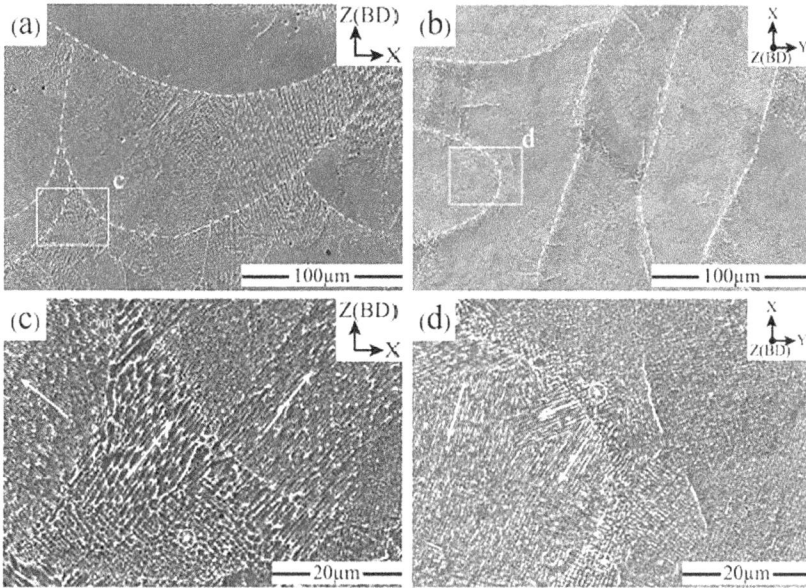

Figure 10.2 Magnified images of etched LPBF-built IN718 samples in (a) the vertical direction and (b) horizontal direction. The areas enclosed in the box in (a) and (b) are magnified in (c) and (d). Yellow dashed lines represent the melt boundaries and the yellow arrows and target symbols indicate the growth direction of cellular dendrites (Deng et al., 2018).

of the standard double ageing treatment on LPBF-built IN718. After a prolonged soaking time, during the simulated HIP process, the microstructure transitioned from heterogeneous columnar grains to homogeneous recrystallised grains with MC-type carbide precipitates. This leads to a decrease in micro-hardness. The ageing treatment resulted in an increase in both hardness values and Young's modulus values, possibly due to the formation of γ'' and γ' precipitates in the Ni-matrix. Annealing twins are also observed for longer soaking periods. These annealing twins were found to be the dominant strengthening mechanisms in the LPBF built samples that were either stress-relieved and hot isostatic pressed (HIP) or solution treated and aged (Jiang et al., 2019).

10.4.3 LDED of IN718

Parametric investigation of LDED-built IN718 samples revealed that microstructure and residual stress are affected by an overlap ratio along with density or porosity. The overlapped region shows relatively higher residual stresses and fluctuation in its values. The texture of the LDED samples is affected by the scanning strategy used (unidirectional or bidirectional scanning). Studies have reported unidirectional and bidirectional scanning

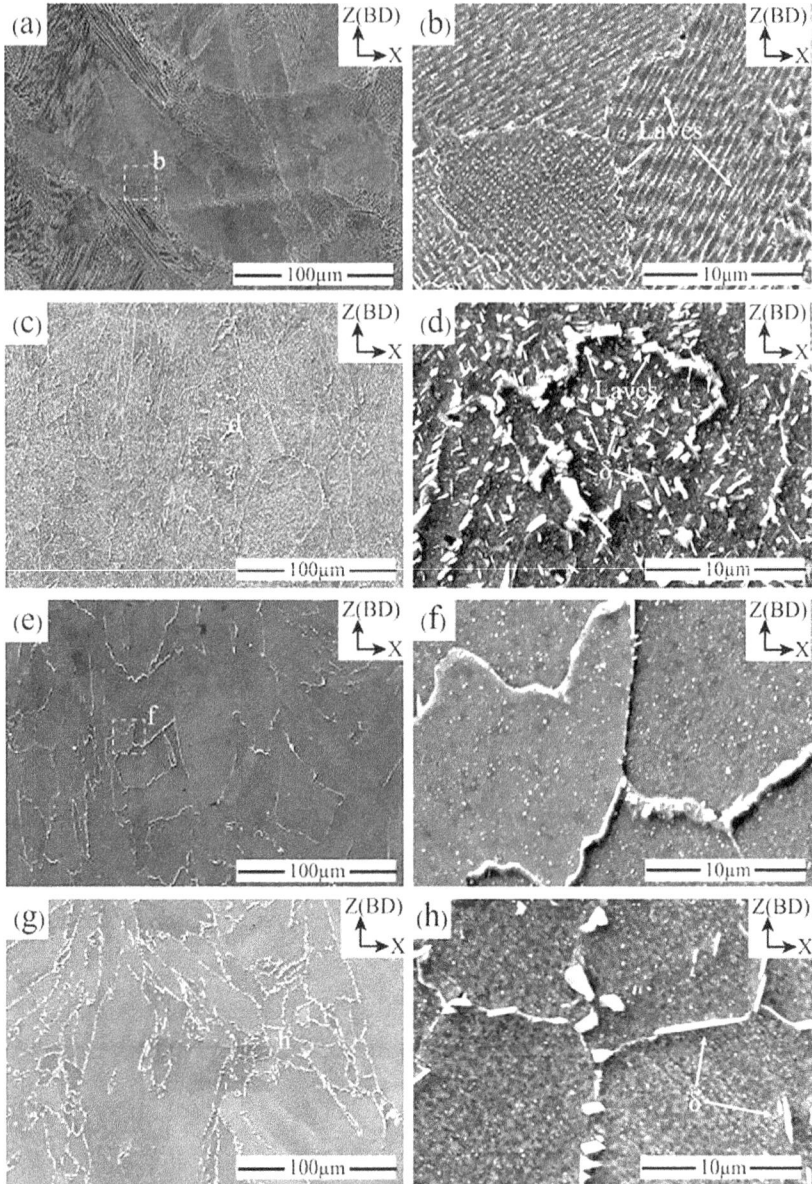

Figure 10.3 Magnified images of heat-treated LPBF-built IN718 samples in different heat-treated conditions. These conditions are (a) direct aged, (b) solution annealed and aged, (c) homogenised and aged and (d) homogenised, solution annealed and aged. (b), (d), (f) and (h) represent the magnified areas as represented in (a), (c), (e) and (g) respectively (Deng et al., 2018).

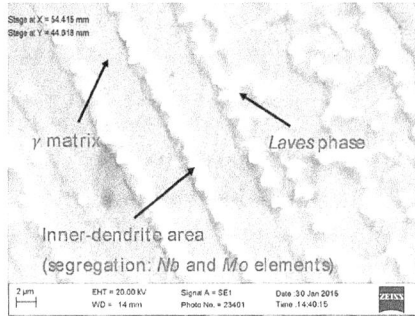

Figure 10.4 SEM micrograph of IN718. Precipitations and dendrites are marked (Zhong et al., 2016).

developing a fibre texture and rotated cube texture in the deposit, respectively (Cao et al., 2013; Dinda et al., 2012; Liu et al., 2011).

X-ray diffraction (XRD) of the LDED-built IN718 sample studies reveal that they consist of an FCC Ni-based austenite phase. Higher magnification scanning electron microscopy (SEM) images show the presence of eutectic phases of γ+Laves phases. Figure 10.4 shows the segregations of Laves phases in the γ matrix. Energy dispersive spectroscopy (EDS) in the interdendritic region reveals the presence of Nb and Mo (Liu et al., 2013; Zhong et al., 2016).

Table 10.2 presents a comparison of the mechanical properties of LPBF, LDED and wrought IN718. "Full heat treatment" refers to solution

Table 10.2 Mechanical Properties of LPBF, LDED and Wrought IN718

	Yield Stress	Ultimate Tensile Stress	ε %	References
Wrought IN718 (full heat treatment)	1125	1365	20	Blackwell (2005)
LPBF-built	~800	~1100	~30	Deng et al. (2018)
LPBF-built and direct aged	~1300	~1500	~15	Deng et al. (2018)
LPBF-built and solution annealed	~1200	~1500	~22.5	Deng et al. (2018)
LPBF-built, homogenised and aged	~1200	~1400	~20	Deng et al. (2018)
LDED-built	650	1000	38	Blackwell (2005)
LDED-built and aged	1204	1393	13	Blackwell (2005)
LDED-built and full heat treated	1257	1436	13	Blackwell (2005)

Table 10.3 Composition of Inconel 625 (wt.%)

Ni	Cr	Mo	Fe	Nb	Co	Si	Ti	Al	C
≥ 58	20–23	8–10	≤ 5	3.15–4.15	≤ 1.0	≤ 0.5	≤ 0.4	≤ 0.4	≤ 1.0

treatment (980°C for 1 h and air cooling) followed by a double ageing cycle (720°C for 8 h with furnace cooling + 620°C for 8 h followed by air cooling).

10.4.4 Inconel 625 (IN625)

IN625 is a superalloy that is solid solution strengthened. It is characterised by its high-temperature strength, excellent fabricability and weldability. It also possesses outstanding corrosion resistance. It is commercially available in two grades according to ASTM B443. Grade 1 of IN625 has a relatively smaller grain size and therefore has higher corrosion resistance, higher corrosion-fatigue resistance strength, higher tensile strength and higher resistance to chloride-ion stress-corrosion cracking and finds application in the marine sector. Grade 2 of IN625 has relatively higher creep resistance strength, higher ductility and primarily finds application in the aerospace sector for building components like aircraft ducting systems, engine exhaust systems and combustion systems. Table 10.3 presents the composition of IN625 (Marchese et al., 2018).

10.4.5 LPBF of IN625

From parametric investigations it is observed that increasing the laser power or decreasing the scan speed (i.e., increasing the energy density) results in an increase of the contact angle in the melt pool and stronger texture in the grains, apart from an increase in the melt pool dimensions (Tian et al., 2020). The choice of scanning strategy especially affects the texture. Unidirectional and bidirectional scanning strategies have been shown to produce grains textured at 60° to the vertical and along the vertical direction, respectively, due to the asymmetry between the heat flow direction and build direction (Li et al., 2015). Nayak et al. parametrically investigated the LPBF of IN625 at a higher layer thickness of 100 μm and found that the optimally dense samples were obtained at an LED of 240 J/mm³ (Nayak et al., 2020a).

The SEM of the as-built samples detect precipitates like γ' and γ'' phases. Some MC carbides and Laves phases are detected, apart from the FCC-Ni γ phase, which is preferentially textured along <100>. The existence of brittle Laves phases in the intergranular region in as-built samples results in its creep properties being poorer than those of the aerospace standard forgings. At the molten pool boundary, γ' (Ni_3Nb) is detected (Tian et al., 2020). Some of the studies also indicate the detection of carbon–oxygen inclusions.

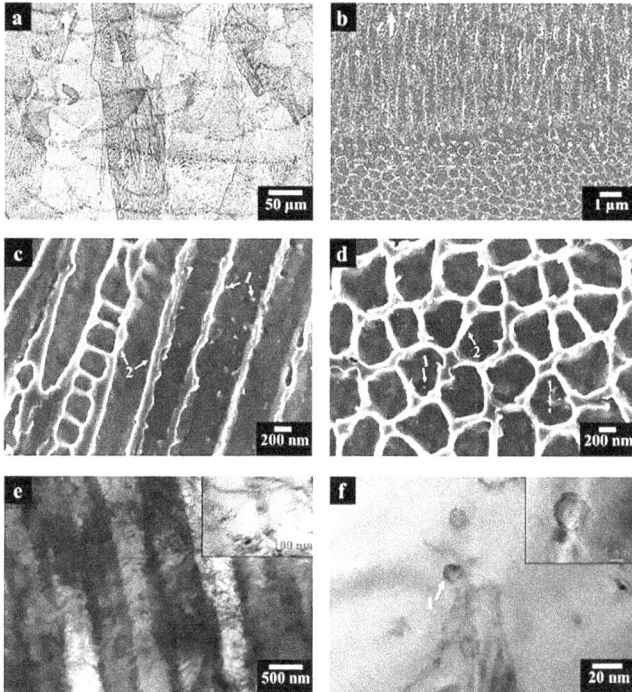

Figure 10.5 The microstructure of an LPBF-built IN625 cross-section as observed under (a) an optical microscope. Melt-pool features and columnar dendrites can be observed: (b) field emission scanning electron microscope (FESEM) for columnar and cellular primary dendrites; (c–d) FESEM images where the elemental Nb, MC carbides rich in Nb, and the areas rich in Mo (in the inter-dendritic regions) are represented by arrows 1 and 2 respectively for (c) columnar dendrites and (d) cellular dendrites; (e) bright field images of transmission electron microscopy (TEM) that show the interdendritic areas of columnar dendrites having high dislocation density; (f) carbide cores with MC carbides rich in Nb, where the inset shows the coherency of the carbide with the matrix (Marchese et al., 2018).

Figure 10.5 shows the microstructure and different phases in the LPBF-built IN625 as observed by Marchese et al., 2018.

The effect of different post-treatments—like annealing (stress relief and recrystallisation), solution treatment and HIPing on LPBF-built IN625 samples—have also been studied in the literature. Annealing at different elevated temperatures resulted in different microstructures and phases. Annealing at 700°C showed no change in the as-built microstructure; annealing at 1000°C resulted in the homogenisation of solute elements and the formation of fine grains; annealing at 1150°C resulted in significant grain growth, although the carbides at the grain boundaries persisted and coarsened. High-temperature annealing resulted in recrystallisation, dissolution of precipitates with high Z-contrast and precipitation of MC carbides.

The fine microstructure and solid solution strengthened the nature of IN625, which resulted in very high hardness (343 HV) in the LPBF form, that is higher than that obtained by traditional forging. Annealing at 700 to 800°C resulted in the precipitation of the δ-phase. When LPBF-built IN625 samples were annealed above 1000°C, dissolution of the δ phase into the matrix was observed (Fang et al., 2018; Kreitcberg et al., 2017).

10.4.6 LDED of IN625

The as-built microstructure of LDED-built IN625 consists of a nickel-based FCC γ phase along with MC, M_6C and $M_{23}C_6$ carbides. The MC carbides rich in Nb are seen in the interdendritic region that might have been formed due to rapid solidification, whereas the M_6C and $M_{23}C_6$ carbides might have been formed due to exposure to the elevated temperature during the LDED process (Rombouts et al., 2012).

Rastering patterns are found not to affect the mechanical properties of LDED-built IN625 (Paul et al., 2007). The impact energy of LDED-built IN625 samples have been reported to increase after solution treatment (Puppala et al., 2014). As the temperature was increased, the wear rate of the LPBF-built IN625 samples was decreased (Feng et al., 2017). The room temperature corrosion properties of the LDED-built samples were found to match the corresponding wrought samples (Tuominen et al., 2003). Dinda et al. (2009) also reported that the as-built dendrites showed no change for annealing below 1000°C. Annealing at 1100°C showed the appearance of new grains between the deposition layer and the substrate (Dinda et al., 2009). Table 10.4 presents the mechanical properties of LPBF, LDED and wrought IN625.

10.5 HASTELLOY-X(HX)

HX (UNS N06002) is a nickel-chromium iron-molybdenum matrix-stiffened high-temperature superalloy. It shows excellent resistance to oxidation and fairly retains strength up to 1200°C, and has good fabricability

Table 10.4 Mechanical Properties of LPBF, LDED and Wrought IN625

	Yield Stress	Ultimate Tensile Stress	ε %	Reference
Wrought	517	930	40	Yadroitsev et al. (2007)
LPBF built	743	1043	31.4	Tian et al. (2020)
LPBF built + annealing	386	910	54.4	Tian et al. (2020)
LDED-built	571	915	49	Paul et al. (2007)
LDED-built + hot-finished and annealed	395	824	51	Paul et al. (2007)

Table 10.5 Chemical Composition of HX (wt.%)

Ni	Cr	Fe	Mo	Co	W	Si	C	O	P	S
Balance	21.20	17.6	8.8	2.0	0.7	0.2	0.06	0.02	0.002	0.002

and weldability. It is used: in combustion chambers and tailpipes in aircraft and ground-based gas turbines; for fans, industrial furnace support members, and in engineering components for nuclear applications (hightemp-metals, 2015). Table 10.5 shows the chemical composition of Hastelloy-X (Sanchez-Mata et al., 2020).

10.5.1 LPBF of HX

Parametric optimisation for the LPBF of HX is indispensable. Very high or very low scanning speeds would result in a lack of fusion and gas porosity, respectively. The dense samples are also found to follow the Hall–Petch relationship (Esmaeilizadeh et al., 2020). LPBF-built as-built HX samples have a microstructure with columnar grains and equiaxed grains, along the building direction and perpendicular to the building direction, respectively. The fine columnar and cellular structures have dimensions less than 1 µm due to high cooling rates. Sub-micrometric face centred cubic (FCC) M_nC_m, Mo-rich $M_{12}C$ and M_6C carbides rich in Mo are also formed and are relatively larger along the grain boundaries (>500 nm) and in the interdendritic areas compared to carbides within dendritic cores (<100 nm). Mo-rich carbides apart from residual stresses are found to be the primary reason for cracking (Figure 10.6) (Marchese et al., 2019).

Montero-Sistiaga (2020) investigated the elevated temperature tensile properties of LPBF-built HX samples at 650, 750 and 850°C in an as-built condition and heat-treated condition (at 1177 and 800°C). At 750°C, tensile strain significantly decreased due to the precipitation of carbides and Mo-rich intermetallic phases. The lowest carbide formation, yielding the highest increase in tensile ductility, was observed for the sample heat-treated at 1177°C due to the relatively higher homogeneous starting microstructure (Montero-Sistiaga et al., 2020).

The HIP of LPBF-built HX resulted in the closing of cracks and defects, barring a few gas-filled micro-pores. HIP promoted recrystallisation and resulted in the reduction of yield strength and ultimete tensile strength by 130 and 60 MPa, respectively. It also improved the fatigue life, by reducing stress concentration and releasing residual stress. It is observed that the micro-cracks in the as-built samples occur along high angle grain boundaries. The hot cracking mechanism is definitely due to the pressure drop in the interdendritic liquid between the dendrite tip and root (Han et al., 2018).

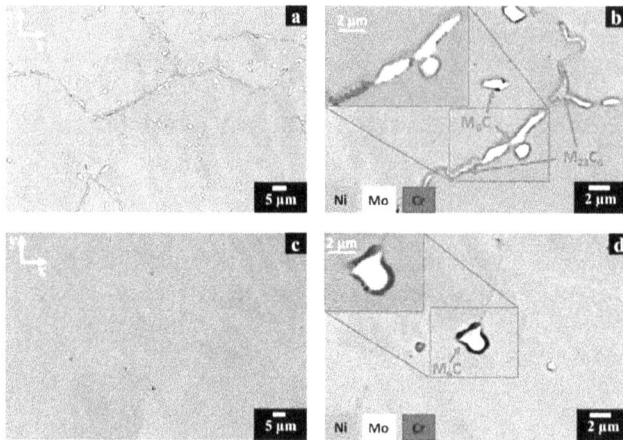

Figure 10.6 Post-processed HX samples using back scattered electron (BSE) and SEM images where (a–b) show the EDS maps detecting the Mo-rich M_6C and Cr-rich $M_{23}C_6$ carbides; (c–d) show the EDS mapping of post-solution annealed HX samples detecting Mo-rich M_6C carbides (Marchese et al., 2019).

10.5.2 LPBF of HX

Jinoop et al. (2019a) parametrically investigated the LDED of HX and obtained a process window for crack-free deposition at the maximum deposition rate. Crystallite sizes were calculated and found to be 25.56 nm. Tensile residual stresses were observed on the deposited surface with a maximum stress of 252 MPa (Jinoop et al., 2019a). Heat treatment increased the energy storage capacity by 1.55 (see Figure 10.7) and reduced the average roughness (Jinoop et al., 2019b). Further heat treatment of LDED-built HX samples also resulted in improved fatigue life, and in a 1.72 times increase in the wear rate and an increase in delamination (Jinoop et al., 2019b). Table 10.6 presents the mechanical properties of HX built using different production techniques

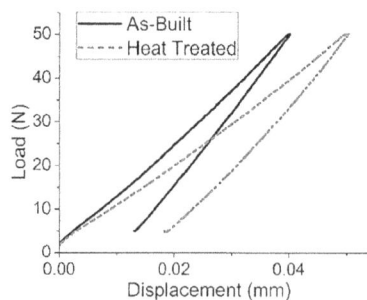

Figure 10.7 Single cycle ball indentation test on as-built HX samples and post-heat-treated HX samples (Jinoop et al., 2019b).

Table 10.6 Mechanical Properties of HX Built using Different Production Techniques

	YS	UTS	ε %	References
Plate 3/8–2 in thick, heat treated at 1177°C and rapidly cooled	339	743	51	hightempmetals (2015)
LPBF-built	480	620	40	Han et al. (2018)
LPBF-built + HIP	350	560	41	Han et al. (2018)
LDED-built	478	765	–	Jinoop et al. (2019a)
Heat treated	393	630	–	Jinoop et al. (2019a)

10.6 WASPALOY

Waspaloy (UNS N07001) is an age-hardenable Ni-superalloy that displays excellent strength at elevated temperature and has good corrosion resistance, especially to oxidation, for service temperatures up to 650°C (for critical rotating applications) and 870°C (for less demanding applications) (Special Metals, 2004). Currently, it is used in compressor and rotor discs, spacers, seals, rings and casings, fasteners and other miscellaneous engine hardware, airframe assemblies and missile systems. It shows good oxidation resistance under frequent thermal cycling conditions and when exposed to temperatures up to 1038°C. The best corrosion performance of the material is seen in its solution treated state. It is used as material for the unpainted rear fairing around exhaust nozzles, and is regularly subjected to temperatures around 1000°C. Waspaloy's composition is given in Table 10.7 (Mumtaz et al., 2008; Special Metals, 2004).

10.6.1 LPBF of Waspaloy

Parametric investigation of LPBF-built Waspaloy samples with a pulse laser revealed that the highest density samples are obtained for pulse width, pulse energy, repetition rate and scan speed of 5 ms, 9 J, 10 Hz and 168 mm/min respectively. Further increasing the energy resulted in large material vaporisation (Mumtaz et al., 2008). Mumtaz and Hopkinson (2007) fabricated a functionally graded material (FGM) of Waspaloy consisting of gradually varying a zirconia (0–10%) composition using LPBF with a high power Nd:YAG laser. An average porosity of 0.34% was observed that changed gradually between layers without any major interface defects (Mumtaz & Hopkinson, 2007).

Table 10.7 Composition of Waspaloy (wt.%)

Ni	Co	Cr	Mo	Fe	Si	Mn	Al	Ti	Cu
54	13.5	19.5	4.2	2.0	0.7	1.0	1.4	3.0	0.5

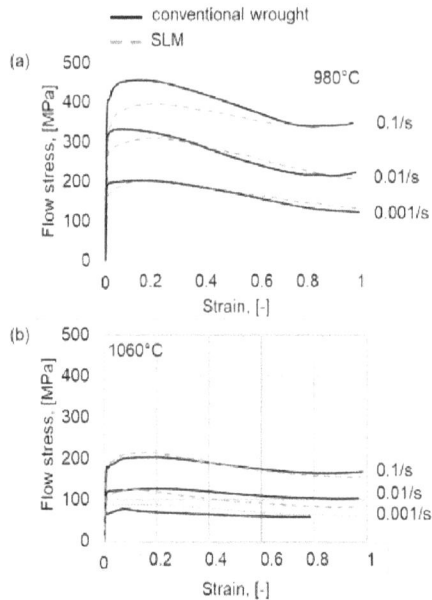

Figure 10.8 Flow curves for (a) LPBF/SLM-built and (b) conventionally wrought Waspaloy (Jedynak et al., 2020).

A similarity was observed in the hot working behaviour of as-built and wrought samples of Waspaloy. Both have similar processing maps and their efficiency peak occurs around 1060°C and strain rates range from 0.008 to 0.02 s⁻¹ (see Figure 10.8) (Jedynak et al., 2020).

10.6.2 LPBF of Waspaloy

Parametric investigations with a pulse laser revealed that low duty cycles led to the formation of highly columnar microstructures and those produced with high duty cycles led to the formation of lower aspect ratio grains. A duty cycle showed little effect on the micro-texture. Changing the pulse length markedly altered the (0 0 1) plane angle, but showed little effect on the preferred orientation's intensity (Moat et al., 2009).

10.7 CM247LC

CM247LC is an Ni-based superalloy that has a high volume of γ' in a fully heat-treated condition. It finds applications in poly-crystalline (PX) or directionally solidified (DS) forms. Its values of mechanical properties at elevated temperatures, oxidation resistance and corrosion resistance are optimum and similar to conventional cast alloys, therefore making it appealing for the LAM of gas turbine components, delivering high-performance. Because of its high γ' content and complex chemistry, the alloy is difficult to manufacture. Its high content of Ti + Al makes it difficult to weld due to increased susceptibility to strain-age (reheat) cracking (SAC) (Boswell et al., 2019).

10.7.1 LPBF of CM247LC

Its poor weldability is also reflected in LPBF as it is prone to micro-cracking during LPBF. Short ageing treatments around the phase transition temperatures (450–600°C) results in cracking density increasing steeply, primarily by a combination of cracking mechanisms (ductility dip and strain age) attributed to the change in the microstructure, hardness and elastic modulus (Boswell et al., 2019). Table 10.8 shows the composition of CM247LC (Wang et al., 2017).

LPBF-built CM247LC samples have been studied with different build orientations, parameter sets and post-processing methods. It was observed that the material's high temperature creep resistance is primarily affected by the build direction. Grains were observed to grow epitaxially in the build direction, resulting in strong anisotropic behaviour (Hilal et al., 2019).

SEM and TEM investigations on the as-built samples revealed cells separated by a γ'/γ eutectic, Hf/Ti/Ta/W-rich precipitates and high densities of dislocation (Wang et al., 2017). Fine primary MC carbides were observed in the intercellular regions (Divya et al., 2016).

Post-standard heat treatments of LPBF-built samples resulted in mechanical properties that are comparable to the cast material (Wang et al., 2017). The progressive recovery of the microstructure and recrystallisation of the microstructure were observed for heat treatments taking place below and over 1230°C. Below 1230°C, no change in microstructure or texture was observed. Above 1230°C, as-deposited cellular colonies coalesced, γ' precipitates and refractory metal-rich particles redistributed, but the texture remained very strong. Above 1240°C, apart from recrystallisation, γ' precipitates became finer and more uniform. The carbide growth distribution did not change above recrystallisation. Post-recrystallisation, only a small remnant of anisotropy remained, which is related to the reduced elastic anisotropy (Muñoz-Moreno et al., 2016).

Table 10.8 Typical Composition of CM247LC (wt.%)

Ni	Cr	Co	Mo	W	Ta	Ti	Al	Hf	C	B	Zr
Balance	8.31	9.15	0.54	9.4	3.2	0.73	5.62	1.28	0.07	0.02	0.01

Table 10.9 Mechanical Properties of CM247LC Built Using
            Different Manufacturing Routes

	YS	UTS	ε %	References
Cast and heat treated	~800	~900	~11	Wang et al. (2017)
LPBF-built	792	~1000	~5.67	Wang et al. (2017)
HIP and heat treated	~900	~1100	~9	Wang et al. (2017)

10.7.2 LPBF of IN625

Parametric investigation of the LDED of CM247LC revealed that the cracking in the LDED-built samples occurred during solidification and was the result of the elemental segregation of alloying elements residing at the grain and other solidification boundaries leading to the formation of a low melting point eutectic composition liquid. It was observed that a cross-hatch toolpath produced the lowest cracking response due to competing stress fields (McNutt, 2015). Table 10.9 presents the mechanical properties of CM247LC, built using different manufacturing routes.

10.8 RECENT TRENDS IN THE LAM OF NI-SUPERALLOYS

Yuanbo T. Tang in 2021 designed a new γ/γ' Ni-based superalloy using computational approaches. This new alloy shows relatively less solidification and solid-state cracking during LAM and shows superior processability and mechanical behaviour when compared with CM247LC and IN939 (Tang et al., 2021). Efforts are being made towards combining LAM systems and other manufacturing systems to build hybrid additive manufacturing systems. Brown et al. (2018) combined an LPBF system with a machining setup. IN718 structures were fabricated and investigated for their surface integrity. It was observed that the microhardness was dependent on the scan direction and use of coolant in subsequent milling and that the hybrid LPBF-milling route can significantly improve the surface integrity of the built sample (Brown et al., 2018). Li et al. (2019) built a bimetallic of IN718 and SS316 using hybrid manufacturing involving LDED and thermal milling. No diffusion was observed between these two metals although the strength of the bimetal was found to be less than the individual alloy's strength (Li et al., 2019). LAM systems are incorporating *in situ* systems to monitor the process so that the parameters can be adjusted to obtain components with the desired properties. Cabeza et al. (2020) monitored strain *in situ* during the LDED of IN718. The degree to which the lattice parameter evolution was affected by thermal, phase and stress-related contributions during processing for representative regions was addressed. These represent the melt pool region, near the melt pool region and the far-field region (Cabeza et al., 2020).

Some of the recent case studies showing the application of LAM-built Ni-superalloys for aerospace applications are:

1. Rosswag Engineering, a German metallurgical company, qualified Waspaloy for high stress, high-temperature applications such as gas turbines for its impressive creep resistance to be manufactured through LPBF. The company also believes that the material's qualification has opened new possibilities for functionally optimised parts. This can be attributed to the material's excellent creep resistance that makes it functional even in extreme temperature and stress conditions (Serotoglu, 2020).

2. The Fraunhofer Institute for Production Technology (IPT) Aachen, Germany devised a new method called Express Wire Coil Cladding (EW2C) for treating the surface of shafts. This is in contrast to the DED wire fed process where the material is constantly fed and joined in the form of wire. Here the wire is first pushed in the form of spirals to the desired locations in the shaft and then welded there using a high powered laser. The wire spirals are placed under tension on the shaft to avoid slipping during the laser process. The EW2C is a significant innovation in the sense that it is used to repair worn-out shafts as they are the primary transmitters of power and torque. In the test series, it was observed that it was possible to wind an IN718 wire of 1.2 mm diameter onto a steel shaft of 35 mm diameter in less than a minute. Research is being conducted in the direction of improving the processing speed by adjusting the laser focal spot diameter so that several spiral coils could be irradiated and melted simultaneously (Fraunhofer, 2020).

3. LPBF is used by MTU Aero Engines to build borescope bosses for PurePower® for powering A320neo. It is to be used with a low-pressure turbine for a turbofan engine. The bosses allow for the inspection of the turbofan using a borescope. The bosses are a part of the turbine case through which the turbine can be inspected. This is the first time that low-pressure turbines in turbofan engines have been equipped with LPBF-built borescopes. According to Airbus, this is a step towards reducing aircraft fuel consumption by 15% through its short and medium-haul aircraft, the A320neo family (Chater, 2015; EOS, 2020).

4. CellCore GmbH used LPBF to build a monolithic thrust chamber for a rocket engine using IN718. Contrary to the conventional techniques, the deployment of LPBF led to minimal post-processing, resulting in minimal tool wear, integration of multiple parts and internal ducts directly into the part, and enhanced cooling and stability due to

innovative lattice structures. Conventionally the cooling ducts were built in the blanks by countersinking operation and then, through multiple working steps, they were sealed, which is cost-intensive and required nearly six months for fabrication. With LPBF, the entire component is built in five days at lower cost (*Rocket Propulsion Engine Built with Selective Laser Melting Technology from SLM Solutions*, 2019; SLM Solutions Group AG, 2021).

5. The metallic parts of CubeSats or miniature cubic satellite buses are built using IN718 on a Concept Laser M2 Cusing based LPBF system. The CubeSats are built lighter and stiffer using integrated lattice structures. Fabricating CubeSats through LPBF caused a 35% lowering of mass compared to the predicted mass in the original CubeSat design (Scott, 2018).

10.9 FUTURE SCOPE

The LPBF of nickel superalloys holds great possibilities for the development of innovative components in the aerospace sector. LPBF can be used to build aerospace components with embedded sensors for online health monitoring. There are few attempts to embed sensors during AM processes (Jung et al., 2020; Petrat et al., 2018). Research is also being conducted into the development of composites using LPBF for enhancing the properties of nickel superalloys for aerospace applications. The type and the concentration of the reinforcement material will determine the properties of the composite components (Han et al., 2020; Kim et al., 2017). Newer nickel superalloys are also being researched and developed for components that can sustain high temperature and pressure in aerospace service conditions (Xu et al., 2020). The welding of LPBF-built nickel superalloys is also being researched for the purpose of joining large size aerospace components that cannot be built in a single step using LPBF (Jokisch et al., 2019).

10.10 CONCLUSIONS

The aerospace sector involves a lot of components that have complex shapes and at the same time need to sustain strength and avoid wear at very-high-temperature conditions. Due to this the LAM of Ni-superalloys has emerged as an exciting field for research. This chapter has dealt with the LAM of some commercially popular Ni-superalloys and their relevance to aerospace application. The chapter has briefly touched upon the basic concepts related to LAM, a brief description of Ni-superalloys, the literature related to LAM of different superalloys that includes both LPBF and LDED, recent trends in the LAM of Ni-superalloys and case studies related to the LAM of Ni-superalloys in the aerospace sector.

ACKNOWLEDGEMENT

S. K. Nayak and A. N. Jinoop acknowledge financial support from RRCAT, DAE, Govt. of India and HBNI, Mumbai, India.

REFERENCES

Additive Manufacturing-General Principles-Terminology. (2018). ISO/ASTM. https://www.iso.org/obp/ui/#iso:std:iso-astm:52900:dis:ed-2:v1:en

Blackwell, P. L. (2005). The mechanical and microstructural characteristics of laser-deposited IN718. *Journal of Materials Processing Technology*, *170*(1), 240–246. https://doi.org/10.1016/j.jmatprotec.2005.05.005

Boswell, J. H., Clark, D., Li, W., & Attallah, M. M. (2019). Cracking during thermal post-processing of laser powder bed fabricated CM247LC Ni-superalloy. *Materials & Design*, *174*, 107793. https://doi.org/10.1016/j.matdes.2019.107793

Brown, D., Li, C., Liu, Z. Y., Fang, X. Y., & Guo, Y. B. (2018). Surface integrity of Inconel 718 by hybrid selective laser melting and milling. *Virtual and Physical Prototyping*, *13*(1), 26–31. https://doi.org/10.1080/17452759.2017.1392681

Cabeza, S., Özcan, B., Cormier, J., Pirling, T., Polenz, S., Marquardt, F., Hansen, T. C., López, E., Vilalta-Clemente, A., & Leyens, C. (2020). Strain monitoring during laser metal deposition of Inconel 718 by neutron diffraction. In S. Tin, M. Hardy, J. Clews, J. Cormier, Q. Feng, J. Marcin, C. O'Brien, & A. Suzuki (Eds.), *Superalloys 2020* (pp. 1033–1045). Springer International Publishing. https://doi.org/10.1007/978-3-030-51834-9_101

Cao, J., Liu, F., Lin, X., Huang, C., Chen, J., & Huang, W. (2013). Effect of overlap rate on recrystallization behaviors of Laser Solid Formed Inconel 718 superalloy. *Optics & Laser Technology*, *45*, 228–235. https://doi.org/10.1016/j.optlastec.2012.06.043

Chater, J. (2015). *Nickel alloy use in additive manufacturing*. https://www.stainless-steel-world.net/pdf/Nickel_alloy_use_in_additive_manufacturing.pdf

Deng, D., Peng, R. L., Brodin, H., & Moverare, J. (2018). Microstructure and mechanical properties of Inconel 718 produced by selective laser melting: Sample orientation dependence and effects of post heat treatments. *Materials Science and Engineering: A*, *713*, 294–306. https://doi.org/10.1016/j.msea.2017.12.043

Dinda, G. P., Dasgupta, A. K., & Mazumder, J. (2009). Laser aided direct metal deposition of Inconel 625 superalloy: Microstructural evolution and thermal stability. *Materials Science and Engineering: A*, *509*(1), 98–104. https://doi.org/10.1016/j.msea.2009.01.009

Dinda, G. P., Dasgupta, A. K., & Mazumder, J. (2012). Texture control during laser deposition of nickel-based superalloy. *Scripta Materialia*, *67*(5), 503–506. https://doi.org/10.1016/j.scriptamat.2012.06.014

Divya, V. D., Muñoz-Moreno, R., Messé, O. M. D. M., Barnard, J. S., Baker, S., Illston, T., & Stone, H. J. (2016). Microstructure of selective laser melted CM247LC nickel-based superalloy and its evolution through heat treatment. *Materials Characterization*, *114*, 62–74. https://doi.org/10.1016/j.matchar.2016.02.004

Donachie, M. J., & Donachie, S. J. (2002). *Superalloys: A technical guide* (2nd ed). ASM International.

Dunbar, A. J., Denlinger, E. R., Heigel, J., Michaleris, P., Guerrier, P., Martukanitz, R., & Simpson, T. W. (2016). Development of experimental method for in situ distortion and temperature measurements during the laser powder bed fusion additive manufacturing process. *Additive Manufacturing*, *12*, 25–30. https://doi. org/10.1016/j.addma.2016.04.007

EOS. (2020). *An Intelligent Strategy for Achieving Excellence: MTU Relies on Additive Manufacturing for Its Series Component Production*. https://www. eos.info/en/3d-printing-examples-applications/all-3d-printing-applications/ mtu-aerospace-with-3d-printing-series-component-production.

Esmaeilizadeh, R., Keshavarzkermani, A., Ali, U., Mahmoodkhani, Y., Behravesh, B., Jahed, H., Bonakdar, A., & Toyserkani, E. (2020). Customizing mechanical properties of additively manufactured Hastelloy X parts by adjusting laser scanning speed. *Journal of Alloys and Compounds*, *812*, 152097. https://doi.org/10.1016/j. jallcom.2019.152097

Fang, X. Y., Li, H. Q., Wang, M., Li, C., & Guo, Y. B. (2018). Characterization of texture and grain boundary character distributions of selective laser melted Inconel 625 alloy. *Materials Characterization*, *143*, 182–190. https://doi.org/10.1016/j. matchar.2018.02.008

Fayed, E. M., Shahriari, D., Saadati, M., Brailovski, V., Jahazi, M., & Medraj, M. (2020). Influence of homogenization and solution treatments time on the microstructure and hardness of Inconel 718 fabricated by laser powder bed fusion process. *Materials*, *13*(11), 2574. https://doi.org/10.3390/ma13112574

Feng, K., Chen, Y., Deng, P., Li, Y., Zhao, H., Lu, F., Li, R., Huang, J., & Li, Z. (2017). Improved high-temperature hardness and wear resistance of Inconel 625 coatings fabricated by laser cladding. *Journal of Materials Processing Technology*, *243*, 82–91. https://doi.org/10.1016/j.jmatprotec.2016.12.001

Fraunhofer, Fraunhofer. (2020). *Fraunhofer IPT develops Express Wire Coil Cladding AM process for surface treatment of shafts*. https://www.metal-am. com/fraunhofer-ipt-develops-express-wire-coil-cladding-am-process-for-surface- treatment-of-shafts/

Han, Q., Gu, Y., Huang, J., Wang, L., Low, K. W. Q., Feng, Q., Yin, Y., & Setchi, R. (2020). Selective laser melting of Hastelloy X nanocomposite: Effects of TiC reinforcement on crack elimination and strength improvement. *Composites Part B: Engineering*, *202*, 108442. https://doi.org/10.1016/j.compositesb.2020.108442

Han, Q., Mertens, R., Montero-Sistiaga, M. L., Yang, S., Setchi, R., Vanmeensel, K., Van Hooreweder, B., Evans, S. L., & Fan, H. (2018). Laser powder bed fusion of Hastelloy X: Effects of hot isostatic pressing and the hot cracking mechanism. *Materials Science and Engineering: A*, *732*, 228–239. https://doi.org/10.1016/j. msea.2018.07.008

hightempmetals. (2015). *HASTELLOY X TECHNICAL DATA*. https://www. hightempmetals.com/techdata/hitempHastXdata.php

Hilal, H., Lancaster, R., Jeffs, S., Boswell, J., Stapleton, D., & Baxter, G. (2019). The influence of process parameters and build orientation on the creep behaviour of a laser powder bed fused Ni-based superalloy for aerospace applications. *Materials*, *12*(9), 1390. https://doi.org/10.3390/ma12091390

INCONEL alloy 718 (INCONEL Alloy 718). (2020). [INCONEL® alloy 718]. www.specialmetals.com

Jedynak, A., Sviridov, A., Bambach, M., Beckers, D., & Graf, G. (2020). On the potential of using selective laser melting for the fast development of forging alloys at the example of Waspaloy. *Procedia Manufacturing*, 47, 1149–1153. https://doi. org/10.1016/j.promfg.2020.04.138

Jiang, R., Mostafaei, A., Pauza, J., Kantzos, C., & Rollett, A. D. (2019). Varied heat treatments and properties of laser powder bed printed Inconel 718. *Materials Science and Engineering: A*, 755, 170–180. https://doi.org/10.1016/j. msea.2019.03.103

Jinoop, A. N., Denny, J., Paul, C. P., Ganesh Kumar, J., & Bindra, K. S. (2019a). Effect of post heat-treatment on the microstructure and mechanical properties of Hastelloy-X structures manufactured by laser based Directed Energy Deposition. *Journal of Alloys and Compounds*, 797, 399–412. https://doi.org/10.1016/j. jallcom.2019.05.050

Jinoop, A. N., Paul, C. P., & Bindra, K. S. (2019b). Laser-assisted directed energy deposition of nickel super alloys: A review. *Proceedings of the Institution of Mechanical Engineers, Part L: Journal of Materials: Design and Applications*, 233(11), 2376–2400. https://doi.org/10.1177/1464420719852658

Jokisch, T., Marko, A., Gook, S., Üstündag, Ö., Gumenyuk, A., & Rethmeier, M. (2019). Laser welding of SLM-manufactured tubes made of IN625 and IN718. *Materials*, 12(18). https://doi.org/10.3390/ma12182967

Jung, I. D., Lee, M. S., Lee, J., Sung, H., Choe, J., Son, H. J., Yun, J., Kim, K., Kim, M., Lee, S. W., Yang, S., Moon, S. K., Kim, K. T., & Yu, J.-H. (2020). Embedding sensors using selective laser melting for self-cognitive metal parts. *Additive Manufacturing*, 33, 101151. https://doi.org/10.1016/j.addma.2020.101151

Kim, S. H., Shin, G.-H., Kim, B.-K., Kim, K. T., Yang, D.-Y., Aranas, C., Choi, J.-P., & Yu, J.-H. (2017). Thermo-mechanical improvement of Inconel 718 using ex situ boron nitride-reinforced composites processed by laser powder bed fusion. *Scientific Reports*, 7(1), 14359. https://doi.org/10.1038/s41598-017-14713-1

Kreitcberg, A., Brailovski, V., & Turenne, S. (2017). Elevated temperature mechanical behavior of IN625 alloy processed by laser powder-bed fusion. *Materials Science and Engineering: A*, 700, 540–553. https://doi.org/10.1016/j.msea.2017.06.045

Li, P., Gong, Y., Xu, Y., Qi, Y., Sun, Y., & Zhang, H. (2019). Inconel-steel functionally bimetal materials by hybrid directed energy deposition and thermal milling: Microstructure and mechanical properties. *Archives of Civil and Mechanical Engineering*, 19(3), 820–831. https://doi.org/10.1016/j.acme.2019.03.002

Li, S., Wei, Q., Shi, Y., Zhu, Z., & Zhang, D. (2015). Microstructure characteristics of Inconel 625 superalloy manufactured by selective laser melting. *Journal of Materials Science & Technology*, 31(9), 946–952. https://doi.org/10.1016/j. jmst.2014.09.020

Liu, F., Lin, X., Huang, C., Song, M., Yang, G., Chen, J., & Huang, W. (2011). The effect of laser scanning path on microstructures and mechanical properties of laser solid formed nickel-base superalloy Inconel 718. *Journal of Alloys and Compounds*, 509(13), 4505–4509. https://doi.org/10.1016/j.jallcom.2010.11.176

Liu, F., Lin, X., Leng, H., Cao, J., Liu, Q., Huang, C., & Huang, W. (2013). Microstructural changes in a laser solid forming Inconel 718 superalloy thin wall in the deposition direction. *Optics & Laser Technology*, 45, 330–335. https://doi. org/10.1016/j.optlastec.2012.06.028

Marchese, G., Bassini, E., Aversa, A., Lombardi, M., Ugues, D., Fino, P., & Biamino, S. (2019). Microstructural evolution of post-processed hastelloy X alloy fabricated by laser powder bed fusion. *Materials*, 12(3), 486. https://doi.org/10.3390/ ma12030486

Marchese, G., Lorusso, M., Parizia, S., Bassini, E., Lee, J.-W., Calignano, F., Manfredi, D., Terner, M., Hong, H.-U., Ugues, D., Lombardi, M., & Biamino, S. (2018). Influence of heat treatments on microstructure evolution and mechanical properties of Inconel 625 processed by laser powder bed fusion. *Materials Science and Engineering: A, 729*, 64–75. https://doi.org/10.1016/j.msea.2018.05.044

McNutt, P. A. (2015). *An investigation of cracking in laser metal deposited nickel superalloy CM247LC* [EngD Thesis, University of Birmingham]. https://etheses.bham.ac.uk//id/eprint/6394/5/McNutt15EngD.pdf

Moat, R. J., Pinkerton, A. J., Li, L., Withers, P. J., & Preuss, M. (2009). Crystallographic texture and microstructure of pulsed diode laser-deposited Waspaloy. *Acta Materialia, 57*(4), 1220–1229. https://doi.org/10.1016/j.actamat.2008.11.004

Montero-Sistiaga, M. L., Liu, Z., Bautmans, L., Nardone, S., Ji, G., Kruth, J.-P., Van Humbeeck, J., & Vanmeensel, K. (2020). Effect of temperature on the microstructure and tensile properties of micro-crack free hastelloy X produced by selective laser melting. *Additive Manufacturing, 31*, 100995. https://doi.org/10.1016/j.addma.2019.100995

Mumtaz, K. A., Erasenthiran, P., & Hopkinson, N. (2008). High density selective laser melting of Waspaloy®. *Journal of Materials Processing Technology, 195*(1), 77–87. https://doi.org/10.1016/j.jmatprotec.2007.04.117

Mumtaz, K. A., & Hopkinson, N. (2007). Laser melting functionally graded composition of Waspaloy® and Zirconia powders. *Journal of Materials Science, 42*(18), 7647–7656. https://doi.org/10.1007/s10853-007-1661-3

Muñoz-Moreno, R., Divya, V. D., Driver, S. L., Messé, O. M. D. M., Illston, T., Baker, S., Carpenter, M. A., & Stone, H. J. (2016). Effect of heat treatment on the microstructure, texture and elastic anisotropy of the nickel-based superalloy CM247LC processed by selective laser melting. *Materials Science and Engineering: A, 674*, 529–539. https://doi.org/10.1016/j.msea.2016.06.075

Nayak, S. K., Mishra, S. K., Jinoop, A. N., Paul, C. P., & Bindra, K. S. (2020a). Experimental studies on laser additive manufacturing of Inconel-625 structures using powder bed fusion at 100 μm layer thickness. *Journal of Materials Engineering and Performance, 29*(11), 7636–7647. https://doi.org/10.1007/s11665-020-05215-9

Nayak, S. K., Mishra, S. K., Paul, C. P., Jinoop, A. N., & Bindra, K. S. (2020b). Effect of energy density on laser powder bed fusion built single tracks and thin wall structures with 100 μm preplaced powder layer thickness. *Optics & Laser Technology, 125*, 106016. https://doi.org/10.1016/j.optlastec.2019.106016

Paul, C. P., Ganesh, P., Mishra, S. K., Bhargava, P., Negi, J., & Nath, A. K. (2007). Investigating laser rapid manufacturing for Inconel-625 components. *Optics & Laser Technology, 39*(4), 800–805. https://doi.org/10.1016/j.optlastec.2006.01.008

Paul, C. P., Jinoop, A. N., Nayak, S. K., & Paul, A. C. (2020). Laser additive manufacturing in Industry 4.0: overview, applications, and scenario in developing economies. In K. R. Balasubramanian & V. Senthilkumar (Eds.), *Additive Manufacturing Applications for Metals and Composites:* (pp. 271–295). IGI Global. https://doi.org/10.4018/978-1-7998-4054-1

Petrat, T., Kersting, R., Graf, B., & Rethmeier, M. (2018). Embedding electronics into additive manufactured components using laser metal deposition and selective laser melting. *Procedia CIRP, 74*, 168–171. https://doi.org/10.1016/j.procir.2018.08.071

Puppala, G., Moitra, A., Sathyanarayanan, S., Kaul, R., Sasikala, G., Prasad, R. C., & Kukreja, L. M. (2014). Evaluation of fracture toughness and impact toughness of laser rapid manufactured Inconel-625 structures and their co-relation. *Materials & Design*, *59*, 509–515. https://doi.org/10.1016/j.matdes.2014.03.013

Reed, R. C. (2006). *The Superalloys: Fundamentals and Applications*. Cambridge University Press.

Rocket propulsion engine built with selective laser melting technology from SLM Solutions. (2019, September 30). https://www.aerospacemanufacturinganddesign. com/article/rocket-propulsion-engine-selective-laser-melting-slm-solutions/

Rombouts, M., Maes, G., Mertens, M., & Hendrix, W. (2012). Laser metal deposition of Inconel 625: Microstructure and mechanical properties. *Journal of Laser Applications*, *24*(5), 052007. https://doi.org/10.2351/1.4757717

Sanchez-Mata, O., Muñiz-Lerma, J. A., Wang, X., Atabay, S. E., Attarian Shandiz, M., & Brochu, M. (2020). Microstructure and mechanical properties at room and elevated temperature of crack-free Hastelloy X fabricated by laser powder bed fusion. *Materials Science and Engineering: A*, *780*, 139177. https://doi.org/10.1016/j.msea.2020.139177

Scott, C. (2018, August 29). *3D Printing Lattice Structures for a More Lightweight CubeSat Bus*. 3Dprint.Com. https://3dprint.com/223717/lattice-structures-cubesat-bus/

Serotoglu, K. (2020). *ROSSWAG QUALIFIES ANOTHER NICKEL-BASED SUPERALLOY FOR AM – WASPALOY (ROSSWAG QUALIFIES ANOTHER NICKEL-BASED SUPERALLOY FOR AM – WASPALOY)*. https://3dprinting industry.com/news/rosswag-qualifies-another-nickel-based-superalloy-for-am-waspaloy-172182/

SLM Solutions Group AG. (2021). *Case Study: Monolithic Thrust Chamber*. SLM-Solutions. https://www.slm-solutions.com/fileadmin/Content/Case_Studies/Case Study_CellCore_ThrustChamber_web.pdf

Special Metals. (2004). *Waspaloy* (Waspaloy). https://www.specialmetals.com/assets/ smc/documents/alloys/other/waspaloy.pdf

Tang, Y. T., Panwisawas, C., Ghoussoub, J. N., Gong, Y., Clark, J. W. G., Németh, A. A. N., McCartney, D. G., & Reed, R. C. (2021). Alloys-by-design: Application to new superalloys for additive manufacturing. *Acta Materialia*, *202*, 417–436. https://doi.org/10.1016/j.actamat.2020.09.023

Tian, Z., Zhang, C., Wang, D., Liu, W., Fang, X., Wellmann, D., Zhao, Y., & Tian, Y. (2020). A Review on Laser Powder Bed Fusion of Inconel 625 Nickel-Based Alloy. *Applied Sciences*, *10*(1), 81. https://doi.org/10.3390/app10010081

Tuominen, J., Vuoristo, P., Mäntylä, T., Latokartano, J., Vihinen, J., & Andersson, P. H. (2003). Microstructure and corrosion behavior of high power diode laser deposited Inconel 625 coatings. *Journal of Laser Applications*, *15*(1), 55–61. https://doi.org/10.2351/1.1536652

Wang, X., Carter, L. N., Pang, B., Attallah, M. M., & Loretto, M. H. (2017). Microstructure and yield strength of SLM-fabricated CM247LC Ni-Superalloy. *Acta Materialia*, *128*, 87–95. https://doi.org/10.1016/j.actamat.2017.02.007

Xu, J., Gruber, H., Lin Peng, R., & Moverare, J. (2020). A Novel γ′-Strengthened Nickel-Based Superalloy for Laser Powder Bed Fusion. *Materials*, *13*(21), 4930. https://doi.org/10.3390/ma13214930

Yadav, S., Jinoop, A. N., Sinha, N., Paul, C. P., & Bindra, K. S. (2020). Parametric investigation and characterization of laser directed energy deposited copper-nickel graded layers. *The International Journal of Advanced Manufacturing Technology*, *108*(11–12), 3779–3791. https://doi.org/10.1007/s00170-020-05644-9

Yadroitsev, I., Thivillon, L., Bertrand, Ph., & Smurov, I. (2007). Strategy of manufacturing components with designed internal structure by selective laser melting of metallic powder. *Applied Surface Science*, *254*(4), 980–983. https://doi.org/10.1016/j.apsusc.2007.08.046

Zhong, C., Gasser, A., Kittel, J., Wissenbach, K., & Poprawe, R. (2016). Improvement of material performance of Inconel 718 formed by high deposition-rate laser metal deposition. *Materials & Design*, *98*, 128–134. https://doi.org/10.1016/j.matdes.2016.03.006

Chapter 11

Impact of Enabling Factors on the Adoption of Additive Manufacturing in the Automotive Industry

Kshitij Sharma, Maitrik Shah, Shivendru Mathur, Neha Choudhary, and Varun Sharma

IIT Roorkee, Haridwar, India

CONTENTS

11.1 INTRODUCTION

Additive manufacturing (AM) is a non-conventional manufacturing process that involves the layer by layer deposition of material to make a product/part as opposed to traditional subtractive methods that focus on removing material to get the desired shape and size. AM is a transformative approach to industrial production used to produce components with great precision using a variety of materials like plastic, metal and composites. AM is a general term that encompasses various techniques like 3-D printing, rapid prototyping (RP), direct digital manufacturing (DDM), layered manufacturing and additive fabrication (Abdulhameed et al., 2019). Common to all these methods is the use of computer-aided-design (CAD) software or 3D object scanners to direct hardware for achieving the desired result, thereby skipping the various traditional manufacturing steps (Attaran, 2017).

AM is an upcoming technology with the potential to revolutionise manufacturing and which has found applications in various sectors like aerospace, automotive, medical and sports (Attaran, 2017). It is a cost-effective, precise and safe way of manufacturing components. Since it is still in its nascent stages, even within the aforementioned sectors, its usage has been limited to making prototypes, custom-made products and parts involving complicated designs (Devi et al., 2019; Quinlan et al., 2017). Since the steps involved in manufacturing using AM are very different from traditional methods, its use promises to alter the existing supply chain that is designed for traditional methods (Devi et al., 2019).

There are many benefits of using AM, like reduced material wastage (Böckin & Tillman, 2019; Priarone et al., 2017), reduced manufacturing steps and safety in production. Since this process simply converts CAD files to components directly, there is less need to have a large inventory of work in progress or finished goods and it can be replaced by a digital inventory (Singamneni et al., 2019). This allows a larger number of manufacturing units to exist in a decentralised manner, closer to the consumer, reducing the lead time (Attaran, 2017; Khajavi et al., 2014).

Specifically, in the automotive industry, the use of AM has been initiated by luxury brands like BMW, Audi and Bugatti to allow for the personalisation of certain parts like dashboards. Now the use has become more widespread with companies also using it to produce mechanical components, spare parts (Attaran, 2017; Khajavi et al., 2014) and electronic components in certain models. These companies have already reported an increase in cost efficiency for certain complex components and have increased the use of AM in many precision products or even regular parts to reduce the weight of the component (Böckin & Tillman, 2019).

11.2 RESEARCH MOTIVATION

A growth in the usage of AM has been observed in the industrial area in recent times. The global market for AM is expected to rise from an estimated value of USD3.99 billion in 2018 to USD11.56 billion by 2026 and have a compound annual growth rate of 14.20% (Bajwa, 2021).

One of the first industries to apprehend the benefits of AM was the automotive industry. AM, being disruptive, gives freedom to the designer where he can prepare rapid prototypes with complex design, which otherwise would not have been possible through subtractive manufacturing processes (Gibson et al., 2010). This new technology provides automotive engineers with innovation privileges (Zhai et al., 2014) and helps in the optimisation of the supply chain. This technology also leads to the higher efficiency of the testing, manufacturing and assembling processes of the production cycle (Delic & Eyers, 2020).

The benefits of AM have huge potential to cater to the requirements of the automotive sector in developing countries like India. The electric vehicle (EV) market alone is expected to grow at a compounded annual growth rate (CAGR) of 44% from 2020 to 2027 and is expected to hit 6.34 million units of annual sales by 2027 as EV manufacturing incorporates the extensive use of AM (Data Bridge Market Research, 2021).

Furthermore, The Union Cabinet, chaired by the Prime Minister, Shri Narendra Modi, has given its approval to introduce the Production-Linked Incentive (PLI) Scheme into the automobile and auto components sectors to enhance India's manufacturing capabilities under the banner of Atma Nirbhar Bharat (Data Bridge Market Research, 2021).

Despite various potential benefits, the widespread implementation of AM is still not visible in the Indian market. To make this feasible and incentivise the benefits of AM, the identification of enabling factors and promoting the adoption of AM in the automobile industry needs to be looked into. Thus, the present chapter analyses the effect of the key enablers on the adoption of AM in the automotive industry.

11.3 LITERATURE REVIEW

There have been various research papers that have analysed the various aspects of AM and its impact on various sectors. Özceylan et al. (2017) analysed the logistics of integrating AM in the traditional supply chain network (TSCN), its effect on the lead time and the potential for associated customer increase. Beiderbeck et al. (2018) studied the supply chain further and listed the potential benefits of AM for automotive spare parts distribution and how it would affect after-market sales. AM is an immature technology and companies will face problems in developing long-term sustainable models for manufacturing. Agrawal & Vinodh (2019) and Ford and Despeisse (2016) studied the problems faced by AM in this regard to provide an overview of innovation, business models and configuration of supply chains.

Böckin and Tillman (2019) and Gebler et al. (2014) studied the environmental aspect of implementing AM by analysing how the property of reduced material usage through optimisation would lead to lower CO_2 emissions and higher energy efficiency. Chen et al. (2015) analysed the sustainability of cheap 3D printing technology in terms of economic and environmental aspects, showing that digitisation of various aspects of production can have a significant impact on present business models.

Sonar et al. (2020) used interpretive structural modelling (ISM) cross-impact matrix multiplication applied to classification (MICMAC) analysis to find the interrelationships between the factors affecting AM implementation in the automotive sector of India. Dwivedi et al. (2017) took a different approach by analysing the barriers faced by the automotive sector and used fuzzy ISM to conclude that government support and production technology limitations were the major driving factors. Rahim and Maidin (2013) used an analytic hierarchical process (AHP) to analyse the problems faced by AM implementation in the Malaysian automotive industry, where investment cost resulted as the highest ranking problem. For selecting the best AM method for metal parts manufacturing in the West, Sobota et al. (2020) interviewed experts by using the best-worst method to obtain the data for analysis, concluding that market demand would be most suited for the selection of a process. No study was found which addressed the impacts and relationships of enablers for AM application in India, and this will be the primary focus of this research.

11.3.1 Enablers

For a product to be competitive it needs to be up to date with industry standards for new design models, technical advancements, aesthetics and safety. Rapid prototyping provided by AM allows manufacturers to implement these changes easily and on time to meet market demands (Gibson, 2017). Traditional manufacturing has disadvantages in this regard as the whole process is too rigid for major improvements and design implementations. AM provides ease of customization for a model that designers can implement without disrupting the whole assembly line. The designers' freedom is achieved by the 3D printer's ability to manufacture very complex designs to meet certain customer requirements which would have been unfeasible otherwise (Ford & Despeisse, 2016).

AM techniques being used currently have the potential to be more economical than conventional manufacturing processes for small to medium batches. The tooling cost is minimal in most of the processes (Eyers & Potter, 2015). For large batches, traditional machining requires numerous tools and machines to manufacture the multiple components which amounts to a significant cost, while AM machines can cut down the costs by manufacturing multiple parts single-handedly. The cost of manufacturing equipment for AM is still expensive but it is bound to decrease in the coming years as the technology becomes more adopted by manufacturers, making it a very efficient process for large-scale industries in the coming future (Ford & Despeisse, 2016).

Traditional techniques require separate manufacturing for each component. 3D printers can manufacture multiple parts simultaneously and with greater accuracy, thereby reducing time, defects and quality problems compared to conventional processes. AM allows easy redesigns to achieve industry-standard strength with low material volume: known as part consolidation. Life cycle analysis has shown the possibility of a significant amount of manufacturing cost reduction by implementing AM, which reveals the importance of enablers, like part weight reduction and shorter assembly times. Savings by AM worldwide are projected to be in the region of USD113–370 billion by 2025 (Gebler et al., 2014).

Since the components manufactured through AM use less material and can be redesigned for reduced weight, the automobiles manufactured are more fuel-efficient, which is more visible in carriers like trucks as they have a greater number of components that can be 3D printed. Naturally, AM wastes less material than subtractive manufacturing as it uses layer by layer deposition to finish a product, eliminating or minimising the machining required while using less material simultaneously (Böckin & Tillman, 2019; Chen et al., 2015).

The spare part requirements for an industry this large necessitates massive warehousing for the storage and proper methods of distribution. The losses encountered in poor inventory management can be easily rectified by keeping a digital inventory (Beiderbeck et al., 2018). Digitisation is one of the most defining factors for AM implementation in regards to its economic and supply chain impact (Chen et al., 2015). Using AM to manufacture the after-sales spare parts provides the benefits of on-demand production and lower inventory costs, and also decreases the lead time by a significant amount. Studies have calculated that increasing the number of computer numeric control (CNC) machines in traditional supply chain networks can decrease the lead time from 20.3 to 2.44 days, allowing brands to serve more customers in a short period (Özceylan et al., 2017).

Unlike traditional subtractive manufacturing, most of the AM techniques use live sensors/cameras to monitor the processing to provide a proper quality inspection. Techniques like closed-feedback loops and image processing are being implemented to reduce defects in the manufacturing in some AM methods, like material jetting (Kim et al., 2018). 3D printers are much safer than the conventional manufacturing machines used when taking into account the physical hazards. Lightweight parts with adequate strength enhance product safety as well (Baumers et al., 2011).

After a thorough examination, 13 enablers shown in Table 11.1 were decided on, with the reasons and references mentioned.

11.3.2 Research Gaps and Objectives

AM is growing in different areas and is rapidly reshaping existing supply chains. However, only a few researchers have analysed the influence of AM adoption in industry (Böckin & Tillman, 2019; Delic & Eyers, 2020; Leal et al., 2017). The previous literature is mostly confined to the conceptual evaluation of the problems associated with AM supply chains (Durach et al., 2017; Haghighat Khajavi et al., 2020; Tasé Velázquez et al., 2020). Though there have been studies related to the barriers to the adoption of AM in the automotive area (Dwivedi et al., 2017), there is no work that critically analyses the enablers of AM adoption. Thus, there is a need for research

Table 11.1 List of Enablers

Enabler	Description	References
1. Tooling cost reduction	AM processes themselves require fewer tools compared to conventional processes. But in the automotive industry where certain products are still required to be manufactured conventionally (owing to large quantities required) the tools and casts used can still be 3D printed. Thus in these cases also tooling costs are reduced since if there is a change in a tool or cast design it can be incorporated easily and when a single tool or cast breaks it is easier to replace that single part with a new 3D printed part rather than manufacture conventionally.	Eyers & Potter (2015) Durão et al. (2019)
2. Reduction in lead time	The supply chain will be modified in the case of AM where manufacturing is done closer to the consumer thus reducing the lead time.	Özceylan et al. (2017) Beiderbeck et al. (2018) Petrovic et al. (2011) Bandyopadhyay & Heer (2018)
3. Use of digital inventory	The conventional inventory can be replaced by a digital inventory, thus only requiring the maintenance of a database for the design of the products. These are manufactured only as per requirement, thereby reducing the cost incurred to maintain the inventory. This is especially useful for spare parts in the automotive industry.	Özceylan et al. (2017) Holmström et al. (2010) Singamneni et al. (2019)
4. Less material wastage	AM stands for processes where the material is added to manufacture a product and thus only a requisite amount of raw material is used as opposed to conventional subtractive manufacturing. This reduces material wastage and scrap generation.	Chen et al. (2015) Özceylan et al. (2017) Baumers et al. (2011) Zhai et al. (2014) Böckin & Tillman (2019) Priarone et al. (2017)
5. Lower prototyping cost	While a product is in the design stage the number of design changes is very high and making a prototype to account for every change is a difficult task, but one made easy by the use of AM. Since time and effort to make a prototype are reduced, better testing can be done at each stage using the same.	Gibson (2017) Jamshidinia & Sadek (2015) Hannibal & Knight (2018)

(Continued)

Table 11.1 (Conitnued) List of Enablers

Enabler	Description	References
6. Shorter assembly line and reduced workforce	Instead of manufacturing different parts and then assembling them, with the help of AM, a product can be manufactured in a single step, thus reducing the assembly line needed for that product. This in turn leads to a reduction in the labour required for the process. This is especially true for the automotive sector where long assembly lines are fairly common for most models.	Gebler et al. (2014) Raymond (2005) Baumers et al. (2011) Attaran (2017)
7. Increased energy efficiency	Due to a reduction in the number of steps in manufacturing processes, the energy required for fabrication will decrease.	Böckin & Tillman (2019) Gebler et al. (2014) Walachowicz et al. (2017) Shahrubudin et al. (2019) Attaran (2017)
8. Part weight reduction	Complex designs including lattice structures that reduce weight without compromising on strength can be better manufactured by 3D printing. Also, newer composite materials are constantly being developed that further improve the physical properties of a product. Part weight is a critical factor in the automotive industry, especially for competitive manufacturing.	Böckin & Tillman (2019) Gebler et al. (2014) Zhai et al. (2014)
9. Mass customisation/ personalisation	User feedback could be easily incorporated into the design process. Customisation is very easy in the case of AM.	Chen et al. (2015) Hannibal & Knight (2018) Delic & Eyers (2020)
10. Quality control	Since AM involves minimal human intervention and is a fairly reliable manufacturing process, the percentage of defective products is much less, easing the process of quality control.	Kim et al. (2018) Delgado Camacho et al. (2018) Shahrubudin et al. (2019)
11. Design flexibility	Altering a design is an easy process in AM, and a product can be manufactured for a range of properties by changing the manufacturing process slightly.	Ford & Despeisse (2016) Özceylan et al. (2017) Gibson (2017) Zhai et al. (2014) Quinlan et al. (2017)

(Continued)

Table 11.1 (Conitnued) List of Enablers

Enabler	Description	References
12. Increased self-reliance	In recent times, governments throughout the world are emphasising self-reliance. In AM the dependency on suppliers will reduce as a lot of products will be manufactured in-house.	N/A
13. Safety	In the context of physical hazards, 3D printers are much safer than any conventional manufacturing.	Baumers et al. (2011) Agrawal & Vinodh (2019)

on the industrial application of, and the deciding factors related to, AM adoption in the automotive industry. The study of the nature of the interactions between enablers and the ranking of enablers could provide significant insights into the market understanding of AM.

11.4 RESEARCH METHOD

There are various decision-making techniques, some of which are listed in Table 11.2.

The ISM MICMAC method is used by many in their research to provide a hierarchical order of the critical factors relating to a problem through digraphs (Devi K et al., 2021; Palaniappan et al., 2020). The data is collected in binary form and without any weights assigned to each enabling factor. The TISM method involves one extra step to describe how each factor affects the other (Agrawal & Vinodh, 2019). The AHP model is used to

Table 11.2 List of Research Models

Author	Research Topic	Research Model
Rahim & Maidin (2013)	AM implementation in the Malaysian automotive industry	AHP
Palaniappan et al. (2020)	AM influence in the food sector	ISM-MICMAC
Agrawal & Vinodh (2019)	Analysis of factors influencing sustainable AM	TISM
Yeh & Chen (2018)	Success factors to 3D printing adoption	Fuzzy AHP
Dwivedi et al. (2017)	Analysis of barriers to implementing AM in the Indian automotive sector	Fuzzy ISM
Anand & Vinodh (2018)	Ranking AM processes for microfabrication	Fuzzy AHP-TOPSIS

develop final rankings of enablers on the basis of their respective importance to the industry while fuzzy AHP is used in situations with unreliable/insufficient information (Peko et al., 2018; Rahim & Maidin, 2013). The analytic network process (ANP) is a more general and cluster based approach when compared to AHP as it uses a network analysis providing more detailed interdependence between the factors (Yazgan et al., 2010), the drawback being the large number of comparisons required to obtain the results. The combined ISM-ANP approach incorporates the results obtained from ISM to forward the calculations in the ANP part, hence reducing the pairwise comparisons while also assigning weights to each factor (Digalwar et al., 2020).

11.5 METHODOLOGY

This analysis aims to study the impact of various enabling factors for the use of AM in the automotive industry. This is accomplished in two steps that involve two very effective mathematical tools, namely ISM and ANP. These methods involve collecting data from a group of experts, processing the data to quantify the opinions furnished by them, and then drawing appropriate conclusions. Since both these tools are essentially data-driven the collection of data is an essential step. For this purpose, a group of six experts from various backgrounds and experience were identified and contacted for data collection (Table 11.3).

The group of experts was contacted and the relevant data regarding the 13 enablers were collected after multiple rounds of to and fro communication. The pool of experts represents a group of individuals with a diverse set of experience, opinions and understanding of AM and/or the automotive industry in general. To extract an unbiased opinion we contacted experts from both industrial and academic backgrounds. The academic experts have a deep understanding of the concept of AM while the industrial experts better understand the practical implications of using it in industry. The data

Table 11.3 Expert Description

	Academia/ Industry	Designation	Experience
1	Academia	Assistant professor	5–10 years
2	Academia	Research scholar	5–10 years
3	Academia	Research scholar	<5 years
4	Academia	Assistant professor	<5 years
5	Industry	Plant head at auto ancillary company	>10 years
6	Industry	Chief financial officer (CFO) at auto ancillary company	>10 years

collection process was completed twice, once while implementing each one of the ISM and ANP methods. First, data pertaining to the ISM method was collected from the experts, and based on that conclusions about the inter-relationships between the enablers were established. The digraph was con-structed and the MICMAC analysis was completed. Then the ANP network diagram was constructed using the data collected in the ISM step. Next the data pertaining to the ANP analysis were collected in the second round and conclusions from the ANP method were deduced. Then the results from both methods were combined to form the final hierarchical ranking. Subsequently, the outcomes of the research were discussed and some con-structive conclusions presented.

11.6 INTERPRETIVE STRUCTURAL MODELLING (ISM)

ISM is a very useful mathematical tool, first proposed by John N. Warfield in 1974. It is an approach that graphically represents a complex system of elements or factors that influence a specific issue or problem. The purpose of ISM is to solve complex problems by identifying the major factors involved and their interrelationships. In this analysis, it will help us identify interre-lationships amongst the various enablers and identify which enabler plays a more important role than the others. After following the steps of the meth-odology, a digraph is derived that shows the dependency and driving abil-ity of the factors. The MICMAC chart constructed from the data collected also helps summarise the driving power and dependability of the various enabling factors. The steps of the ISM process are:

1. All the important parameters that can influence the given study are identified and data regarding their interdependence is collected from the pool of experts. Next, a contextual relationship between the iden-tified variables is established and a structural self-interaction matrix (SSIM) is constructed to represent them.
2. An initial reachability matrix (IRM) is derived from the SSIM by expressing the relationships in binary form.
3. The transitivity rule is checked for each pair of factors and a final reachability matrix (FRM) is prepared that represents all the relation-ships between the variables.
4. Level partitioning of the variables is done to reflect their hierarchy based on their driving power and dependence.
5. Based on the level partitioning, a digraph is constructed representing the interdependence of all the enablers.
6. A binary interaction matrix is constructed to represent the connec-tions in the digraph in binary form.
7. MICMAC analysis is done and a plot between the driving power and dependence of all the enablers is made.

11.7 ANALYTIC NETWORK PROCESS (ANP)

Multi-criteria decision analysis is a sub-discipline of operations research that explicitly evaluates multiple conflicting criteria in decision making. It involves choosing amongst the alternatives present for a particular problem involving multiple criteria. ANP and AHP are two methods used to solve such problems. Both are powerful synthesis methodologies used to combine judgements and data and effectively rank all the options available. ANP is a more general form of AHP and it is used in this study since we have a complex network of enablers and not a hierarchical structure. The ANP method involves constructing the network diagram using the data from the IRM obtained in the ISM step. After the network diagram is made the relevant data is entered and the relative importance of all the enablers is calculated. The steps involved in ANP are:

1. The goals, criteria and sub-criteria are identified and the enablers are classified into clusters to help in the construction of the network.
2. The network diagram is constructed representing all the enablers and clusters and the connections are formed using the IRM obtained during the ISM step.
3. Based on the network diagram, pairwise comparison between the enablers is done to ascertain their relative importance and data is collected from the experts in this step. The unweighted and weighted supermatrix is constructed based on the data entered.
4. The supermatrix is then normalised to generate the limit matrix representing the relative importance of all the factors. The ANP ranking is established based on these values with the relative weights represented in percentage form.
5. The hierarchy from the ISM method and the relative importance obtained from the ANP method are combined to obtain the final rankings of all the enablers.

11.8 APPLICATION AND RESULTS

11.8.1 ISM Application

This section describes the application of ISM to develop the interrelationships among the identified enablers.

Step 1: Development of SSIM
The SSIM is developed to represent the relationships among the listed enablers from the collected data (Table 11.4). The V, A, X and O notations are:
- *V*- Factor (i) causes/affects factor (j)
- *A*- Factor (j) causes/affects factor (i)
- *X*- Factor (i) and (j) cause/affect each other
- *O*- Factors have no connection.

Table 11.4 Structural Self-Interaction Matrix

	A	B	C	D	E	F	G	H	I	J	K	L	M
Tool Cost Reduction (A)		A	O	A	X	X	O	A	O	A	A	O	A
Reduction in Lead time (B)			O	X	X	A	V	A	V	X	X	O	A
Use of Digital Inventory (C)				A	O	A	O	O	X	V	A	O	O
Less Material Wastage (D)					A	X	V	A	A	X	X	A	O
Lower Prototyping Cost (E)						A	A	A	A	X	V	O	O
Shorter Assembly Line (F)							X	A	A	A	X	O	V
Increased Energy Efficiency (G)								O	A	O	A	O	A
Part Weight Reduction (H)									A	X	A	O	A
Mass Customization (I)										X	A	A	A
Quality Control (J)											A	A	A
Design Flexibility (K)												O	V
Increased Self-Reliance (L)													A
Safety (M)													

Step 2: Development of Initial Reachability Matrix

The IRM (Table 11.5) is created from the SSIM (Table 11.5). This is a binary matrix obtained by replacing the VAXO notation with binary values using the rules:

- V entry has a binary value of 1 for (i,j) and 0 for (j, i)
- A entry has a binary value of 0 for (i,j) and 1 for (j, i)
- X entry has a binary value of 1 for both (i,j) and (j, i)
- O entry has a binary value of 0 for (i,j) and 1 for (j, i)

Step 3: Development of Final Reachability Matrix

The FRM is obtained by applying the transitivity rule to the IRM (Table 11.6). The transitivity rule is applied to check for indirect relationships between the enablers as they might be significant for identifying the hidden impacts of certain factors. The transitivity refers to whether a first factor is influencing another second factor, and whether this second factor is influencing a third factor; and if so then the first factor must be influencing the third factor also.

Table 11.5 Initial Reachability Matrix

	A	B	C	D	E	F	G	H	I	J	K	L	M
Tool Cost Reduction (A)	1	0	0	0	1	1	0	0	1	0	1	0	0
Reduction in Lead time (B)	1	1	0	1	1	0	1	0	1	1	1	0	0
Use of Digital Inventory (C)	0	0	1	0	0	0	0	0	1	1	0	0	0
Less Material Wastage (D)	1	1	1	1	0	1	1	1	0	1	1	0	0
Lower Prototyping Cost (E)	1	1	0	1	1	0	0	0	0	1	1	0	0
Shorter Assembly Line (F)	1	1	1	1	1	1	1	0	0	0	1	0	1
Increased Energy Efficiency (G)	0	0	1	0	1	1	1	0	0	0	0	0	0
Part Weight Reduction (H)	1	1	0	1	1	1	0	1	0	1	0	0	1
Mass Customisation (I)	0	0	1	1	1	1	1	1	1	1	0	0	0
Quality Control (J)	1	1	0	1	1	1	0	1	1	1	0	0	0
Design Flexibility (K)	1	1	1	1	0	1	1	1	1	1	1	0	1
Increased Self-Reliance (L)	0	1	0	1	0	0	0	0	1	1	0	1	0
Safety (M)	1	1	0	0	0	0	1	1	1	1	0	1	1

Table 11.6 Final Reachability Matrix

	A	B	C	D	E	F	G	H	I	J	K	L	M	Driving Power
A	0	0	0	0	1	1	0	0	0	0	0	0	0	2
B	1	0	0	1	1	0	1	0	1	1	1	0	0	7
C	0	1*	0	1*	0	0	1*	0	1	1	0	0	0	5
D	1	1	1	0	0	1	1	0	0	1	1	0	0	7
E	1	1	0	1	0	0	1*	0	1*	1	1	0	0	7
F	1	1	1	1	1	0	1	0	1*	1*	1	0	1	10
G	0	0	0	0	1	1	0	0	0	0	0	0	0	2
H	1	1	0	1	1	1	0	0	1*	1	1*	0	1	9
I	0	1*	1	1	1	1	1	1	0	1	0	0	0	8
J	1	1	0	1	1	1	0	1	1	0	0	0	0	7
K	1	1	1	1	1	1	1	1	1	1	0	0	1	11
L	0	1	0	1	0	0	0	0	1	1	0	0	0	4
M	1	1	0	0	0	0	1	0	1	1	0	1	0	6
Dependence Power	8	10	4	9	8	7	8	3	9	10	5	1	3	

After checking the transitivity rules, a final reachability matrix is developed. The changed values are written with an asterisk (1*).

Step 4: **Division of Levels or Level Partitioning**

After the FRM has been created, the reachability and antecedent sets for each enabler is found. The reachability set consists of the factors themselves and others that are reachable from that factor. The antecedent set consists of factors itself and others which may reach it. The interaction of these two sets is established, and the enablers for which reachability and intersection set are similar, marked as the highest level of the ISM hierarchy. The second level is partitioned after removing the top-level factors, and this process continues until all factors have been partitioned. Finally, complete level partitioning is achieved (Table 11.7).

Step 5: **Formation of Digraph**

Subsequently, enablers at different levels are used to create a digraph. The factors placed at the highest level are the top-level factor in the digraph, and subsequently the other factors are placed at different levels in the form of nodes as obtained in Table 11.7. The factors (nodes) that have the same level are placed side by side. Each factor is looked over to establish the relationship with the factors at the same level and the immediate next level. A solid arrow is used to indicate a direct relationship, a double-headed arrow for a bidirectional relation and a dashed arrow for a transitivity relation among the enablers. The digraph is finally converted into an ISM model by replacing nodes with the proper name of the enablers as shown in Figure 11.1. The factors at the bottom level are the most influencing

Table 11.7 Level Partitioning Matrix

	Reachability Set	Antecedent Set	Intersection Set	Level
1	1,5,6	1,2,4,5,6,8,10,11,13	1,5,6	I
7	5,6,7	2,3,4,5,6,7,9,11,13	5,6,7	I
2	2,4,5,9,10,11	2,3,4,5,6,8,9,10,11,12,13	2,4,5,9,10,11	II
4	2,3,4,6,10,11	2,3,4,5,6,8,9,10,11,12	2,3,4,6,10,11	II
10	2,4,5,6,8,9,10	2,3,4,5,6,8,9,10,11,12,13	2,4,5,6,8,9,10	II
3	3,9	3,6,9,11	3,9	III
5	5,9,11	5,6,8,9,11	5,9,11	III
9	3,5,6,8,9	3,5,6,8,9,11,12,13	3,5,6,8,9	III
12	12	12,13	12	IV
13	13	6,8,11,13	13	V
6	6,11	6,8,11	6,11	VI
11	6,8,11	6,8,11	6,8,11	VI
8	8	8	8	VII

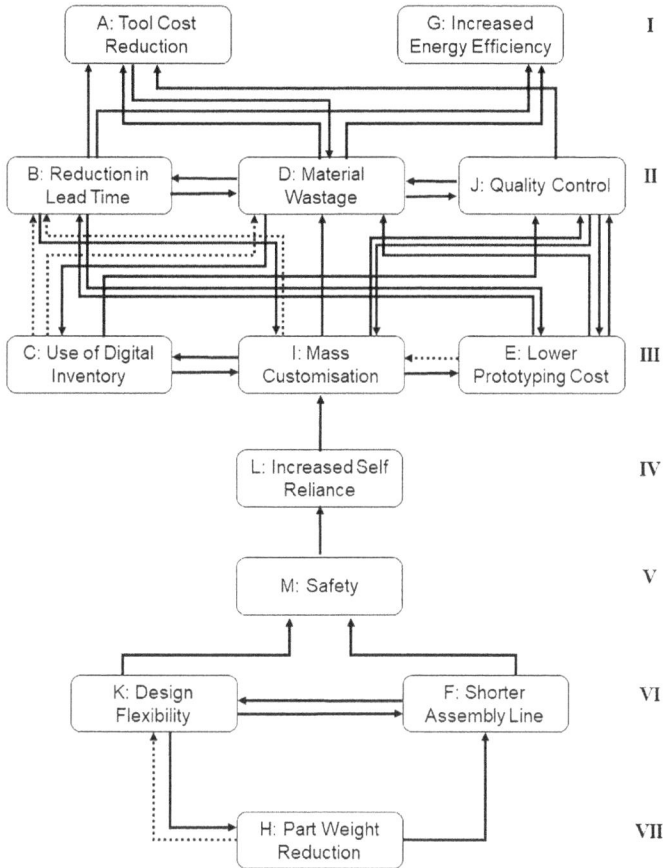

Figure 11.1 ISM model.

factors. Priority decreases as we move further up the model. The factors at the top depend on the lower level factors. The flow of the direction of influence is shown by the arrows in the ISM model.

Step 6: **Creation of Binary Interaction Matrix**

A binary interaction matrix is constructed by representing the connections indicated in the digraph in the form of a matrix (Table 11.8).

Step 7: **Cross-Impact Matrix Multiplication Applied to Classification (MICMAC)**

MICMAC is used to analyse the enablers based on the driving power and dependence power of each enabler obtained from the FRM (Table 11.6). They are classified into four sections based on the relative values of the driving power and dependence (Figure 11.2):

1. **Linkage region:** Enablers belonging to this region have strong driving and dependence power. They strongly drive other

Table 11.8 Binary Interaction Matrix

	A	B	C	D	E	F	G	H	I	J	K	L	M
A	1	0	0	0	0	0	0	0	0	0	0	0	0
B	1	1	0	1	1	0	1	0	1	1	0	0	0
C	0	1*	1	1*	0	0	0	0	1	1	0	0	0
D	1	1	1	1	0	0	1	0	0	1	0	0	0
E	0	1	0	1	1	0	0	0	1*	1	0	0	0
F	0	0	0	0	0	1	0	0	0	0	1	0	1
G	0	0	0	0	0	0	1	0	0	0	0	0	0
H	0	0	0	0	0	1	0	1	0	0	1*	0	0
I	0	1*	1	1	1	0	0	0	1	1	0	0	0
J	1	1	0	1	1	0	0	0	1	1	0	0	0
K	0	0	0	0	0	1	0	1	0	0	1	0	1
L	0	0	0	0	0	0	0	0	1	0	0	1	0
M	0	0	0	0	0	0	0	0	0	0	0	1	1

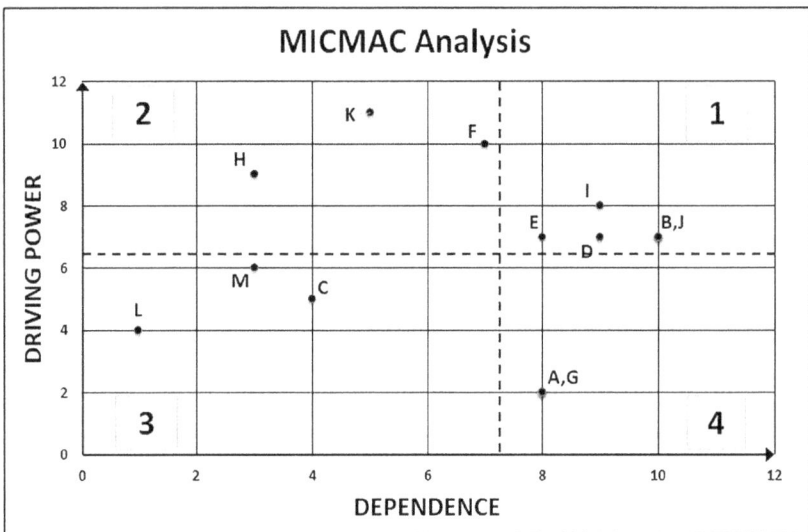

Figure 11.2 MICMAC analysis.

enablers as well as having strong feedback on themselves. Reduction in lead time (B), reduced material wastage (D), lower prototyping cost (E), mass customisation (I) and quality control (J) are in this region.

2. **Independent region:** Enablers belonging to this region have strong driving power with weak dependency. They strongly

drive other enablers but are not easily influenced by others. Design flexibility (K), shorter assembly line (F) and part weight reduction (H) are in this region.

3. **Autonomous region:** Enablers belonging to this region have weak driving power and dependency. They have less influence and are not easily influenced either. Use of digital inventory (C), increased self-reliance (L) and safety (M) are in this region.

4. **Dependent region:** Enablers belonging to this region have weak driving power with strong dependence power. They are easily influenced by other enablers but they don't have much effect on others. Tooling cost reduction (A) and increased energy efficiency (G) are in this region.

11.8.2 ANP Application

The ANP network model (refer to Figure 11.3) is formed from the contextual relationship obtained from the ISM matrix to show the importance of all enablers, which is difficult to present in the ISM method.

Step 1: Distribution of Enablers into Clusters
The enablers are classified into four clusters: design, manufacturing, supply chain and environment (Table 11.9). The design cluster consists of enablers like design flexibility, lower prototyping cost, mass customisation and part weight reduction, which are important at the design phase of a product. The manufacturing cluster consists of enablers like quality control, tooling cost reduction and manufacturer's safety, which are related to manufacturing activities

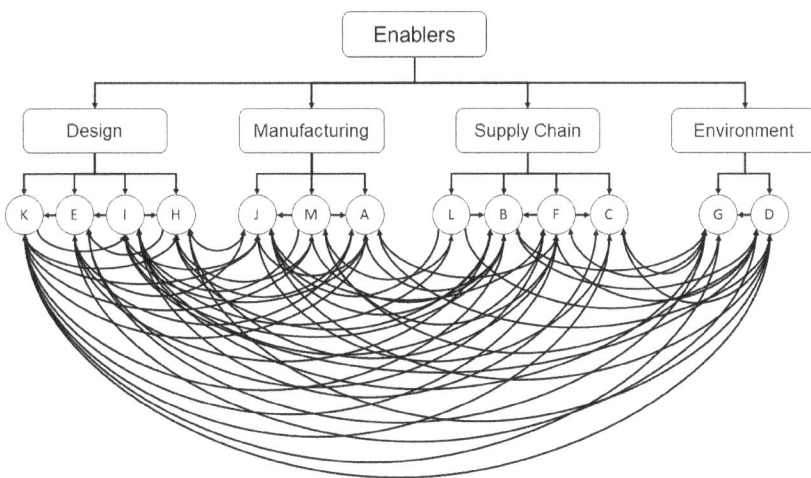

Figure 11.3 ANP model

Table 11.9 Cluster Formation

Design (P)	Manufacturing (Q)	Supply Chain (R)	Environment(s)
Design flexibility	Quality control	Increased self reliance	Increased energy efficiency
Lower prototyping cost	Tooling cost reduction	Reduction in lead time	Less material wastage
Mass customisation	Manufacturer's safety	Shorter assembly line	
Part weight reduction		Use of digital inventory	

in product development. Similarly, the supply chain cluster contains enablers influencing the supply chain of the product, such as increased self-reliance, reduction in lead time, shorter assembly line and the use of a digital inventory. The environment cluster captures the impact on the environment in the process of product development through enablers like increased energy efficiency and less material wastage.

Step 2:   **Formation of the ANP Model**

Based on the IRM (Table 11.4), an ANP model (Figure 11.3) is generated, indicating the connections between the different enablers belonging to different clusters.

Step 3:   **Formation of Weighted and Unweighted Supermatrix**

After the collection of data, a cluster weight matrix (Table 11.10) and a supermatrix are developed. The unweighted supermatrix is obtained by a pairwise comparison of the enablers (Table 11.11). The unweighted supermatrix is normalised by multiplying its elements by the corresponding weights of clusters to generate a weighted supermatrix (Table 11.12).

Step 4:   **Formation of Limit Matrix**

The limiting matrix is obtained by raising the weighted supermatrix to powers by multiplying itself until the numbers in the columns reach the same value (Table 11.13). These values represent

Table 11.10 Cluster Weight Matrix for ANP

Cluster Matrix	Design	Environmental	Manufacturing	Supply Chain
Design	0.21	0.25	0.25	0.25
Environmental	0.25	0.25	0.25	0.25
Manufacturing	030	0.25	0.25	0.25
Supply Chain	0.25	0.25	0.25	0.25

Table 11.11 Unweighted Supermatrix for ANP

Clusters	Enablers	Design				Environment		Manufacturing			Supply Chain			
		K	E	I	H	G	D	J	M	A	L	B	F	C
Design	K	0.00	1.00	0.00	0.00	0.00	0.50	0.00	0.00	0.33	0.00	0.41	0.50	0.00
	E	0.00	0.00	0.50	1.00	1.00	0.00	0.40	0.00	0.41	0.00	0.26	0.50	0.00
	I	0.33	0.00	0.00	0.00	0.00	0.00	0.40	0.50	0.26	1.00	0.33	0.00	1.00
	H	0.67	0.00	0.50	0.00	0.00	0.50	0.20	0.50	0.00	0.00	0.00	0.00	0.00
Environment	G	0.33	0.00	0.25	0.00	0.00	1.00	0.00	1.00	0.00	0.00	0.33	0.50	0.00
	D	0.67	1.00	0.75	1.00	0.00	0.00	1.00	0.00	0.00	1.00	0.67	0.50	0.00
Manufacturing	J	0.33	0.33	1.00	0.41	0.00	0.50	0.00	0.50	0.00	1.00	0.50	0.00	1.00
	M	0.33	0.00	0.00	0.26	0.00	0.00	0.00	0.00	0.00	0.00	0.00	0.33	0.00
	A	0.33	0.67	0.00	0.33	0.00	0.50	1.00	0.50	0.00	0.00	0.50	0.67	0.00
Supply chain	L	0.00	0.00	0.00	0.00	0.00	0.00	0.00	0.67	0.00	0.00	0.00	0.00	0.00
	B	0.14	1.00	0.00	0.33	0.00	0.33	0.50	0.33	0.00	1.00	0.00	0.50	0.00
	F	0.43	0.00	0.50	0.67	0.50	0.26	0.50	0.00	1.00	0.00	0.00	0.00	0.00
	C	0.43	0.00	0.50	0.00	0.50	0.41	0.00	0.00	0.00	0.00	0.00	0.50	0.00

Table 11.12 Weighted Supermatrix for ANP

Clusters	Enablers	Design				Environment		Manufacturing			Supply chain			
		K	E	I	H	G	D	J	M	A	L	B	F	C
Design	K	0.00	0.21	0.00	0.00	0.00	0.13	0.00	0.00	0.16	0.00	0.14	0.13	0.00
	E	0.00	0.00	0.10	0.21	0.50	0.00	0.10	0.00	0.21	0.00	0.09	0.13	0.00
	I	0.07	0.00	0.00	0.00	0.00	0.00	0.10	0.13	0.13	0.25	0.11	0.00	0.50
	H	0.14	0.00	0.10	0.00	0.00	0.13	0.05	0.13	0.00	0.00	0.00	0.00	0.00
Environment	G	0.08	0.00	0.06	0.00	0.00	0.25	0.00	0.25	0.00	0.00	0.11	0.13	0.00
	D	0.16	0.25	0.18	0.25	0.00	0.00	0.25	0.00	0.00	0.25	0.22	0.13	0.00
Manufacturing	J	0.10	0.10	0.30	0.12	0.00	0.13	0.00	0.13	0.00	0.25	0.17	0.00	0.50
	M	0.10	0.00	0.00	0.08	0.00	0.00	0.00	0.00	0.00	0.00	0.00	0.08	0.00
	A	0.10	0.20	0.00	0.10	0.00	0.13	0.25	0.13	0.00	0.00	0.17	0.17	0.00
Supply chain	L	0.00	0.00	0.00	0.00	0.00	0.00	0.00	0.17	0.00	0.00	0.00	0.00	0.00
	B	0.04	0.25	0.00	0.08	0.00	0.08	0.13	0.08	0.00	0.25	0.00	0.13	0.00
	F	0.11	0.00	0.12	0.16	0.25	0.06	0.13	0.00	0.50	0.00	0.00	0.00	0.00
	C	0.11	0.00	0.12	0.00	0.25	0.10	0.00	0.00	0.00	0.00	0.00	0.13	0.00

Table 11.13 Limit Matrix for ANP

Clusters	Enabler	Design				Environment			Manufacturing			Supply chain		
		K	E	I	H	G	D	J	M	A	L	B	F	C
Design	K	0.08	0.08	0.08	0.08	0.08	0.08	0.08	0.08	0.08	0.08	0.08	0.08	0.08
	E	0.11	0.11	0.11	0.11	0.11	0.11	0.11	0.11	0.11	0.11	0.11	0.11	0.11
	I	0.07	0.07	0.07	0.07	0.07	0.07	0.07	0.07	0.07	0.07	0.07	0.07	0.07
	H	0.04	0.04	0.04	0.04	0.04	0.04	0.04	0.04	0.04	0.04	0.04	0.04	0.04
Environment	G	0.07	0.07	0.07	0.07	0.07	0.07	0.07	0.07	0.07	0.07	0.07	0.07	0.07
	D	0.12	0.12	0.12	0.12	0.12	0.12	0.12	0.12	0.12	0.12	0.12	0.12	0.12
Manufacturing	J	0.11	0.11	0.11	0.11	0.11	0.11	0.11	0.11	0.11	0.11	0.11	0.11	0.11
	M	0.02	0.02	0.02	0.02	0.02	0.02	0.02	0.02	0.02	0.02	0.02	0.02	0.02
	A	0.11	0.11	0.11	0.11	0.11	0.11	0.11	0.11	0.11	0.11	0.11	0.11	0.11
Supply chain	L	0.00	0.00	0.00	0.00	0.00	0.00	0.00	0.00	0.00	0.00	0.00	0.00	0.00
	B	0.07	0.07	0.07	0.07	0.07	0.07	0.07	0.07	0.07	0.07	0.07	0.07	0.07
	F	0.12	0.12	0.12	0.12	0.12	0.12	0.12	0.12	0.12	0.12	0.12	0.12	0.12
	C	0.06	0.06	0.06	0.06	0.06	0.06	0.06	0.06	0.06	0.06	0.06	0.06	0.06

Table 11.14 Relative Weights as per ANP

Ranking	Enabler	ANP Weight
1	Less Material Wastage (D)	0.123
2	Shorter Assembly Line (F)	0.119
3	Tooling Cost Reduction (A)	0.111
4	Quality Control (J)	0.109
5	Lower Prototyping cost (E)	0.107
6	Design Flexibility (K)	0.081
7	Mass Customisation/Personalisation (I)	0.074
8	Reduction in Lead Time (B)	0.074
9	Increased Energy Efficiency (G)	0.071
10	Use of Digital Inventory (C)	0.063
11	Part Weight Reduction (H)	0.043
12	Safety (M)	0.021
13	Increased Self Reliance (L)	0.004

Table 11.15 Final Ranking of Enablers

Enabler	ANP Weight	ISM Level	Final Rating	Normalised	Ranking
Shorter Assembly Line (F)	0.119	6	0.717	22.76%	1
Design Flexibility (K)	0.081	6	0.487	15.46%	2
Lower Prototyping cost (E)	0.107	3	0.321	10.20%	3
Part Weight Reduction (H)	0.043	7	0.298	9.48%	4
Low Material Wastage (D)	0.123	2	0.247	7.83%	5
Mass Customisation/ Personalisation (I)	0.074	3	0.222	7.06%	6
Quality Control (J)	0.109	2	0.218	6.91%	7
Use of Digital Inventory (C)	0.063	3	0.189	6.00%	8
Reduction in Lead Time (B)	0.074	2	0.148	4.70%	9
Tooling Cost Reduction (A)	0.111	1	0.111	3.53%	10
Safety (M)	0.021	5	0.107	3.38%	11
Increased Energy Efficiency (G)	0.071	1	0.071	2.24%	12
Increased Self Reliance (L)	0.004	4	0.014	0.45%	13
Total	**1.000**		**3.149**	**100.00%**	

the relative importance of the variables and based on these values the enablers are ranked (Table 11.14) using ANP data only.

Step 5: **Ranking of Enablers**

The final weights were obtained by combining the data from the ISM and ANP and the consequent final rankings (Table 11.15).

The ANP weights are multiplied by the ISM hierarchy level for each enabler to estimate the final ratings which are then normalised to estimate the final ranking. This study utilises the combined effect of ISM and ANP weights to analyse the adoption of AM.

11.9 DISCUSSION

Conclusions drawn from the ISM analysis (Figure 11.1) show that Part Weight Reduction (H) (Böckin & Tillman, 2019) is the most influential enabler as it is classified in level 7 and present at the base of the digraph and influences all the other factors present above it. It directly influences Design Flexibility (K) (Devi et al., 2019) and Shorter Assembly Line (F) (Attaran, 2017) present just above at level 6, both of which are also amongst the most influential enablers. This can also be inferred from the MICMAC chart (Figure 11.2) which shows that Part Weight Reduction (H), Design Flexibility (K) and Shorter Assembly Line (F) lie in the independent region (Quadrant 2) having high driving powers and low dependences, thereby having a large influence on the other enablers present. At the other extreme, we have Tooling Cost Reduction (A) (Tosello et al., 2019) and Increased Energy Efficiency (G) (Attaran, 2017) at the top of the digraph at level 1. Both these enablers are influenced by all the enablers in the numerically higher levels present below them. This is again verified by the MICMAC chart where both these enablers are present in the dependent region (Quadrant 4), having high dependence and low driving powers. All the other enablers lie somewhere in between in the digraph and have either both high dependence and high driving power together, or both low dependence and low driving power.

The ISM tool, however, only represents the interrelationships and relative importance in a qualitative manner. ANP helps to quantify these relations. For ease of calculation, the enablers were divided into four clusters: Design (P), Manufacturing (Q), Supply Chain (R) and Environment (S). In the ANP method, the relative importance of the clusters as a whole and individual enablers are both calculated. The ranking of the clusters obtained was Q > P > R = S (Table 11.10) and the ranking of the enablers obtained was D > F > A > J > E > K > I > B > G > C > H > M > L (Table 11.14). The ranking of the enablers has certain contradictions concerning the result obtained from the ISM process. In this analysis, Less Material Wastage (D) (Priarone et al., 2017) and Shorter Assembly Line (F) are identified as the two most important enablers, while Increased Self Reliance (L) and Safety (M) are obtained as the least important enablers. There are multiple reasons for such deviations. The ISM method focuses more on the relative values of the driving power and dependence for all enablers while ANP values the total interrelationships of the variables more. Since different parameters are being evaluated, certain discrepancies are expected. This has been observed in the previous literature as well, where a lower level ISM factor having high dependence is ranked highest in the ANP analysis owing to the interrelationships it possesses (Choudhary et al., 2021). Also in the ANP method, the relative weight of the cluster to which an enabler belongs plays a role in determining its final weight while this is not the case for ISM.

Therefore to address this contradiction a combined final ranking is prepared that takes into account the findings from both methods and that is closest to the perfect rating. The final weight is obtained by multiplying the ISM level with the ANP weight and this value is normalised to obtain the final rating. The final rating obtained ranks Shorter Assembly Line (F) and Design Flexibility (K) as the most influential enablers, while Increased Energy Efficiency (G) and Increased Self Reliance (L) are obtained as the least influential enablers. The overall ranking obtained is F > K > E > H > D > I > J > C > B > A > M > G > L (Table 11.15).

Shorter Assembly Line is obtained as the most influential enabler after the two methods are collectively applied. This result is along expected lines as the most evident and disruptive change that AM can bring about is for the assembly line and supply chain in general. This is because, using AM, multiple steps in manufacturing can be combined into a single AM process that will do away with the need for large-scale machinery and reduce the moving parts involved. This will alter and shorten the manufacturing process and the assembly line, as a result (Attaran, 2017). Design Flexibility is the second most influential enabler. This is because in the automotive industry innovation and new designs are highly appreciated and most companies also develop concept cars that may not be launched on the market but highlight a company's vision for the future and its endeavour towards a breakthrough innovation. Thus, all these experiments that are carried out in the design phase can easily be tested by making simple prototypes of the new parts. Also, since certain complex designs cannot be manufactured by traditional methods, in such cases AM also comes in handy and thus the flexibility in the design process made possible by AM proves to be a very important enabling factor. On the other hand, Increased Self-Reliance is the least influential factor which can be attributed to the fact that just by improving the manufacturing capability within the country it can't be ensured that the country or industry as a whole will become self-reliant. This is because other factors, such as the availability of raw material or skilled labour, are also contributing factors. The second least influential factor is Increased Energy Efficiency, since the number of steps in manufacturing is reduced and so leads to energy savings. But, this is not completely true since traditional methods can also give higher output in terms of yield per hour and that also contributes to better energy efficiency (Böckin & Tillman, 2019). Lastly, since the conclusions derived in this study take into account the opinions of only a few individuals, they are bound to be biased and a similar study conducted with a different group of experts might furnish different results.

11.10 MANAGERIAL IMPLICATIONS

This research offers insight for managers who are planning to adopt AM. The results reveal that the enablers of the design cluster have a weightage of around 42.2%. This shows that even though AM is a type of manufacturing process it has significant impacts on the complete product life cycle. So managers looking to streamline their designing process can look into the added benefits associated with AM. Also, some major enabling factors of AM are design flexibility (Zhai et al., 2014) and lower prototyping cost (Jamshidinia & Sadek, 2015), adding up to 25.66% of weightage. These factors play a crucial role in companies that are constantly trying to improve themselves as the cost of innovation decreases (Hannibal & Knight, 2018). One of the biggest enablers of AM is the shorter assembly line (Baumers et al., 2011; Raymond, 2005) which accounts for around 22.76% of total weightage. If the company's assembly line has some inherent inefficiencies, managers can opt for AM to produce certain parts, thereby skipping a large portion of the assembly line as a whole. Part weight reduction and lower material wastage mean that the raw material being procured by the company is well utilised (Zhai et al., 2014). This combined with increased energy efficiency (Shahrubudin et al., 2019; Walachowicz et al., 2017) could be seen as a good initiative on the part of the company to become carbon neutral. Industries providing varieties of customised products that want to satisfy their customer in every aspect possible, from design to delivery, could also look into AM as a viable option, since mass customisation/personalisation (Delic & Eyers, 2020; Hannibal & Knight, 2018), quality control (Delgado Camacho et al., 2018; Kim et al., 2018; Shahrubudin et al., 2019) and reduction in lead time (Bandyopadhyay & Heer, 2018; Petrovic et al., 2011) have a combined weightage of 18.67% in AM adoption as enabling factors. Finally enabling factors like digital inventory (Holmström et al., 2010) and increased self-reliance perfectly resonate with the initiatives of the Government of India, like Digital India and Atma Nirbhar Bharat.

Apart from asking the question related to pairwise comparison, the survey conducted initially also asked the respondents about their opinions on the use of AM in the automotive industry and the existing research being done in this field. The following were some of their views.

AM is a versatile technology, suitable for most applications including automobile applications.

AM will greatly impact the automobile segment. With a printable battery and in-design PCB printing, the future of AM automobiles seems attractive.

The success of a product thoroughly depends on the type of AM opted for; fabrication and the fabricator must also have immense knowledge about the advantages the respective AM provides to the product.

AM in automotives still has time to be used widely due to the per piece cost not matching the conventional manufacturing cost, due to high volume; it is good for the customisation of parts. In research, it's been used widely and a lot of new parts have been made using AM for automotives and put to test in real-life scenarios. The future is certainly bright for AM in automotives once the cost of per piece production comes down.

On the whole, everyone agreed that the future of AM seems bright and with certain improvements done in the technology, it could easily capture a large proportion of the manufacturing segment.

11.11 CONCLUSIONS

This study has aimed to identify the most influential enabling factors for the implementation of AM in the automotive industry. This has been accomplished using the ISM–ANP methodology based on the data collected from a pool of experts belonging to the industry and to academia. The data collected was analysed and the final ranking of the enablers was deduced, considering their interrelationships as well as their ability to influence and get influenced by other enablers. Shorter Assembly Line (J) and Design Flexibility (K) were identified as the most influential enablers while Increased Self-Reliance (L) and Increased Energy Efficiency (G) were the least influential. Although influential enablers should be given preference in decision making, all enablers must be included in the analysis.

Like any other study, this report does not provide the final word when it comes to a large industry like automotives and does have some shortcomings to it. As our approach towards this problem is vastly different from pre-existing research on the topic, we hope this study serves as a pathway for further developments in this field. The enablers in our study were selected from existing research papers based on AM and thus are not exhaustive. Coming to the experts who were a part of this study, their personal opinions on the matter might have led to biased results. A wider pool of participants can be implemented to possibly eliminate this bias. Lastly, while the ISM and ANP methodology is a very dynamic approach to this decision-based problem, our research can be taken forward if the results are used in some specific case studies of AM application in India, while similar studies can also be conducted with the help of other mathematical tools.

REFERENCES

Abdulhameed, O., Al-Ahmari, A., Ameen, W., & Mian, S. H. (2019). Additive manufacturing: Challenges, trends, and applications. *Advances in Mechanical Engineering*, *11*(2), 168781401882288. https://doi.org/10.1177/1687814018822880

Agrawal, R., & Vinodh. (2019). Application of total interpretive structural modelling (TISM) for analysis of factors influencing sustainable additive manufacturing: a case study. *Rapid Prototyping Journal*, *25*(7), 1198–1223. https://doi.org/10.1108/RPJ-06-2018-0152

Anand, M. B., & Vinodh, S.(2018). Application of fuzzy AHP-TOPSIS for ranking additive manufacturing processes for microfabrication. *Rapid Prototyping Journal*, 24(2), 424–435. https://doi.org/10.1108/RPJ-10-2016-0160

Attaran, M. (2017). The rise of 3-D printing: The advantages of additive manufacturing over traditional manufacturing. *Business Horizons*, 60(5), 677–688. https://doi.org/10.1016/j.bushor.2017.05.011

Bajwa, N. (2021). *Automobile Industry in India - Auto Sector Growth Analysis*. March 2020. https://www.investindia.gov.in/sector/automobile.

Bandyopadhyay, A., & Heer, B. (2018). Additive manufacturing of multi-material structures. *Materials Science and Engineering: R: Reports*, 129, 1–16. https://doi.org/10.1016/j.mser.2018.04.001

Baumers, M., Tuck, C., Bourell, D. L., Sreenivasan, R., & Hague, R. (2011). Sustainability of additive manufacturing: measuring the energy consumption of the laser sintering process. *Proceedings of the Institution of Mechanical Engineers, Part B: Journal of Engineering Manufacture*, 225(12), 2228–2239. https://doi.org/10.1177/0954405411406044

Beiderbeck, D., Deradjat, D., & Minshall, T. (2018). The impact of additive manufacturing technologies on industrial spare parts strategies. *Centre for Technology Management Centre for Technology Management Working Paper Series*, 1. https://doi.org/10.17863/CAM.21296

Böckin, D., & Tillman, A.-M. (2019). Environmental assessment of additive manufacturing in the automotive industry. *Journal of Cleaner Production*, 226, 977–987. https://doi.org/10.1016/j.jclepro.2019.04.086

Chen, D., Heyer, S., Ibbotson, S., Salonitis, K., Steingrímsson, J. G., & Thiede, S. (2015). Direct digital manufacturing: definition, evolution, and sustainability implications. *Journal of Cleaner Production*, 107, 615–625. https://doi.org/10.1016/j.jclepro.2015.05.009

Choudhary, N., Kumar, A., Sharma, V., & Kumar, P. (2021). Barriers in adoption of additive manufacturing in medical sector supply chain. *Journal of Advances in Management Research*, 18(5), 637–660. https://doi.org/10.1108/JAMR-12-2020-0341

Data Bridge Market Research. (2021). Global Additive Manufacturing Market Booming Demand Leading To Exponential CAGR Growth By 2028 – The Courier. In *Data Bridge Market Research*. https://www.mccourier.com/global-additive-manufacturing-market-booming-demand-leading-to-exponential-cagr-growth-by-2028/

Delgado Camacho, D., Clayton, P., O'Brien, W. J., Seepersad, C., Juenger, M., Ferron, R., & Salamone, S. (2018). Applications of additive manufacturing in the construction industry – A forward-looking review. *Automation in Construction*, 89, 110–119. https://doi.org/10.1016/j.autcon.2017.12.031

Delic, M., & Eyers, D. R. (2020). The effect of additive manufacturing adoption on supply chain flexibility and performance: An empirical analysis from the automotive industry. *International Journal of Production Economics*, 228, 107689. https://doi.org/10.1016/j.ijpe.2020.107689

Devi, K. S., Paranitharan, K. P., & Agniveesh A. I. (2021). Interpretive framework by analysing the enablers for implementation of Industry 4.0: an ISM approach. *Total Quality Management & Business Excellence*, 32(13–14), 1494–1514. https://doi.org/10.1080/14783363.2020.1735933

Devi, P. A., K.B.L. Prasanna, & Vundela, V. S. (2019). Study of challenges in Additive Manufacturing in the Automotive Industry. *International Journal of Research in Advent Technology*, 7(1), 64–67. https://doi.org/10.32622/ijrat.71201917

Digalwar, A., Raut, R. D., Yadav, V. S., Narkhede, B., Gardas, B. B., & Gotmare, A. (2020). Evaluation of critical constructs for measurement of sustainable supply chain practices in lean-agile firms of Indian origin: A hybrid ISM-ANP approach. *Business Strategy and the Environment*, 29(3), 1575–1596. https://doi.org/10.1002/bse.2455

Durach, C. F., Kurpjuweit, S., & Wagner, S. M. (2017). The impact of additive manufacturing on supply chains. *International Journal of Physical Distribution & Logistics Management*, 47(10), 954–971. https://doi.org/10.1108/IJPDLM-11-2016-0332

Durão, L. F. C. S., Barkoczy, R., Zancul, E., Lee Ho, L., & Bonnard, R. (2019). Optimizing additive manufacturing parameters for the fused deposition modeling technology using a design of experiments. *Progress in Additive Manufacturing*, 4(3), 291–313. https://doi.org/10.1007/s40964-019-00075-9

Dwivedi, G., Srivastava, S. K., & Srivastava, R. K. (2017). Analysis of barriers to implement additive manufacturing technology in the Indian automotive sector. *International Journal of Physical Distribution & Logistics Management*, 47(10), 972–991. https://doi.org/10.1108/IJPDLM-07-2017-0222

Eyers, D. R., & Potter, A. T. (2015). E-commerce channels for additive manufacturing: an exploratory study. *Journal of Manufacturing Technology Management*, 26(3), 390–411. https://doi.org/10.1108/JMTM-08-2013-0102

Ford, S., & Despeisse, M. (2016). Additive manufacturing and sustainability: an exploratory study of the advantages and challenges. *Journal of Cleaner Production*, 137, 1573–1587. https://doi.org/10.1016/j.jclepro.2016.04.150

Gebler, M., Schoot Uiterkamp, A. J. M., & Visser, C. (2014). A global sustainability perspective on 3D printing technologies. *Energy Policy*, 74(C), 158–167. https://doi.org/10.1016/j.enpol.2014.08.033

Gibson, I. (2017). The changing face of additive manufacturing. *Journal of Manufacturing Technology Management*, 28(1), 10–17. https://doi.org/10.1108/JMTM-12-2016-0182

Gibson, I., Rosen, D. W., & Stucker, B. (2010). Additive Manufacturing Technologies. In *Springer*. Springer US. https://doi.org/10.1007/978-1-4419-1120-9

Haghighat Khajavi, S., Flores Ituarte, I., Jaribion, A., An, J., Chee Kai, C., & Holmstrom, J. (2020). Impact of Additive Manufacturing on Supply Chain Complexity. *Proceedings of the Annual Hawaii International Conference on System Sciences*, 2020-Janua, 4505–4514. https://doi.org/10.24251/HICSS.2020.551

Hannibal, M., & Knight, G. (2018). Additive manufacturing and the global factory: Disruptive technologies and the location of international business. In *International Business Review* (Vol. 27, Issue 6, pp. 1116–1127). https://doi.org/10.1016/j.ibusrev.2018.04.003

Holmström, J., Partanen, J., Tuomi, J., & Walter, M. (2010). Rapid manufacturing in the spare parts supply chain. *Journal of Manufacturing Technology Management*, 21(6), 687–697. https://doi.org/10.1108/17410381011063996

Jamshidinia, M., & Sadek, A. A. (2015). Additive Manufacturing of Steel Alloys Using Laser Powder-Bed Fusion. *Advanced Materials and Processes*, 173(1), 20–25.

Khajavi, S. H., Partanen, J., & Holmström, J. (2014). Additive manufacturing in the spare parts supply chain. In *Computers in Industry* (Vol. 65, Issue 1, pp. 50–63). https://doi.org/10.1016/j.compind.2013.07.008

Kim, H., Lin, Y., & Tseng, T.-L. B. (2018). A review on quality control in additive manufacturing. *Rapid Prototyping Journal*, *24*(3), 645–669. https://doi.org/10.1108/RPJ-03-2017-0048

Leal, R., Barreiros, F. M., Alves, L., Romeiro, F., Vasco, J. C., Santos, M., & Marto, C. (2017). Additive manufacturing tooling for the automotive industry. *The International Journal of Advanced Manufacturing Technology*, *92*(5–8), 1671–1676. https://doi.org/10.1007/s00170-017-0239-8

Özceylan, E., Çetinkaya, C., Demirel, N., & Sabırlıoğlu, O. (2017). Impacts of Additive Manufacturing on Supply Chain Flow: A Simulation Approach in Healthcare Industry. *Logistics*, *2*(1), 1. https://doi.org/10.3390/logistics2010001

Palaniappan, A., Vinodh, S., & Ranganathan, R. (2020). Analysis of factors influencing AM application in food sector using ISM. *Journal of Modelling in Management*, *15*(3), 919–932. https://doi.org/10.1108/JM2-11-2018-0190

Peko, I., Gjeldum, N., & Bilić, B. (2018). Application of AHP, Fuzzy AHP and PROMETHEE Method in Solving Additive Manufacturing Process Selection Problem. *Tehnicki Vjesnik - Technical Gazette*, *25*(2), 453–461. https://doi.org/10.17559/TV-20170124092906

Petrovic, V., Vicente Haro Gonzalez, J., Jordá Ferrando, O., Delgado Gordillo, J., Ramón Blasco Puchades, J., & Portolés Griñan, L. (2011). Additive layered manufacturing: sectors of industrial application shown through case studies. *International Journal of Production Research*, *49*(4), 1061–1079. https://doi.org/10.1080/00207540903479786

Priarone, P. C., Ingarao, G., di Lorenzo, R., & Settineri, L. (2017). Influence of Material-Related Aspects of Additive and Subtractive Ti-6Al-4V Manufacturing on Energy Demand and Carbon Dioxide Emissions. *Journal of Industrial Ecology*, *21*(S1), S191–S202. https://doi.org/10.1111/jiec.12523

Quinlan, H. E., Hasan, T., Jaddou, J., & Hart, A. J. (2017). Industrial and Consumer Uses of Additive Manufacturing: A Discussion of Capabilities, Trajectories, and Challenges. *Journal of Industrial Ecology*, *21*(S1), S15–S20. https://doi.org/10.1111/jiec.12609

Rahim, S. L., & Maidin, S. (2013). Feasibility Study of Additive Manufacturing Technology Implementation in Malaysian Automotive Industry Using Analytic Hierarchy Process. *Applied Mechanics and Materials*, *465–466*, 715–719. https://doi.org/10.4028/www.scientific.net/AMM.465-466.715

Raymond, L. (2005). Operations management and advanced manufacturing technologies in SMEs. *Journal of Manufacturing Technology Management*, *16*(8), 936–955. https://doi.org/10.1108/17410380510627898

Shahrubudin, N., Lee, T. C., & Ramlan, R. (2019). An Overview on 3D Printing Technology: Technological, Materials, and Applications. *Procedia Manufacturing*, *35*, 1286–1296. https://doi.org/10.1016/j.promfg.2019.06.089

Singamneni, S., Yifan, L.V., Hewitt, A., Chalk, R., Thomas, W., & Jordison, D. (2019). Additive Manufacturing for the Aircraft Industry: A Review. *Journal of Aeronautics & Aerospace Engineering*, *08*(01). https://doi.org/10.35248/2168-9792.19.8.215

Sobota, V. C. M., van de Kaa, G., Luomaranta, T., Martinsuo, M., & Ortt, J. R. (2020). Factors for metal additive manufacturing technology selection. *Journal of Manufacturing Technology Management*, *32*(9), 26–47. https://doi.org/10.1108/JMTM-12-2019-0448

Sonar, H., Khanzode, V., & Akarte, M. (2020). Investigating additive manufacturing implementation factors using integrated ISM-MICMAC approach. *Rapid*

Prototyping Journal, 26(10), 1837–1851. https://doi.org/10.1108/RPJ-02-2020-0038

Tasé Velázquez, D. R., Tadeu Simon, A., Luís Helleno, A. L., & Hernández Mastrapa, L. (2020). Implications of additive manufacturing on supply chain and Logistics. *Independent Journal of Management & Production, 11*(4), 1323. https://doi.org/10.14807/ijmp.v11i4.1136

Tosello, G., Charalambis, A., Kerbache, L., Mischkot, M., Pedersen, D. B., Calaon, M., & Hansen, H. N. (2019). Value chain and production cost optimization by integrating additive manufacturing in injection molding process chain. In *International Journal of Advanced Manufacturing Technology* (Vol. 100, Issues 1–4). https://doi.org/10.1007/s00170-018-2762-7

Walachowicz, F., Bernsdorf, I., Papenfuss, U., Zeller, C., Graichen, A., Navrotsky, V., Rajvanshi, N., & Kiener, C. (2017). Comparative Energy, Resource and Recycling Lifecycle Analysis of the Industrial Repair Process of Gas Turbine Burners Using Conventional Machining and Additive Manufacturing. *Journal of Industrial Ecology, 21*(S1), S203–S215. https://doi.org/10.1111/jiec.12637

Yazgan, H. R., Boran, S., & Goztepe, K. (2010). Selection of dispatching rules in FMS: ANP model based on BOCR with choquet integral. *The International Journal of Advanced Manufacturing Technology, 49*(5–8), 785–801. https://doi.org/10.1007/s00170-009-2416-x

Yeh, C.C., & Chen, Y.F. (2018). Critical success factors for adoption of 3D printing. *Technological Forecasting and Social Change, 132*, 209–216. https://doi.org/10.1016/j.techfore.2018.02.003

Zhai, Y., Lados, D. A., & LaGoy, J. L. (2014). Additive Manufacturing: Making Imagination the Major Limitation. *JOM, 66*(5), 808–816. https://doi.org/10.1007/s11837-014-0886-2

Chapter 12

Thermal Analysis and the Melt Flow Behavior of Ethylene Vinyl Acetate for Additive Manufacturing

Vivek Dhimole and Prashant K. Jain
PDPM IIITDM Jabalpur, Jabalpur, India

Narendra Kumar
Dr. B. R. Ambedkar National Institute of Technology Jalandhar,
Jalandhar, India

CONTENTS

12.1 INTRODUCTION

Additive Manufacturing (AM) describes the technologies that create 3D parts in layer by layer fashion [1,2]. The process of fabricating a part begins with a computer aided design (CAD) model preparation which is then followed by various steps such as standard tessellation, slicing, toolpath generation and part fabrication [3,4]. Sometimes, post-processing is also performed on the fabricated part, if required [5]. Various AM processes are available in the commercial market and can be classified as solid, liquid and powder form, based on the form of raw material. Fused deposition modeling (FDM) is the solid-based AM process in which material is extruded from a nozzle onto a movable platform to build 3D parts [6]. The melt flow behavior of the material and temperature field during the deposition plays a vital role in the fabrication of the parts and affects the part quality [7,8]. Since melt flow behavior itself is affected by the various parameters, researchers have analyzed the behavior of FDM systems through simulation as well as by experimental methods. Costa et al. [9] suggested an analytical solution for the heat transfer between the filaments using MATLAB. They also covered the effect of process parameters during the material deposition process. Zhang and Chou developed a 3D finite element model using element activations for the FDM process to simulate mechanical as well as thermal behavior [10]. In addition, the study shows simulations of residual stresses, part distortions and tool path effects on part fabrication. Ramanath et al. presented the melt flow behavior of poly-ε-caprolactone (PCL) in an FDM process along with the mathematical modeling and finite element modeling (FEM) analysis to analyze the effect of the velocity, pressure, temperature distribution, melt temperature and rheology on the part's quality [11]. Melt flow behavior in the terms of temperature and pressure distribution are shown in Figure 12.1.

Saari et al. presented a thermal simulation of the mini-extruder, developed for the fabrication of elastomeric parts [13]. Nikzad et al. also presented work related to the simulation of melt flow in a channel for an ABS–iron composite material in FDM [14]. Simulation was done for the melt flow to characterize the thermal, mechanical and rheological properties. Parameters such as temperature, velocity and pressure drop were also investigated and their simulation results are shown in Figure 12.2 [14].

Saude et al. also presented melt flow behavior in the isothermal condition for a polymer matrix [15]. They conducted the simulation study on the pressure and velocity of a melt flow channel for various materials such as ABS, PP, PLA, ABS–copper and ABS–iron. Zhou et al. presented a numerical simulation study to observe the temperature field in the FDM process [16]. They presented the simulation results for the temperature evolution and the formation of the modeled part. Results were investigated by using continuous media theory-based finite element analysis. In another study, a 3D transient thermal FEM model was developed by Ji and Zhou [16]. Results

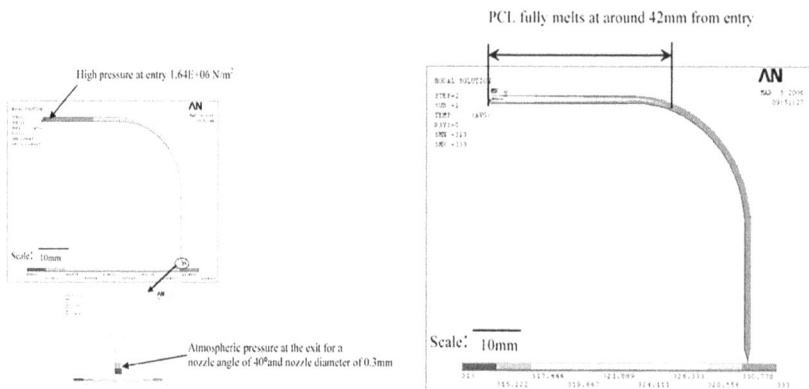

Figure 12.1 Temperature and pressure distribution plot for PCL [12].

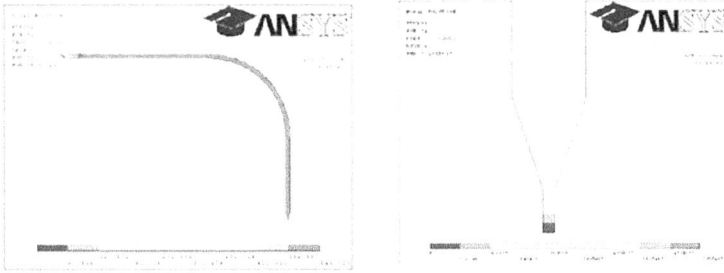

Figure 12.2 Temperature and pressure distribution of an ABS–iron composite within a liquefier.

showed that the temperature distribution field was like an ellipse and the highest temperature gradient was found near the edges of the fabricated part. Ramu and Yadava studied the thermal stress distribution within a metallic layer based on a 2D temperature and thermal stress model. The temperature and thermal stress distributions were calculated for a single scanned layer formed using selective laser sintering (SLS) [17].

The above literature reveals that most researchers have performed FEM simulation on standard materials and the commercial FDM process. Elastomeric materials have not yet been explored as candidate materials for FDM systems due to the various challenges in the processing of the flexible filament, that is buckling due to low column strength. Appropriate filament dimensions and properties are the other constraints in the exploration of elastomeric materials. Instead of considering the filament form of materials, the direct use of pellets may eliminate these problems. A customized material deposition tool (MDT) has already been developed which processes the materials in pellet form. The advantage of this developed MDT is that many other materials in pellet form can be explored to fabricate parts, which cannot be processed on commercial FDM machines [18–24]. In this work ethylene vinyl acetate (EVA), a thermoplastic elastomer, has been considered as the candidate material. The melt flow behavior of an EVA material should be known to the user before the experiment. Detailed FEM analysis of the indigenous developed MDT system and its melt flow behavior have been carried out using ANSYS software. Results are presented in terms of velocity, temperature and pressure inside the barrel and nozzle. In addition, the effect of the temperature gradient and thermal stress between two successive rasters have been analyzed.

12.2 MATERIAL AND METHODS

EVA is a thermoplastic elastomer, which changes properties with temperature. Viscosity depends on the shear rate and temperature. Therefore, the viscosity behavior of EVA can be presented by using the power and

Table 12.1 Properties Considered for FEM Simulation

Properties	Value
Thermal conductivity (EVA)	0.34 W/m.K
Surface tension (EVA)	0.05 N/m
Specific heat capacity (EVA)	1400 J/Kg.K
Density (EVA)	950 Kg/m³
Angular screw velocity	0.833 RPS
Volumetric flow rate (EVA)	6.75E–9 m³/Sec
Meshing method	Quad (8 or 9 nodes per element)/Tri(6 nodes per element)

Arrhenius laws. Table 12.1 shows the properties taken into consideration for performing the simulation.

In ANSYS, a polyflow module is used to conduct this simulation, which allows the use of this condition.

The power law contains two parameters: K (Pascal*sec) and n (dimensionless).

$$\left(F\gamma\right) = K\gamma^{n-1} \tag{12.1}$$

When n < 1, the fluid is pseudo-plastic or shear thinning and so:

$$K = 635\,\text{Pa}^*\text{Sn and } n = 0.65$$

Polyflow follows the Newton–Raphson method, but in this problem, Picard iteration (fixed-point iteration or successive substitution) has been used to solve, due to the larger radius of convergence than the Newton–Raphson method.

If flow is non-isothermal, temperature dependencies of viscosity must be taken into account with the shear rate dependence. The viscosity law is [7]:

$$\eta = H\left(T\right)\cdot\eta_0\left(\gamma\right) \tag{12.2}$$

Here, H *(T)* = Arrhenius relation

$\eta_0\left(\gamma\right)$ = the viscosity law at a reference temperature $T\alpha$.

The Arrhenius relation is

$$H\left(T\right) = \exp\left[\alpha\left(\frac{1}{T - T_0} - \frac{1}{T_\alpha - T_0}\right)\right] \tag{12.3}$$

Here α = activation energy/R = 20928 (the activation energy for EVA is 174 KJ/mole and R is the universal gas constant); T = 383 K; and T_a is the reference temperature for which $H(T)$ = 1.

For mechanical analysis, the governing equations are taken, including thermal strain, as follows [10]:

$$\frac{\partial \sigma_x}{\partial x} + \frac{\partial \tau_{yx}}{\partial y} + \frac{\partial \tau_{zx}}{\partial z} = 0$$

$$\frac{\partial \tau_{xy}}{\partial x} + \frac{\partial \sigma_y}{\partial y} + \frac{\partial \tau_{zy}}{\partial z} = 0 \qquad (12.4)$$

$$\frac{\partial \tau_{zx}}{\partial x} + \frac{\partial \sigma_{zy}}{\partial y} + \frac{\partial \sigma_z}{\partial z} = 0$$

The thermal strain is:

$$\varepsilon = \alpha \left(T - T_{ref} \right) \qquad (12.5)$$

where α = the coefficient of thermal expansion and

T_{ref} = the reference temperature.

The boundary conditions at the outer surface are convection with a heat convection coefficient and a *temperature according to:*

$$Q = h \left(T - T_{amb} \right) \qquad (12.6)$$

where h = the heat transfer coefficient and

T_{amb} = the ambient temperature.

MDT has been designed and developed to use material in pellet form. MDT was mounted on the computer numerical control (CNC) milling machine to conduct the preliminary experiments. The components of MDT are fabricated using mild steel. The advantage of the developed system is that any polymeric material in pellet form can be explored and its suitability can be checked for additive manufacturing of the parts. The main components of MDT include the barrel, screw, nozzle, funnel, band heater and thermocouple. MDT works on the screw extrusion principle where the pellets are fed from the top of the barrel with the help of the funnel. A small clearance of 0.25 mm is given between the screw and barrel. The band heater provides heat to the barrel, and the material receives the heat from the barrel by conduction as well as from friction between the barrel and pellets. The rotation of the screw creates sufficient pressure on the material which provides continuous extrusion from the nozzle. A schematic representation of MDT is shown in Figure 12.3.

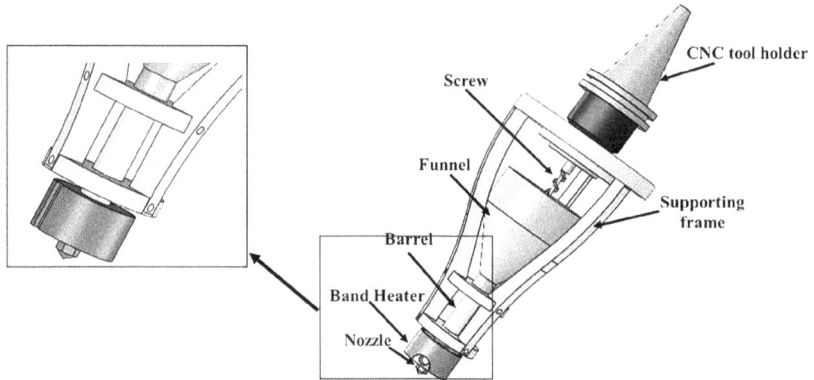

Figure 12.3 Schematic diagram of the MDT system.

A designed MDT works on the screw extrusion principle. Since a screw rotates and carries material from the top to the bottom of the barrel, it has some angular velocity. A Poiseullie and Couette equation is used to find out the volumetric flow rate.

The formula for Q (flow) is shown below about the flight axis [25]:

$$Q = Wh^3 / 12\mu(p_1 - p_2 / l) + (Whrw\cos\theta) / 2 \qquad (12.7)$$

where
W = successive distance between two screw helixes
H = distance between the screw and barrel
r = the screw's radius
$(p_1 - p_2)$ = pressure difference between the inlet and outlet
w = angular velocity of the screw
θ = angle between the axis and screw turn (helix)

The heat transfer coefficient has a significant role for thermal simulation; it is calculated by the Rayleigh number and the Nusselt number:

$$Ra = 9.81\beta\Delta T(L)^3 / v\alpha$$
$$Nu = 0.59(Ra)^{0.25} \qquad (12.8)$$

The heat transfer coefficient h is:

$$h = Nu^*k / L \qquad (12.9)$$

Here, Ra = the Rayleigh number

Nu = the Nusselt number

$\overline{\Delta T}$ = the temperature difference the body and atmosphere

$\beta = \dfrac{1}{\Delta T}$

L = the characteristic length, which is (area/perimeter)

$v = k$inematic viscosity

α = thermal diffusivity.

FEM analysis was performed on the entire MDT system and the various parameters analyzed through simulation. The melt flow behavior was assessed in the barrel and nozzle respectively. Also the free extrusion and thermal gradient between the two rasters were simulated. The appropriate material properties and laws necessary for the simulation studies have already been discussed.

12.3 RESULTS AND DISCUSSION

12.3.1 Thermal Analysis of the Material Deposition Tool System

Thermal analysis of the MDT system was also carried out to obtain the temperature distribution along with the barrel and supporting frame as shown in Figure 12.4. Uniform meshing was done for this simulation and was refined at the barrel and nozzle. A total of 23,303 nodes and 10,978 elements were taken for this simulation.

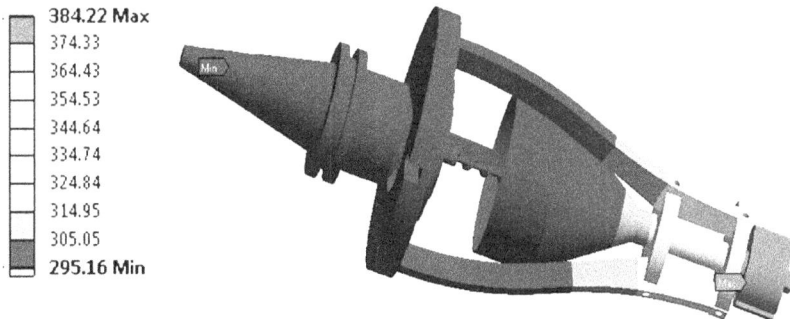

Figure 12.4 Temperature distribution on an MDT system.

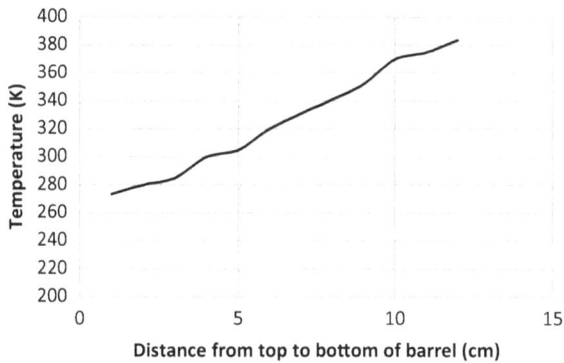

Figure 12.5 Temperature variation along the height of the MDT (top to bottom).

EVA melts at about 358 K and, experimentally, it was found that its melt flow from the nozzle was appropriate for printing parts at 383 K. Consequently the heater temperature was kept at 383 K, and the same temperature was considered to be the heat source temperature. Free natural convection was taken as 30 W/m²K. The temperature distribution results show that the temperature is at a maximum near the heater portion, that is the exit of the nozzle, and is at a minimum on the tool holder of the CNC milling machine as shown in Figure 12.5. It can be concluded from the results that the material will be melted near the nozzle and then extruded. Since the minimum temperature is observed at the tool holder so it will not severely affect the machine due to heat while performing the experiments. Also, the material will not stick at the entrance due to a low-temperature value which enables the smooth feeding of the material from the funnel to the barrel.

12.3.2 Simulation of Melt Flow in the Barrel

The melt flow in the barrel was analyzed by developing the model with axis-symmetric properties which allows us to consider meshing only in the radial direction. As the melt flow behavior was unknown in the barrel, uniform meshing was considered throughout the barrel geometry as shown in Figure 12.6. One hundred and twenty-six nodes and 83 elements were taken for the simulation. Since the pellets were fed into the barrel from the top, a 313 K temperature was obtained at the entrance, based on MDT simulation. At the exit of the barrel, a temperature of 383 K was taken into account as per the thermocouple reading, which could be considered as a suitable temperature for printing parts with EVA.

Figure 12.7 shows the temperature and velocity distribution in the barrel. The results show that the temperature varies from 313 to 383 K along the length of the barrel and that the velocity varies from e–7 to 2e–06 m/sec.

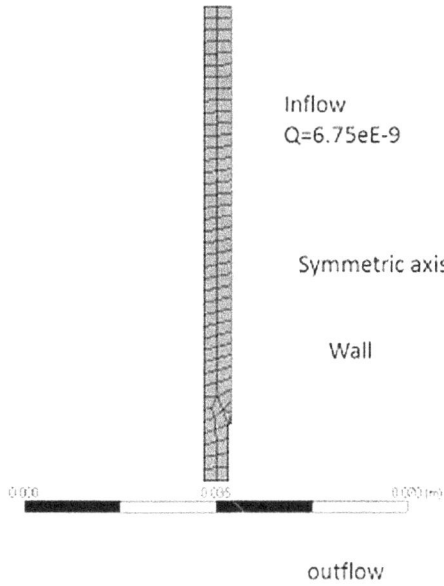

Inflow
Q=6.75eE-9

Symmetric axis

Wall

0.000   0.005   0.001(m)

outflow

Figure 12.6 Schematic diagram of the barrel.

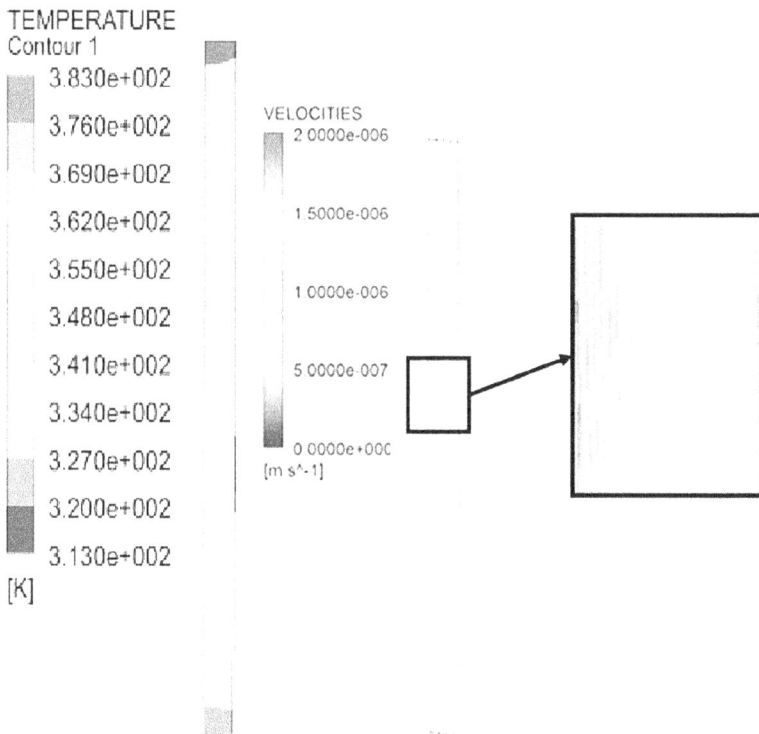

TEMPERATURE
Contour 1
3.830e+002
3.760e+002
3.690e+002
3.620e+002
3.550e+002
3.480e+002
3.410e+002
3.340e+002
3.270e+002
3.200e+002
3.130e+002
[K]

VELOCITIES
2.0000e-006

1.5000e-006

1.0000e-006

5.0000e-007

0.0000e+000
[m s^-1]

Figure 12.7 Temperature and velocity distribution of the barrel.

The temperature distribution result shows a minimum temperature of 313 K at the entrance side, which indicates the material at the entrance will not agglomerate. This allows the smooth feeding of the pellets in the barrel. Gradually the temperature in the barrel is increased which implies a change in the material state from solid to semi-solid.

12.3.3 Simulation of Melt Flow in the Nozzle

A nozzle with a 0.4 mm exit diameter was used for the generation of the simulation data for melt flow in the nozzle. The geometry and boundary conditions were considered as axis symmetric for the round nozzle with respect to the centerline. At the conical section of the nozzle, the mesh was refined in order to get small changes in the flow rate; 177 nodes and 147 elements were taken for this simulation. Figure 12.8 shows the meshed geometry of the nozzle with applied boundary conditions.

Figure 12.9 shows the pressure and velocity distribution in an isothermal flow condition in the nozzle. This shows that pressure and velocity vary from inlet to outlet. A parallel streamline shows that the flow is laminar, velocity is at a maximum and pressure is at a minimum at the nozzle outlet.

Figure 12.10 shows the pressure variation between two points: $(0, 0)$ and $(0.0003, 0.004)$.

12.3.4 Free Extrusion and the Swelling of Melt

The same nozzle of 0.4 mm diameter was used with round geometry for the simulation of the free extrusion and the swelling of the material. As

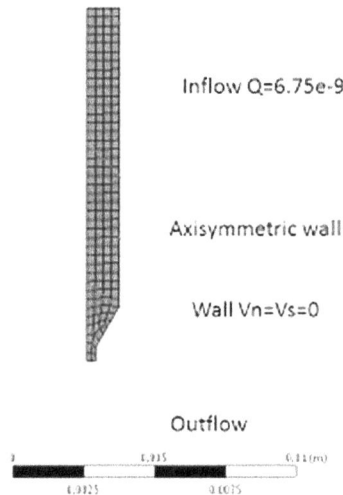

Figure 12.8 Schematic diagram of the nozzle.

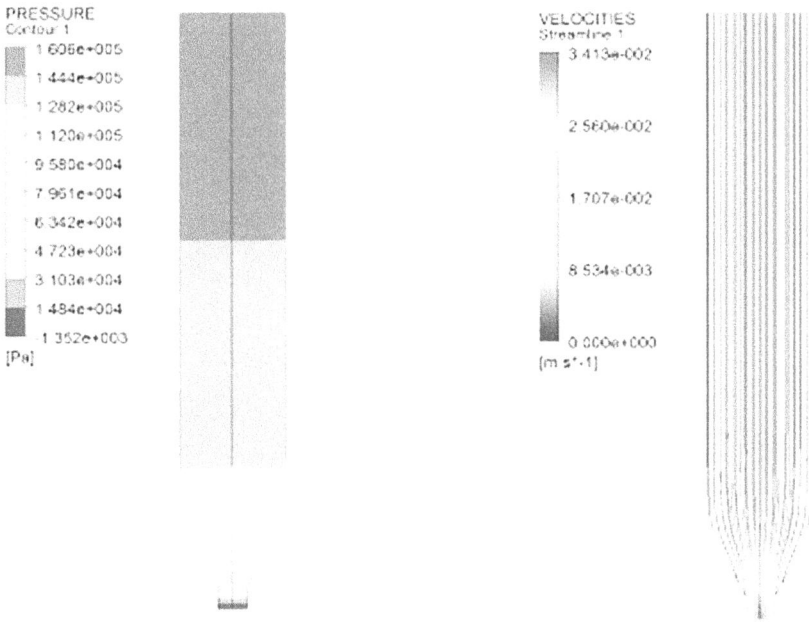

PRESSURE
Contour 1
    1.606e+005
    1.444e+005
    1.282e+005
    1.120e+005
    9.580e+004
    7.961e+004
    6.342e+004
    4.723e+004
    3.103e+004
    1.484e+004
    -1.352e+003
[Pa]

VELOCITIES
Streamline 1
    3.413e-002

    2.560e-002

    1.707e-002

    8.534e-003

    0.000e+000
[m s^-1]

Figure 12.9 Pressure and velocity distribution in the nozzle.

the nozzle is circular, the model considers only radial direction. The time-dependent approach was used for the simulation to obtain the evolution of geometry: 465 nodes and 420 elements were taken for the simulation; the mesh size was refined along the free surface boundary. In Figure 12.11, the upper part shows the end of the nozzle tip and the bottom shows the free surface of the extruded material within the meshed condition.

Figure 12.12 shows the swelling of the free surface of the melt, which has been captured in the form of snapshots at time intervals

When the material exits from the nozzle, the forces exerted by the nozzle wall disappear and the material attempts to come to a condition of equilibrium. Therefore the diameter of the extruded material becomes larger than the nozzle diameter. This phenomenon is known as swell, and can be clearly observed in Figure 12.13. The simulation was stopped at $t = 0.09$ sec for study. The swelling coefficient (S) was also calculated by measuring the ratio of the extruded diameter and the nozzle diameter (S = D2/D1 = 1.08432) at $t = 0.0026$ sec.

Figure 12.13 shows the temperature distribution of the extruded material, which is a quite higher temperature at the exit point. When the extruded filament is exposed in the air the heat transfer rate increases from the material to the air. Therefore, the temperature of the extruded filament gradually decreases.

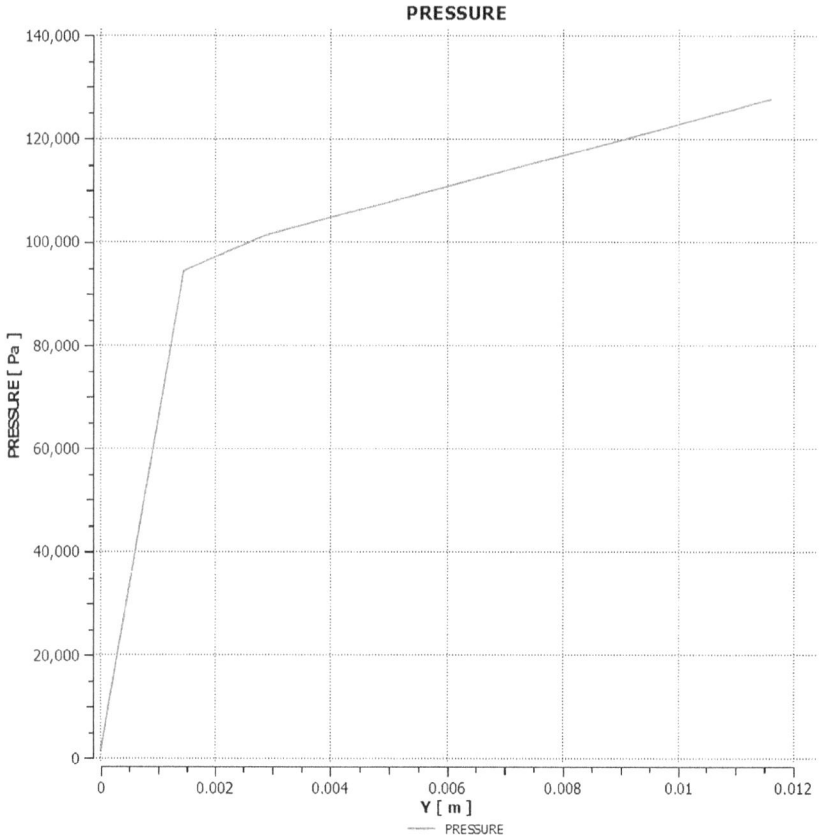

Figure 12.10 Pressure distribution along the Y-axis in the nozzle from bottom to top.

12.3.5 Evolution of Temperature Distribution Along the Rasters

The temperature distribution along the rasters plays a critical role in their bonding. A simulation was carried out in the study to determine the temperature along the deposition path; 9701 nodes and 5930 elements were taken for the simulation. The meshing size was resized in the material deposition zone. Two rasters were considered to simplify the simulation. The EVA material properties were taken as an input to analyze the temperature distribution. Calculations of the convection heat transfer coefficient used for this simulation have been discussed in Section 12.2. The element activation-deactivation technique was used for this simulation. The evolution of the temperature distribution along the deposition path at 1, 5, 8, 9, 12 and 16 sec between two successive rasters of a layer is shown in Figure 12.14.

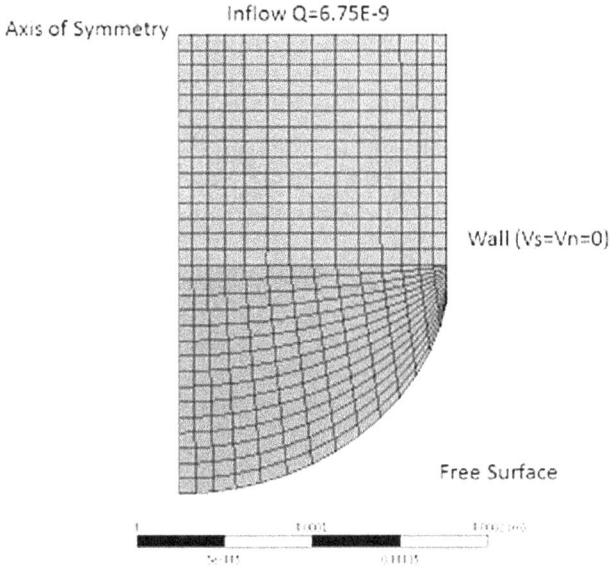

Figure 12.11 Schematic diagram of free extrusion.

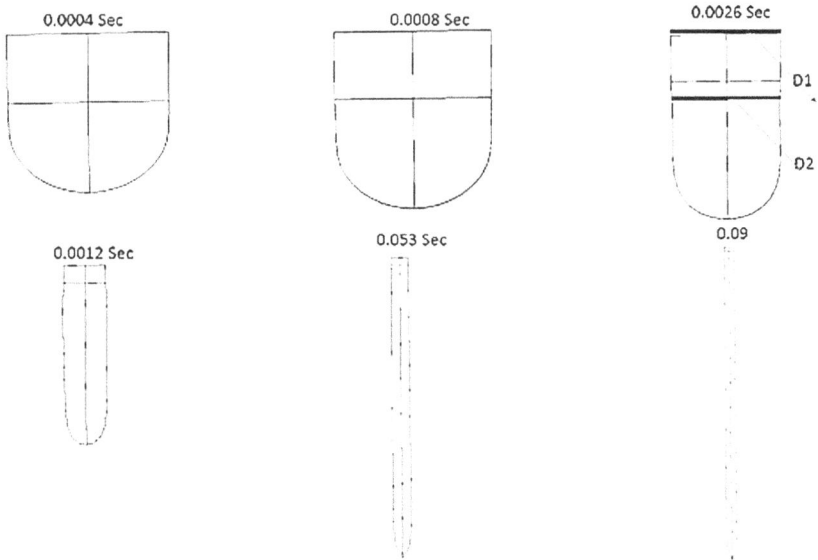

Figure 12.12 Swelling and extrusion results.

Figure 12.15 shows the temperature variation at a point on a single raster. When the nozzle deposits material on the platform, the extruded material temperature decreases gradually with respect to time along with a raster. Moreover, when the nozzle comes back to deposit the adjacent raster, the

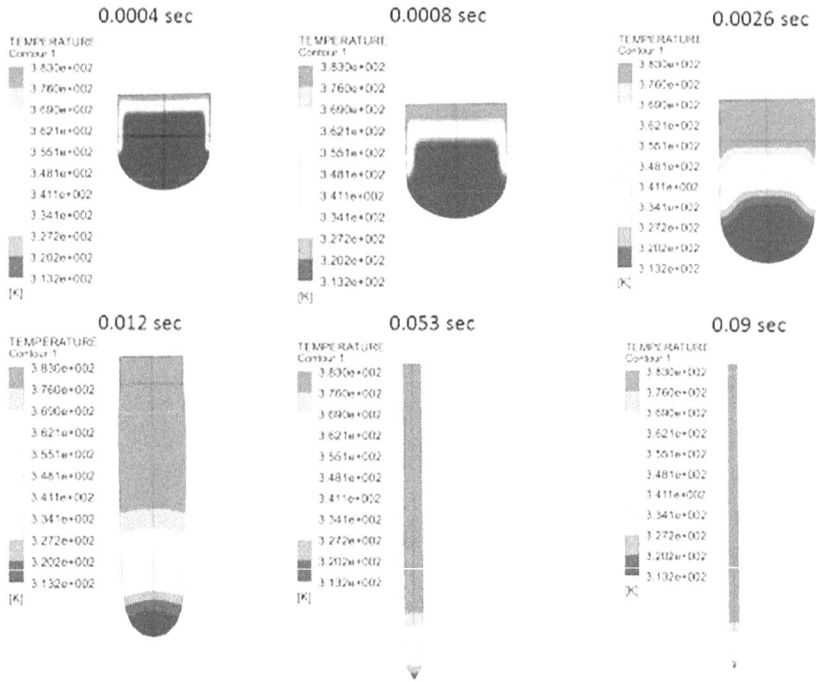

Figure 12.13 Evolution of temperature distribution when material is extruded from the nozzle.

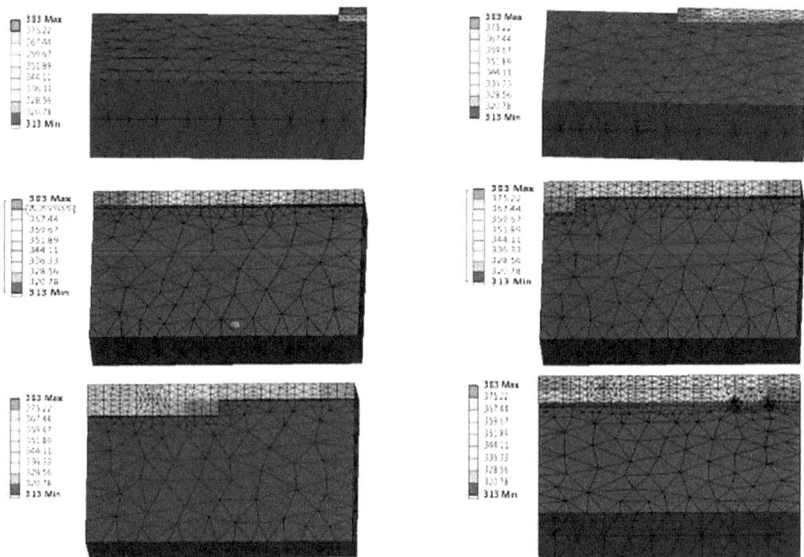

Figure 12.14 Evolution of temperature distribution between the two successive rasters of a layer.

Temperature vs Time

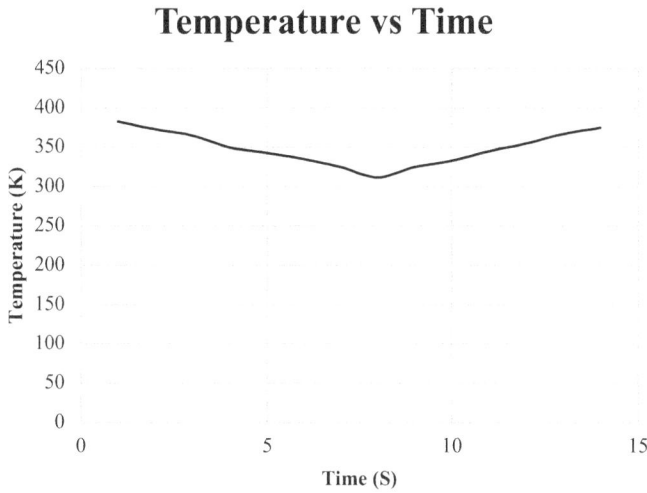

Figure 12.15 Temperature distribution in two consecutive rasters.

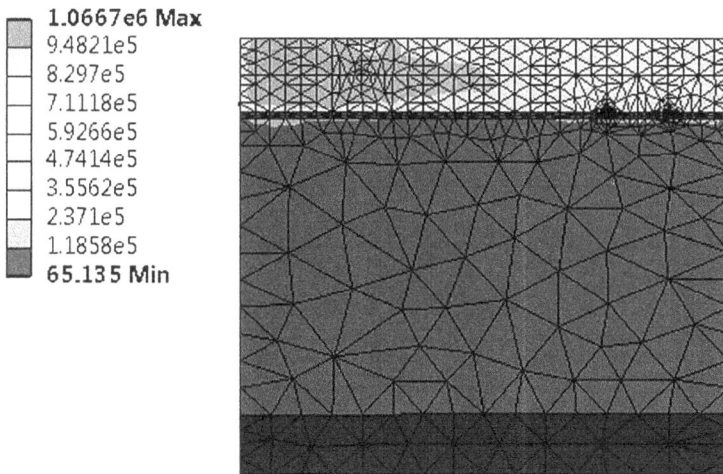

Figure 12.16 Stress distribution between two successive rasters on a layer.

temperature of the first raster again increases due to its contact with the adjacent raster.

Another simulation was carried out to find out the thermal stress distribution between two consecutive rasters, as shown in Figure 12.16. The result shows the thermal stress distribution between the two rasters. The maximum thermal stress developed was 1 MPa. This shows reduced variation in the thermal stress along the seam line of the two rasters. Therefore, the bonding between the two rasters will be good enough.

12.4 CONCLUSION

The finite element analysis of the novel MDT system has been presented in this chapter. The results show the feasibility of the MDT system for AM. The thermal analysis of the MDT system indicates that it has the capability to process the pellet form of the EVA material and can provide the sufficient heat required for the extrusion. An insight into the melt flow behavior of EVA was also presented by performing a simulation of pressure, velocity and temperature. Free extrusion and swelling simulations were further investigated to predict parameters like road gap and layer thickness before the experiment. The temperature field distribution was also analyzed using the element activation and deactivation technique between two rasters to assess bonding along the seam line of the rasters. Based on this study, other materials can be explored on the developed MDT system for AM in future.

REFERENCES

[1] S. B. Selvaraj and S. Singamneni, "Pre-Moisturized beta-hemihydrate for 3D Printed Molds," *Mater. Manuf. Process.*, vol. 31, no. 8, pp. 1102–1112, 2016.
[2] N. Kumar, S. Shaikh, P. K. Jain, and P. Tandon, "Effect of fractal curve based toolpath on part strength in fused deposition modelling," *Int. J. Rapid Manuf.*, vol. 5, no. 2, pp. 186–198, 2015.
[3] C. K. Chua, K. F. Leong, and C. S. Lim, *Rapid Prototyping: Principles and Applications*. World Scientific, 2010.
[4] A. Garg, A. Bhattacharya, and A. Batish, "On surface finish and dimensional accuracy of FDM parts after cold vapor treatment," *Mater. Manuf. Process.*, vol. 6914, no. October, p. 150724213500001, 2015.
[5] M. Taufik and P. K. Jain, "A study of build edge profile for prediction of surface roughness in fused deposition modeling," *J. Manuf. Sci. Eng.*, vol. 138, no. 6, pp. 1–11, 2016.
[6] K. Chockalingam, N. Jawahar, and J. Praveen, "Enhancement of anisotropic strength of fused deposited ABS parts by genetic algorithm," *Mater. Manuf. Process.*, vol. 31, no. 15, 2016.
[7] A. Bellini, Fused Deposition of Ceramics: A Comprehensive Experimental, Analytical and Computational Study of Material Behavior, Fabrication Process and Equipment Design, Drexel University, 2002.
[8] O. A. Mohamed, S. H. Masood, and J. L. Bhowmik, "Experimental investigations of process parameters influence on rheological behavior and dynamic mechanical properties of FDM manufactured parts," *Mater. Manuf. Process.*, vol. 31, no. 15, pp. 1983–1994, 2015.
[9] Costa, S.F., Duarte, F.M. and Covas, J.A., 2008. "Towards modelling of Free Form Extrusion: analytical solution of transient heat transfer," *International Journal of Material Forming*, 1(1), pp. 703–706
[10] Y. Zhang and Y. K. Chou, "Three-dimensional finite element analysis simulations of the fused deposition modelling process," *Proc. Inst. Mech. Eng. Part B-Journal Eng. Manuf.*, vol. 220, no. 72, pp. 1663–1671, Jan. 2006.

[11] H. S. Ramanath, C. K. Chua, K. F. Leong, and K. D. Shah, "Melt flow behaviour of poly-epsilon-caprolactone in fused deposition modelling," *J. Mater. Sci. Mater. Med.*, vol. 19, no. 7, pp. 2541–2550, 2008.

[12] M. Saari, M. Galla, B. Cox, P. Krueger, A. Cohen, and E. Richer, "Additive manufacturing of soft and composite parts from thermoplastic elastomers," *Solid Free. Fabr. Symp. Austin, TX Univ. Texas Austin*, pp. 949–958, 2015.

[13] M. Nikzad, S. H. Masood, I. Sbarski, and A. Groth, "A study of melt flow analysis of Polycarbonate (PC) in fused deposition modelling process," *Int. Conf. Exhib. Sustain. Energy Adv. Mater.*, vol. 14, no. S1, pp. 29–37, 2011.

[14] N. Saude, M. Ibrahim, and M. H. I. Ibrahim, "Melt flow behavior of polymer matrix extrusion for Fused Deposition Modeling (FDM)," *Appl. Mech. Mater.*, vol. 660, pp. 84–88, 2014.

[15] Yang, H. and Zhang, S., "Numerical Simulation for Temperature Field in the Fused Deposition Modeling Process." *Journal of Mechanical Science and Technology*, vol. 32(7), pp. 3337–3344, 2018.

[16] L. B. Ji and T. R. Zhou, "Finite Element Simulation of Temperature Field in Fused Deposition Modeling," *Adv. Mater. Res.*, vol. 97–101, pp. 2585–2588, 2010.

[17] P. M. Ramu and V. Yadva, "Determination of thermal stress distribution in metallic layer during selective laser sintering using finite element method," *International Journal of Manufacturing Technology and Management*,vol. 13, pp. 280–296, 2008.

[18] N. Kumar, P. K. Jain, P. Tandon, and P. M. Pandey, "The effect of process parameters on tensile behavior of 3D printed flexible parts of ethylene vinyl acetate (EVA)," *J. Manuf. Process.*, vol. 35, pp. 317–326, 2018.

[19] N. Kumar, P. K. Jain, P. Tandon, and P. Mohan Pandey, "Experimental investigations on suitability of polypropylene (PP) and ethylene vinyl acetate (EVA) in additive manufacturing," *Mater. Today Proc.*, vol. 5, no. 2, pp. 4118–4127, 2018.

[20] N. Kumar, P. K. Jain, P. Tandon, and P. M. Pandey, "Extrusion-based additive manufacturing process for producing flexible parts," *J. Brazilian Soc. Mech. Sci. Eng.*, vol. 40, no. 3, pp. 1–12, 2018.

[21] N. Kumar, P. K. Jain, P. Tandon, and P. M. Pandey, "Additive manufacturing of flexible electrically conductive polymer composites via CNC-assisted fused layer modeling process," *J. Brazilian Soc. Mech. Sci. Eng.*, vol. 40, no. 4, pp. 1–13, 2018.

[22] N. Kumar and P. K. Jain, 2021. Analysing the influence of raster angle, layer thickness and infill rate on the compressive behaviour of EVA through CNC-assisted fused layer modelling process. *Proceedings of the Institution of Mechanical Engineers, Part C: Journal of Mechanical Engineering Science*, 235(10), pp. 1731–1740, 2021.

[23] S. K. Gawali, N. Kumar, and P. K. Jain, "Investigations on the Development of Heated Build Platform for Additive Manufacturing of Large-Size Parts," In *Manufacturing Engineering* (pp. 1–17). Springer, Singapore. pp. 1–13, 2020.

[24] N. Kumar, P. K. Jain, P. Tandon, and P. M. Pandey, "3D printing of flexible parts using eva material," *Mater. Phys. Mech.*, vol. 37, no. 2, pp. 124–132, 2018.

[25] J. Wilkes, "Chapter 6: Solution of viscous flow problems," *Fluid Mech. Chem. Eng. with Microfluid. CFD*, vol. 2, pp. 272–320, 2005.

Chapter 13

Directed Energy Deposition for Metals

Nitish P. Gokhale and Prateek Kala

Birla Institute of Technology, Pilani, India

CONTENTS

13.1 INTRODUCTION

The metal deposition process can be done by using processes like laser beam melting (LBM), electron beam melting (EBM), binder jetting technology, sheet lamination technology and directed energy deposition (DED). In DED processes, the material in metal wire or metal powder form is added to the substrate in cooperation with melting heat sources such as lasers, electron beams or electric arcs. Hence, it can be termed "a metal deposition process".

DED processes direct energy into a concentrated region to heat the surface of the substrate and simultaneously melt the substrate and filler material to deposit the material on the substrate. In contrast to other powder bed fusion processes, in DED processes the material is not put down on the entire surface but added exactly at the time of melting. DED processes deposit the material in feed stock or powder form. The building process is similar to the extrusion process. The laser beam or electron beam has been used for melting the wire or powder material in commercial DED processes. In this chapter, DED processes used for depositing 3D metallic structures are considered instead of normal laser or plasma-cladding processes.

Many industries have developed DED processes which use a powder feeding system and a laser or electron beam source for metal deposition. Laser engineered net shaping (LENS), direct metal deposition (DMD), laser cladding and electron beam melting are some examples of these processes. The operating procedure is the same as for these processes. However, with a laser power source the powder feeding mechanism is different. The initial operation cost of these systems is high due to the laser and electron beam which are used as power sources, which require special materials in powder form. In addition, the parts obtained through these processes are not totally dense and may require further processing for obtaining denser parts. On the other hand, DED processes like the wire and arc additive manufacturing (WAAM) process have attracted the attention of researchers because they are capable of producing denser metal components with relatively lower costs. The WAAM process uses two major processes: gas metal arc welding (GMAW) and gas tungsten arc welding (GTAW). A number of research attempts have been done on the WAAM process in recent years to fabricate free-form parts. This chapter also focuses on the advantages and limitations of different DED processes in relation to other commercial techniques. The parts with complex geometry can be easily manufactured by these processes and with minimum wastage of material due to its buy to fly ratio.

13.2 CLASSIFICATION OF DED PROCESSES

DED techniques are the processes in which feeding material is directed at the deposition position along with the input energy source. The processes are divided into three types: LENS, electron beam free form fabrication (EBF3) and WAAM. The working principle of these processes is the same, but the type of energy source and physical form of the feeding material is different. Figure 13.1 shows the schematics of LENS based direct energy deposition processes.

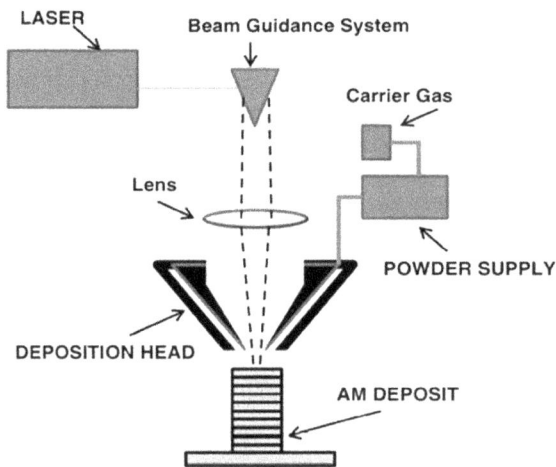

Figure 13.1 Schematic of LENS or laser metal deposition (LMD) setup [1].

In the LENS process, a moving deposition head with a laser beam pathway and metal powder carrier is used for deposition of material. The powder particles can be injected onto the surface and be melted by a laser beam. The molten metal pool is generated using a laser beam and material in powder form that has been added into the molten pool to build the part. The process and path of travel can be controlled in such a way that the complete 3D metallic structure is completed row by row and then layer by layer. As the LENS process utilizes material in powder form the produced parts contain porosity in the structure. The laser metal deposition process can be commercially termed "LENS". In the laser metal deposition (LMD) process, the surface of the substrate or previous layer is melted, and the metal powder is simultaneously dropped to build the metallic parts. The metal powder of materials like nickel-based alloys, such as INCONEL and HASTELLOY, cobalt-based alloys, such as Stellite, carbides and stainless steels, and titanium alloys is fed by a coaxial nozzle. The melting of the surface is done by using an Nd:YAG laser or a CO_2 laser. The molten pool is protected from oxidation by providing argon gas. The LMD process offers a higher build rate and higher build volume compared to powder bed technologies. It also consumes power up to 70 kWh/Kg, like the LBM process, which is comparatively higher than the EBM process. The average layer thickness of LMD processes varies from 40 μm to 1 mm. The different types of LMD processes advance due to continuous development in this field. The build platform remains stationary and the assembly head containing the laser source and powder nozzle is moved as per the requirements of the structure. The LMD process is mainly used to produce metallic parts like turbine blades and gears. The metal powder in the LMD process can be replaced by wire and the laser beam can be replaced by an electron beam to build a structure with higher deposition rates.

NASA first introduced the EBF3 process and extensively used it as a space-compatible technology. In EBF, material in wire form is supplied over the heat source generated by an electron beam. The building environment is kept controlled by a vacuum to ensure the sharp focusing of the electron beam. Figure 13.2 shows a schematic of the EBF3 process.

The WAAM process also uses feeding material in wire form but a welding arc for a heat source. The heating source for the WAAM process can be similar to GTAW, GMAW or plasma welding processes. A laser source is also used rarely, a process called wire-laser additive manufacturing. The filler wire material can be melted to deposit beads on the substrate. The beads are diffused into each other to form a layer. This process is continual in a layer by layer manner until the desired height of the 3D structure is reached. The schematics of the WAAM process is as shown in Figure 13.3. An extensive range of metals and metal alloys like steel, aluminum or titanium can be used in WAAM. The geometrical accuracy of the parts produced by the WAAM method is one of the main issues and requires the performing of

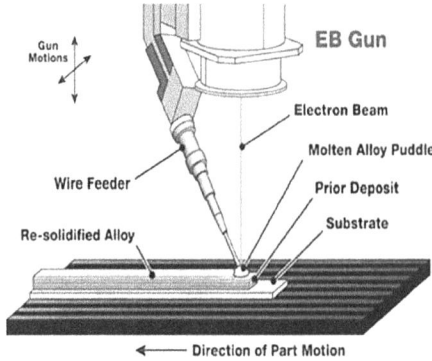

Figure 13.2 EBF3 [2].

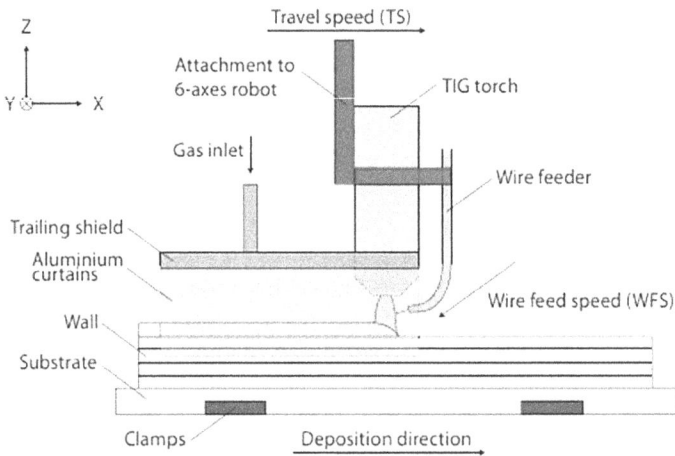

Figure 13.3 Schematic of the WAAM process [3].

some subtractive machining, whereas part density and other mechanical properties are compatible with the casting or forming process. Therefore, unlike powder-based processes, WAAM is able to build large scaled fully functional metallic parts. The buy to fly ratio and operating cost for WAAM are also less than that of the commercial powder-based processes. The WAAM process therefore creates a viable option for a large-scale additive manufacturing process.

13.3 MATERIAL FEEDING

Material in the form of powder and wire has been utilized in DED processes. Both modes of wire feeding have their own limitations.

13.3.1 Wire Feeding

The wire feeding method is more efficient than the powder feeding method due to the reduced wastage of material. The use of wire as a feedstock is efficient for simple as well as complex geometries with a number of thin and thick features. However, for depositing fully dense 3D metallic parts with complex geometry a proper process control is needed. In order to get better dimensional accuracy as well as low porosity mostly wire feeding processes are preferred. It is easier to achieve geometrical accuracy using subtractive machining after deposition of material using wire based processes.

13.3.2 Omni-Directional Wire Feeding

In order to achieve a similar orientation of the wire feeding nozzle in all tool travel directions, the omni-directional wire feeding system has been used. The omni-directional wire feeding mechanism helps to deposit isotropic metallic parts. In the case of WAAM processes, omni-directional wire feeding plays an important role due to the wire being fed from either side of the heating source. Omni-directional wire feeding can be achieved by either providing co-axial feeders from both sides of the WAAM source or by dynamic movement of the feeding nozzle at the time of tool travel. Basically, most of the DED processes are not omni-directional, though omni-directional orientation has been achieved by adjusting the feeding nozzle dynamically.

13.3.3 Powder Feeding

The powder form of any material, like metals or ceramics, is easily available, but the buy to fly ratio for powder based processes is less than that of wire based processes. The amount of powder dispensed from the nozzle is not fully melted in the molten pool. In order to minimize wastage, cycling of the unused powder has to be done.

Powder feeding can be done by applying fluid pressure on the powder. In a first step the gas is supplied to the powder container to properly fluidize the powder. In the next step powder is dispensed as per requirements by dropping the fluid pressure. The focusing of the powder particles on the substrate plate can be done through the coaxial nozzle, four-nozzle feeding or through a single nozzle. The uniqueness of flow in the four-nozzle feeding system helps in depositing a build structure with more consistency in features.

13.4 MATERIALS FOR DED PROCESSES

Attempts have been made to deposit different materials—like titanium and titanium alloys [4], Inconel 625 [5], nickel and copper nickel alloys, stainless steel 300 series, and aluminum alloys 1100, 2318, 2319, 3000 series

[6]—using the WAAM process. Wang et al. [7] fabricated components of alloys like 2219-Al and Inconel 625. The parts of a steel silicon bronze bimetal have also been manufactured by MIG-welding-based additive manufacturing [8]. The functional components of a nickel-aluminum-bronze [9] alloy have been built by a MIG-based additive manufacturing system. The fine surface appearance and better microstructure patterns have been observed in the manufactured part and have been found to have similar properties as that of casted components of the same material, so it has been found to be a good option for producing components for marine applications [9].

The Ti-6Al-4V alloy is mostly used for aerospace applications as it has a high strength to weight ratio and a tensile strength of 1000–1200 MPa which is almost four times greater than that of aluminum (up to 300 MPa). Ti-6Al-4V wires are commercially available in size ranges from 0.9 to 4.0 mm. However, in most studies, researchers have used wire of 1.2 mm as a filler material. This could be because a wire diameter of more than 1.2 mm might not provide better dimensional accuracy. Besides this, a wire diameter less than 1.2 mm might not give the required deposition rate and the wire feeder unit required for feeding the thinner wire demands more initial investment. Most research attempts involving the deposition of Ti-6Al-4V parts using the WAAM process have shown beta columnar grains and a highly coarser microstructure which is indicative of good strength throughout the layer deposited [10]. Baufeld et al. [11] verified the suitability of the TIG welding process for the deposition of a Ti-6Al-4V component by performing a metal deposition process. Some of the researchers have attempted to deposit an NiTi shape memory alloy due to its distinguished super elastic properties. The NiTi alloy with a higher percentage of Ni can be used in applications in the automotive industry, aeronautical industry and for biomedical applications. Zeng et al. [12] studied the mechanical properties and microstructure of the material deposited using an NiTi alloy wire. Columnar grains in the first layer of deposition and equiaxial grains in the topmost layer were observed. The deposited material possesses stable mechanical properties with super elastic ones too.

The selection of filler wire materials will also affect the range of process parameters selected, for example the TIG-based metal deposition of aluminum requires an AC voltage supply which avoids the formation of oxide scales on the deposited surface [10]. The deposition of aluminum parts requires a heat source with alternating current for the automatic removal of the oxides formed in deposition. Williams et al. [10] concluded that the use of a low power heat source also provided a fine and equally spaced grain structure of aluminum. The defect-like voids in the deposited part were prominently found in the aluminum parts. However, these defects have been eliminated by optimization of the process parameters. Aluminum has good thermal conductivity and a lower melting point than steel alloys, so the laser-based additive manufacturing process may be a suitable option rather than arc-based processes.

13.5 INFLUENCE OF PROCESS PARAMETERS

Most of the commercial DED processes have programmed systems to control the process parameters. The user needs to set the process parameters as per requirements. The optimal value of process parameters can be set depending on material and geometric accuracy. However, in the case of the WAAM type of DED process parameter selection needs to be done through experimentation since these processes are in their developing state [13].

The effect of voltage, current, wire feeding speed, torch movement velocity and arc gap was studied by researchers. The quality of deposition is strongly dependent upon the proper process parameter selection [14]. Among these parameters, current has the most dominant effect on the surface appearance of the parts being deposited. Xiong et al. [15] found that parts produced by a 1.2 mm diameter wire of an H08Mn2Si alloy within the current range of 100 A to 180 A showed better surface finish. In another work, a similar research group studied the effect of torch velocity on the surface finish of the deposited parts. They concluded that a higher torch velocity yielded a good surface finish. Besides this, a low wire feeding speed was found to ensure a good surface finish in a MIG-welding-based additive manufacturing process [16].

The adaptive path control method has also been used by some researchers. Xiong et al. [17] utilized this for controlling process parameters. The online data of process parameters have been monitored and simultaneously analyzed in the adaptive process control of the metal deposition process. In this work, an adaptive controller was integrated with an IR camera to measure the distance between the electrode and work piece. This helped in reducing the tedious job of the operator of setting each parameter manually [17]. Youheng et al. [18] developed a closed loop control system to get the optimum layer thickness and torch movement velocity. The error in the optimum layer thickness was minimized to be less than 0.5 mm by this control system. The authors established the process for the deposition of inclined thin walled parts. The angle of inclination was controlled by managing the center distance between adjacent layers and the layer height. They also reported that, at a constant wire feeding speed, as the torch movement velocity increases the maximum inclination angle was obtained, that is an angle greater than 45° [19].

The parts produced by TIG-welding-based WAAM are influenced by process parameters like current, operating voltage, torch movement velocity, wire feeding speed and electrode work piece distance. Among all of these parameters researchers particularly studied the effect of torch movement velocity, current and wire feeding speeds which significantly affect the process output. In most of the research attempts, bead width, deposition height (Figure 13.4a) and spacing between adjacent beads (Figure 13.4b) have been considered as output parameters.

(a)   (b)

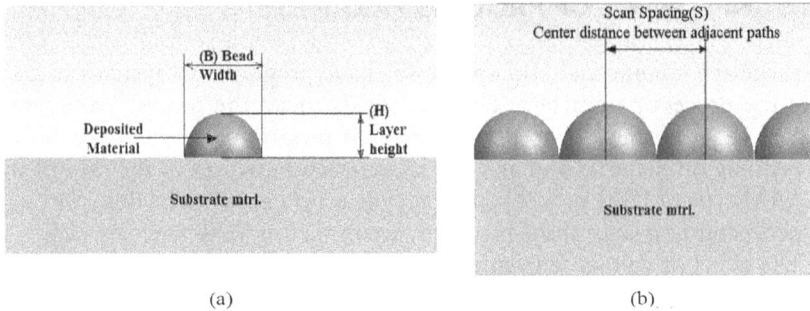

Figure 13.4 Output parameters analyzed after deposition.

Wang et al. [7] studied the effect of TIG-welding-based WAAM process parameters on the morphology and grain structure of the deposited parts of aluminum. The effect of wire feeding speed, current and torch movement velocity on deposition have been studied. The height and width of the deposition decreased with an increase in any one of the above parameters. They suggested that the optimization of the wire feeding angle improves the quality of the layer deposited and allows more complex shapes to be produced with minimum error [7]. Olivares and Díaz [20] carried out a comparison of hot and cold TIG-based-WAAM processes. It was found that the deposition rate and speed of deposition were higher in the hot wire arc method than the cold wire method.

In TIG and MIG-welding-based metal deposition processes the current, voltage, wire feeding speed and torch movement velocity are the parameters that need to be controlled for uniform metal deposition. The same parameters have been considered for controlling the plasma-arc-based metal deposition process. This technique has higher deposition rates and yields a greater functional wall thickness than other wire arc additive manufacturing methods. The plasma welding process has a more concentrated arc than other arc welding processes. However, researchers have reported some issues like oxidation and distortion to some extent. The height of the deposition in each layer has been less than TIG or MIG-welding-based WAAM processes.

Jhavar et al. [21] performed an experimental investigation of the plasma-arc-based metal deposition process for the optimization of process parameters. With the optimization of experimental results they were able to deposit a good quality bead (1.7 mm width and 0.7 mm deposition height), so they concluded that the process can be used for resurfacing [21] and repairing dies or molds. The authors claimed that the developed technology is more beneficial than any other high energy source process [21].

The deposition direction in each subsequently deposited layer is one of the key parameters and affects the quality of the deposition in micro-plasma arc additive manufacturing. In this regard, Aiyiti et al. [22] carried out deposition in such a way that the deposition direction for each subsequent layer was 90°. They reported that the parts so produced were free from cracks and voids [22].

13.6 MECHANICAL PROPERTIES AND MICROSTRUCTURE

The mechanical properties and microstructure of DED processes need to be investigated due to these processes involving a higher temperature range and higher cooling rates such as $105\ ^0$ C/s. DED processes like WAAM processes involve a number of heating and cooling cycles. Since material has been deposited in a layer by layer manner, WAAM processes have a complicated thermal history. These thermal cycles will affect the mechanical properties and microstructure of the deposited structure.

Most of the studies have reported on the analysis of the mechanical properties of the parts produced by WAAM. They have tested the parts deposited using the WAAM process for micro-hardness, microstructure [23], levels of residual stresses induced and tensile strength [24–27]. The welding-based additive manufacturing process has a higher residual stress and it may lead to shorter life [28]. These tests provide an insight into the variations that occur in a prototype due to a change in the process parameters.

In this direction, Liu et al. [8] studied the tensile strength and hardness of the parts produced by the MIG-welding-based additive manufacturing technique. They found only a 7.6 and 4.8% variation in the average and ultimate tensile strength, respectively. The hardness values have also been found to be uniform and only some variation was observed in the upper layers of mild steel parts [10].

Lu et al. [25] examined the microstructure and mechanical properties of parts produced by a MIG-welding-based additive manufacturing technique. The grain size of the parts was found to be 12.469 μm [25] which corresponds to the fine granular ferrite structure. The yield strength in the direction parallel to the building direction ranged from 461 to 618.5 MPa which was found to be higher than the perpendicular direction (519.5–693.5 MPa) [24].

Lin et al. [29] performed two sets of experiments using plasma-based WAAM. In the first set, a cooling system was used for cooling the substrate. They fabricated a thin wall using a Ti-6Al-4V filler wire. They found the Widmanstatten microstructure with large columnar grains in the deposited structures and parts that were free of voids and cracks [29]. In the second set of experiments, thin walled Ti-6Al-4V parts were deposited without cooling the substrate. They found that parts possess good mechanical properties due to an alpha lamellar structure which is equivalent to forged components. In other work, the same authors used a pulsed plasma arc based additive manufacturing process [30] for the thin-walled deposition of Ti-6Al-4V parts. They concluded that the use of a pulsed plasma arc welding process provides improved mechanical properties compared to conventional manufacturing process like casting or welding. A novel compulsively

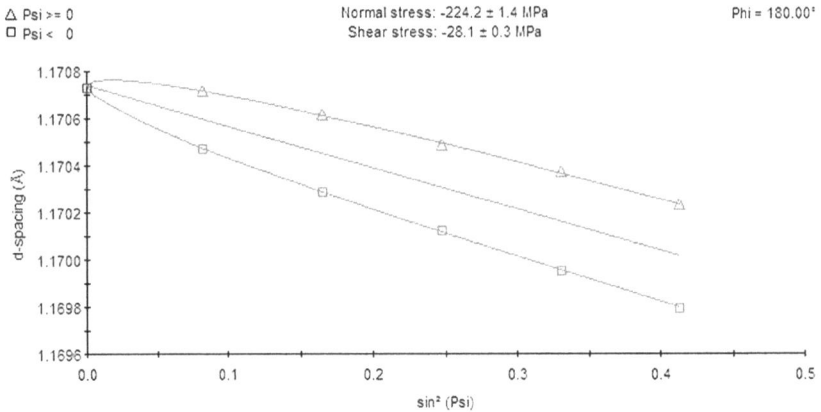

Figure 13.5 Residual stress measurement.

contracted plasma-based additive manufacturing process has been suggested by Liu et al. [31]. The proposed system would help to contract the plasma arc and molten metal droplets falling on the substrate. It would provide a better shielding and concentrated heating zone during the metal deposition process. In addition, the deposited structures possess a homogeneous microstructure, fine grain size and better surface form properties in the deposition. The developed system has shown its effectiveness to produce free form metallic parts with superior mechanical properties.

The residual stress induced in the material deposited using DED processes have been studied. The residual stresses formed due to the non-uniform heating and cooling cycles during metal deposition. The residual stress generated in the thin walled structures deposited using a TIG-based WAAM process was investigated by researchers [32]. In their study, they used a $sin^2\aleph$ method to calculate the residual stress. In this method, the $sin^2\aleph$ value was plotted against the lattice spacing (see Figure 13.5).

The slope of the line as shown in Figure 13.5 gives the residual stress at that point. Overall compressive residual stress was observed across the deposited sample. A solid-state phase transformation has occurred due to this low heat input, which may lead to compressive residual stresses in the core part of the deposited structure.

13.7 ADVANTAGES AND DISADVANTAGES OF DED PROCESSES

DED processes are distinguished from conventional manufacturing processes such as casting, subtractive machining and metal forming processes. The prominent nature of these processes makes them beneficial over conventional manufacturing processes. From an application point of view, these processes give more user-friendly operation and ease of customization with less impact on process complexity and cost. When these processes were used for small batch production, a significant reduction was observed in material wastage, time and cost [2]. The parts having complex geometry and having varying material composition can be easily fabricated with DED processes compared to conventional manufacturing processes. The lower value of the buy to fly ratio is one of the advantages of additive manufacturing processes which helps in the manufacturing of aerospace equipment. The uniqueness of DED processes is that they offer a prototype in the least time and minimize bottlenecks in small volume production. The use of DED based additive manufacturing processes helps in making an environmentally friendly production of goods as they emit less compared to conventional processes like casting or metal forming [33]. The power consumption is also restricted up to 10 kWh/Kg by using the WAAM process.

REFERENCES

[1] W. E. Frazier, "Metal additive manufacturing: A review," *J. Mater. Eng. Perform.*, vol. 23, no. 6, pp. 1917–1928, 2014, doi:10.1007/s11665-014-0958-z.

[2] L. Yang et al., *Additive Manufacturing of Metals: The Technology, Materials, Design and Production*, 2017. Springer.

[3] F. Martina, S. W. Williams, and P. Colegrove, "Design of an empirical process model and algorithm for the Tungsten Inert Gas wire+arc additive manufacture of TI-6AL-4V components," *24th Int. SFF Symp. - An Addit. Manuf. Conf. SFF 2013*, no. August, pp. 697–707, 2013.

[4] S. Liu and Y. C. Shin, "Additive manufacturing of Ti6Al4V alloy: A review," *Mater. Des.*, vol. 164, p. 107552, 2019, doi:10.1016/j.matdes.2018.107552

[5] W. Yangfan, C. Xizhang, and S. Chuanchu, "Microstructure and mechanical properties of Inconel 625 fabricated by wire-arc additive manufacturing," *Surf. Coatings Technol.*, vol. 374, no. May, pp. 116–123, 2019, doi:10.1016/j.surfcoat.2019.05.079

[6] S. Singh, S. Ramakrishna, and R. Singh, "Material issues in additive manufacturing: A review," *J. Manuf. Process.*, vol. 25, pp. 185–200, 2017, doi:10.1016/j.jmapro.2016.11.006

[7] J. F. Wang, Q. J. Sun, H. Wang, J. P. Liu, and J. C. Feng, "Effect of location on microstructure and mechanical properties of additive layer manufactured Inconel 625 using gas tungsten arc welding," *Mater. Sci. Eng. A*, vol. 676, pp. 395–405, 2016, doi:10.1016/j.msea.2016.09.015

[8] L. Liu, Z. Zhuang, F. Liu, and M. Zhu, "Additive manufacturing of steel-bronze bimetal by shaped metal deposition: Interface characteristics and tensile properties," *Int. J. Adv. Manuf. Technol.*, vol. 69, no. 9–12, pp. 2131–2137, 2013, doi:10.1007/s00170-013-5191-7

[9] D. Ding, Z. Pan, S. van Duin, H. Li, and C. Shen, "Fabricating superior NiAl bronze components through wire arc additive manufacturing," *Materials (Basel).*, vol. 9, no. 8, 2016, doi:10.3390/ma9080652

[10] S. W. Williams, F. Martina, A. C. Addison, J. Ding, G. Pardal, and P. Colegrove, "Wire + Arc Additive Manufacturing," *Mater. Sci. Technol.*, vol. 32, no. 7, pp. 641–647, 2016, doi:10.1179/1743284715Y.0000000073

[11] B. Baufeld, O. Van Der Biest, R. Gault, and K. Ridgway, "Manufacturing Ti-6Al-4V components by Shaped Metal Deposition: Microstructure and mechanical properties," *IOP Conf. Ser. Mater. Sci. Eng.*, vol. 26, no. 1, pp. S106–S111, 2011, doi:10.1088/1757-899X/26/1/012001

[12] Z. Zeng et al., "Wire and arc additive manufacturing of a Ni-rich NiTi shape memory alloy: Microstructure and mechanical properties," *Addit. Manuf.*, vol. 32, no. January, 2020, 2020, doi:10.1016/j.addma.2020.101051

[13] S. H. Nikam, N. K. Jain, and M. S. Sawant, "Optimization of parameters of micro-plasma transferred arc additive manufacturing process using real coded genetic algorithm," *Int. J. Adv. Manuf. Technol.*, vol. 106, no. 3–4, pp. 1239–1252, 2020, doi:10.1007/s00170-019-04658-2

[14] M. Liberini et al., "Selection of optimal process parameters for wire arc additive manufacturing," *Procedia CIRP*, vol. 62, pp. 470–474, 2017, doi:10.1016/j.procir.2016.06.124

[15] J. Xiong, Y. Li, R. Li, and Z. Yin, "Influences of process parameters on surface roughness of multi-layer single- pass thin-walled parts in GMAW-based additive manufacturing," *J. Mater. Process. Tech.*, vol. 252, no. February 2017, pp. 128–136, 2017, doi:10.1016/j.jmatprotec.2017.09.020

[16] J. Xiong, G. Zhang, and W. Zhang, "Forming appearance analysis in multi-layer single-pass GMAW-based additive manufacturing," *Int. J. Adv. Manuf. Technol.*, vol. 80, no. 9–12, pp. 1767–1776, 2015, doi:10.1007/s00170-015-7112-4

[17] J. Xiong and G. Zhang, "Adaptive control of deposited height in GMAW-based layer additive manufacturing," *J. Mater. Process. Technol.*, vol. 214, no. 4, pp. 962–968, 2014, doi:10.1016/j.jmatprotec.2013.11.014

[18] F. Youheng, W. Guilan, Z. Haiou, and L. Liye, "Optimization of surface appearance for wire and arc additive manufacturing of Bainite steel," *Int. J. Adv. Manuf. Technol.*, vol. 91, no. 1–4, pp. 301–313, 2017, doi:10.1007/s00170-016-9621-1

[19] J. Xiong, Y. Lei, H. Chen, and G. Zhang, "Fabrication of inclined thin-walled parts in multi-layer single-pass GMAW-based additive manufacturing with flat position deposition," *J. Mater. Process. Technol.*, vol. 240, pp. 397–403, 2017, doi:10.1016/j.jmatprotec.2016.10.019

[20] E. A. G. Olivares and V. M. V. Díaz, "Study of the hot-wire TIG process with AISI-316L filler material, analysing the effect of magnetic arc blow on the dilution of the weld bead," *Weld. Int.*, vol. 32, no. 2, pp. 139–148, 2018, doi:10.1080/09507116.2017.1347327

[21] S. Jhavar, C. P. Paul, and N. K. Jain, "Micro-plasma transferred arc additive manufacturing for die and mold surface remanufacturing," *Jom*, vol. 68, no. 7, pp. 1801–1809, 2016, doi:10.1007/s11837-016-1932-z

[22] W. Aiyiti, W. Zhao, B. Lu, and Y. Tang, "Investigation of the overlapping parameters of MPAW-based rapid prototyping," *Rapid Prototyp. J.*, vol. 12, no. 3, pp. 165–172, 2006, doi:10.1108/13552540610670744

[23] R. Pramod, S. M. Kumar, B. Girinath, A. R. Kannan, N. P. Kumar, and N. S. Shanmugam, "Fabrication, characterisation, and finite element analysis of cold metal transfer–based wire and arc additive–manufactured aluminium alloy 4043 cylinder," *Weld. World*, 2020, doi:10.1007/s40194-020-00970-8

[24] J. D. Spencer, P. M. Dickens, and C. M. Wykes, "Rapid prototyping of metal parts by three dimentional welding," *Mech E J. Eng. Manuf.*, vol. 212, pp. 175–182, 1998, doi:10.1243/0954405981515590

[25] X. Lu, Y. F. Zhou, X. L. Xing, L. Y. Shao, Q. X. Yang, and S. Y. Gao, "Open-source wire and arc additive manufacturing system: formability, microstructures, and mechanical properties," *Int. J. Adv. Manuf. Technol.*, vol. 93, no. 5–8, pp. 2145–2154, 2017, doi:10.1007/s00170-017-0636-z

[26] T. Artaza, A. Suárez, M. Murua, J. C. García, I. Tabernero, and A. Lamikiz, "Wire arc additive manufacturing of Mn4Ni2CrMo steel: Comparison of mechanical and metallographic properties of PAW and GMAW," *Procedia Manuf.*, vol. 41, pp. 1071–1078, 2019, doi:10.1016/j.promfg.2019.10.035

[27] E. Aldalur, F. Veiga, A. Suárez, J. Bilbao, and A. Lamikiz, "High deposition wire arc additive manufacturing of mild steel: Strategies and heat input effect on microstructure and mechanical properties," *J. Manuf. Process.*, vol. 58, no. September, pp. 615–626, 2020, doi:10.1016/j.jmapro.2020.08.060

[28] S. Suryakumar, K. P. Karunakaran, U. Chandrasekhar, and M. A. Somashekara, "A study of the mechanical properties of objects built through weld-deposition," *Proc. Inst. Mech. Eng. Part B J. Eng. Manuf.*, vol. 227, no. 8, pp. 1138–1147, 2013, doi:10.1177/0954405413482122

[29] J. Lin et al., "Microstructural evolution and mechanical property of Ti-6Al-4V wall deposited by continuous plasma arc additive manufacturing without post heat treatment," *J. Mech. Behav. Biomed. Mater.*, vol. 69, no. November 2016, pp. 19–29, 2017, doi:10.1016/j.jmbbm.2016.12.015

[30] J. J. Lin et al., "Microstructural evolution and mechanical properties of Ti-6Al-4V wall deposited by pulsed plasma arc additive manufacturing," *Mater. Des.*, vol. 102, pp. 30–40, 2016, doi:10.1016/j.matdes.2016.04.018

[31] W. Liu, C. Jia, M. Guo, J. Gao, and C. Wu, "Compulsively constricted WAAM with arc plasma and droplets ejected from a narrow space," *Addit. Manuf.*, vol. 27, no. March, pp. 109–117, 2019, doi:10.1016/j.addma.2019.03.003

[32] N. P. Gokhale, P. Kala, V. Sharma, and M. Palla, "Effect of deposition orientations on dimensional and mechanical properties of the thin-walled structure fabricated by tungsten inert gas (TIG) welding-based additive manufacturing process," *J. Mech. Sci. Technol.*, vol. 34, no. 2, pp. 701–709, 2020, doi:10.1007/s12206-020-0115-6

[33] P. C. Priarone, E. Pagone, F. Martina, A. R. Catalano, and L. Settineri, "CIRP Annals - Manufacturing Technology Multi-criteria environmental and economic impact assessment of wire arc additive manufacturing," vol. 69, pp. 2–5, 2020, doi:10.1016/j.cirp.2020.04.010

Chapter 14

An Investigation of Active Cutting Energy for Rough and Finish Turning of Alloy Steel

Raman Kumar, Paramjit Singh Bilga, and Sehijpal Singh

Guru Nanak Dev Engineering College, Ludhiana, India

CONTENTS

14.1 INTRODUCTION

Sustainability has turned out to be a pressing concern in manufacturing units. Sustainable development is defined as the capability to satisfy current requirements without compromising the ability of the developing society to meet its environmental requirements. Sustainable manufacturing is mainly driven by environmental concerns and the escalating population of the world. Therefore, there is a need to balance the environment and the increasing population (Chandel, Kumar, & Kapoor, 2021). The human population will continue rising until 2050; the 7 billion landmark was reached in 2011 and is expected to reach 8 billion individuals by 2024 and 9 billion by 2040 (He, Goodkind, & Kowal, 2016). This rise in population creates further demand for manufactured products. So, an expansion of a population can be seen as an excellent opportunity for manufacturing units since it builds demand for manufactured products. This demand leads to greater use of electrical power/energy, thus, higher fuel costs, a higher levy and higher carbon emissions. Mechanical machining is an important aspect of the manufacturing industry, and it mainly entails cutting metals employing various techniques. This process consumes a substantial amount of electrical energy. Worldwide machine tool manufacturing trade is a USD68.6 billion business, and 37% of global energy is consumed in manufacturing processes, yet only minimal attention is paid towards discrete product manufacturing (Diaz-Elsayed, Dornfeld, & Horvath, 2015). As a result, researchers have started work on developing systematic energy-saving methods. The significant research studies related to the optimization and modeling of power consumption are discussed here.

The response surface methodology (RSM) technique was applied to develop a power consumption model and the Taguchi technique to scrutinize cutting variables and the cutting environment during computer numerical control (CNC) turning (Aggarwal, Singh, Kumar, & Singh, 2008). The RSM models for the responses were also presented (Aggarwal et al., 2008). RSM was applied to develop machining power models for turning EN-31 steel. The second-order models predicted better than the first-order ones (Abhang and Hameedullah, 2010). The cutting power, Ra and tool wear were analyzed on a CNC lathe with carbide tools, using cutting speeds of 160 and 250 m/min., without cutting fluid at a steady feed and depth of cut while turning (Rosa, Diniz, Andrade, & Guesser, 2010). The regression equations were established for Ra, cutting and feed forces while turning stainless steel: the cutting variables and fluid were considered using a carbide tool. The cutting fluid was found to lower forces and enhance surface finish (Cetin, Ozcelik, Kuram, & Demirbas, 2011).

The Taguchi method was applied while turning hardened steel with a coated tool. The multiple linear regression models were projected for power consumption alongside surface roughness, specific cutting force, tool wear and machining force in cutting variables (Suresh, Basavarajappa, & Samuel, 2012). The experiments were conducted on a CNC lathe to machine an extruded aluminum shaft and the Taguchi method was applied to optimize the cutting parameters—speed, feed rate and depth of cut. It was reported that the feed rate (49%) was most significant, followed by the depth of cut (45%), for reducing energy consumption (Mouleeswaran, Babu, & Jothi Prakash, 2012). The nose radius had an insignificant effect on power consumption. The feed rate variation did not affect the power during the CNC turning of aluminum alloy employing a carbide insert (Bhushan, 2013). The cutting parameters were optimized during rough turning while considering the power consumed, energy consumed, cutting power and output parameters. Out of the total energy consumed by all the experiments, less than 50% was consumed in the cutting process. The feed rate was the highest contributor in minimizing total energy consumption and minimizing Ra (Camposeco-Negrete, 2013).

The effect of cutting speed and feed rate was investigated when turning AISI 1045 steel with a coated carbide tool. The high-speed turning experiments were performed on Nitronic 33 steel alloy under diverse cutting environments and cutting speeds; the different responses were investigated. Ra's optimum conditions—cutting force, tool life and power—was achieved at 90 m/min. with a dry cutting environment (Balogun, Edem, Bonney, Ezeugwu, and Mativenga, 2015a). The power consumption of the three machine tools was investigated. They reported that with the variation of cutting and machining loads, the electrical energy for machining processes varies from 0.0 to 48.1% and needs to be optimized. The multiple responses were optimized while rough turning aluminum at a constant material removal volume (Balogun, Edem, and Mativenga, 2015b). The machining parameters—cutting speed, feed rate and depth of cut and corner radius—were optimized while turning AISI 1018 steel to reduce energy utilization at a steady MRR. The results revealed that the same amount of material was removed at different combinations of cutting parameters. The quantity of energy expenditure varied from 80.75 to 141.77 kJ (Camposeco-Negrete, de Dios Calderón Nájera, & Miranda-Valenzuela, 2016). The turning operation was performed on alloy steel to reduce active power/energy and energy efficiency, and the results revealed a substantial reduction in energy (Bilga, Singh, & Kumar, 2016); active power was modeled by RSM (Kumar, Bilga, & Singh, 2018) and the active cutting power was optimized by the Taguchi method (Kumar, Bilga, & Singh, 2020). Three methods of assigning weights to the energy allied output parameters, Ra and MRR were utilized in turning alloy steel. Weight assignment plays an imperative part in optimization (Kumar, Bilga, & Singh, 2017). An empirical model of power consumption was developed while turning aluminum using RSM, and results revealed that the developed model could be utilized within a 7% error (Garg, Garg, & Sangwan, 2018). Minimum cutting force and power was achieved during the turning process's mathematical analysis while maximizing MRR (Wakjira, Altenbach, & Ramulu, 2020). The active power consumption during the drilling of ST52.3 alloy steel was optimized, and regression models were built using RSM (Sidhu, Singh, & Kumar, 2021b). Enhancing the energy efficiency of machining processes can help to make production more sustainable and it is critical to study, assess and optimize the amount of energy used during machining processes (Sidhu, Singh, Kumar, Pimenov, & Giasin, 2021). As compared to conventional machining methods, ecological, financial and technological feasibility may be improved by clean technology machining procedures; the results culminate in sustainability and interoperability (Karim et al., 2021).

This literature review reveals that most of the research associated with machining responses, that is surface roughness, machining cost and productivity, had been done. Still, little attention has been given to the power/energy consumption response. The turning operation is broadly utilized in most manufacturing units and consumes much energy while being performed. Therefore, reducing the energy consumption of this operation can successfully assist manufacturing units in achieving sustainable manufacturing. Active power (kW) is the true power to perform the work and is responsible for electricity bills. A better utilization of active energy can be made with parametric optimization. The empirical models of active energy consumption can predict active power consumption within model development constraints. Therefore, optimizing the machining or manufacturing operation is suggested to be a better choice than changing the existing setup that includes more cost and optimization, leading to enhanced monetary and communal sustainability (Kumar, Singh, Bilga, and Jatin, 2021b; Kumar, Singh, Sidhu, and Pruncu, 2021a; Sidhu, Singh, & Kumar, 2021a). So, active energy consumption is optimized for the rough as well as the finish CNC turning of EN 353 alloy steel; the experiments were designed with Taguchi's L_{27} orthogonal array while varying nose radius, cutting speed, feed rate and depth of cut along with their interactions; RSM was used to model the responses with Minitab software. EN 353 alloy steel is mostly used to manufacture automobile spare parts. The coated tungsten carbide inserts were used for rough turning, and cermet inserts were used for finish turning.

14.2 MATERIALS AND METHODS

14.2.1 The Taguchi Method

The Taguchi concept of quality regulation is an engineering methodology for minimizing faults and failures in produced items. The Taguchi method uses an orthogonal array (OA) design, which results in a substantially lower "variance" for the experiment when the control parameters are "optimized". The S/N (signal to noise) ratio and OA are widely utilized. "The nominal-the-better", "the larger-the-better" and "the smaller-the-better" are the main quality features. The minimum value ACE is desired, so the smaller-the-better category of the S/N ratio is utilized in the current study. The S/N ratios may be calculated using Minitab software (Montgomery, 2017; Ross & Ross, 1988; Roy, 2001). The L_{27} (3^4) OA of Taguchi's design of experiments with four parameters was considered. The OA has three interactions (A*B, A*C and A*D) and 27 experiments with parameters at three levels.

14.2.2 Response Surface Methodology

RSM was anticipated by Box and Wilson (1951) in the early 1950s and has received extensive consideration because of its excellent empirical execution in modeling. A mathematical and statistical process framework offers better empirically fitting models between input parameters and output responses. It is vital to build, enhance, simplify the process and describe the region of interest. R-squared, that is, the coefficient of determination and analysis of variance (ANOVA), is utilized to check an established model's fitness. R-squared signifies the deviation in the selected regression model's response (Montgomery, 2017; Ross & Ross, 1988; Roy, 2001). A mathematical model of ACE is established for rough and finish turning.

14.2.3 Workpiece, Cutting Inserts, Input Parameters and Their Levels

Rough and finish turning cutting information is presented in Table 14.1 as supplied by the manufacturer of cutting tools, but the trial of experiments is also given deliberation in an assortment of input parameters. Table 14.1 also depicts workpiece and tool material information. Table 14.2 shows the information of selected cutting parameters. The cutting speed (A) is calculated, as shown in Eq. (14.14.1):

$$A = \frac{\pi dN}{1000} \tag{14.1}$$

where d is the work piece diameter and N is the number of revolutions per minute.

Table 14.1 Technical Details of Work Piece, Tool Material and Cutting Conditions

Standard Cutting Conditions	**Rough turning**: depth of cut (mm): 1–5, feed (mm/rev.): 0.2–0.5, cutting speed (m/min.): 120–250
	Finish turning: depth of cut (mm): 0.1–1, feed (mm/rev.): 0.08–0.3, cutting speed (m/min.): 150–300
	Percentage composition: C: 0.10~0.2, Mn: 0.50~1, Si: 0.50~1, P: 0.018~0.02, Cr: 0.75~1.25, Mo: 0.08~0.15, Ni: 1~1.5 and the rest is Fe
Work Piece Material: EN 353 Alloy Steel	**Dimensions**: total length: 98 mm; turning length: 68 mm, diameter: 36.8 mm (finish turning), 44 mm (rough turning)
	Wet Cutting Environment
	Rough cut: T 9025 coated tungsten carbide (TM chip breaker)
	Finish cut: GT 9530 coated cermet grade (TSF chip breaker)
	ISO coding: TNMG 160404-08-12
Tungaloy Make, Cutting Inserts	**Shape**: Equilateral Triangle; Relief Angle: 0°, Cutting Edge Length: 16 mm, Corner Radius: 0.4, 0.8 and 1.8 mm, Accuracy Tolerances of Corner Height: ± 0.18; Thickness: ± 0.13; Inscribed Circle Diameter: ± 0.05

Table 14.2 Input Parameters/Levels for Rough and Finish Turning Operations

Parameters	Units	Symbol	Rough Turning Operation Level			Finish Turning Operation Level		
			1	*2*	*3*	*1*	*2*	*3*
Cutting Speed	m/min.	A	165.79	207.24	248.69	179.11	236.88	294.66
Feed Rate	mm/ rev.	B	0.2	0.25	0.3	0.1	0.15	0.2
Depth of Cut	mm	C	1	1.4	1.8	0.2	0.4	0.6
Nose Radius	mm	D	0.4	0.8	1.2	0.4	0.8	1.2

Table 14.3 Details of Equipment and Instrument

CNC Turning Machine; Stallion 100 HS (HMT) make	Power Rating: 5.5 kW; Speed Range: 100~4500 RPM; Turning Diameter: up to 250 mm; Tail Stock Travel:120 mm and Rapid Traverse:18 mm/min.
Clamp-on Power Logger	Current Range: 50 mA–500 A with accuracy ± 0.3% rdg. ± 0.1% f.s. + clamp sensor accuracy
(PW 3360-20); HIOKI Make	Voltage: 5–1000 V with accuracy ± 0.3% rdg. ± 0.1% f.s. Active Power: 300 W–9 MW with accuracy ± 0.3% rdg. ± 0.1% f.s. + clamp sensor accuracy

14.2.4 Instrument and Equipment

The CNC turning machine was utilized for experimentation and the clamp-on power logger was used to measure active power; the details are shown in Table 14.3.

14.3 EXPERIMENTAL PROCEDURE OBSERVATIONS AND CALCULATIONS FOR TURNING OPERATIONS

An EN 353 alloy steel workpiece was furnished with an alike diameter of 44 mm for a rough turning operation of length 98 mm. The rough turning operation was completed first, and the same workpieces were prepared with a 36.8 mm diameter for a finish turning operation. The cutting parameters were changed following the L-27 design matrix. A clamp-on power logger was mounted on the tool's power distribution cables to quantify active power usage. The power logger was set to monitor active power once every second, and the average scores were used to conduct further studies. Following the design matrix, the same method was repeated when the first experiment was completed. Thus, each experiment was carried out three times, with the average active power levels being used for further study. ITW India, BioCool-80 semi-synthetic metal working fluid (oil) was used with 100 liters of water and 5 liters of oil. The calculated ACE of the rough and finish turning operations as per the design matrix and S/N ratios are shown in Table 14.4. The ACE is the difference between the machine's total active power consumption during the turning operation and the active power consumption without the turning operation.

Table 14.4 L-27 Design Matrix, Experimental Calculations and S/N Ratios

					Rough Turning		Finish Turning	
Exp. no.	A	B	C	D	ACE (Wh)	S/N Ratio	ACE (Wh)	S/N Ratio
1	1	1	1	1	1.457	−3.269	0.358	8.922
2	1	1	2	2	1.842	−5.306	1.018	−0.155
3	1	1	3	3	2.149	−6.645	1.155	−1.252
4	1	2	1	2	0.946	0.482	0.373	8.566
5	1	2	2	3	0.773	2.236	0.594	4.524
6	1	2	3	1	1.936	−5.738	1.168	−1.349
7	1	3	1	3	0.827	1.650	0.254	11.903
8	1	3	2	1	1.129	−1.054	1.174	−1.393
9	1	3	3	2	1.672	−4.465	1.261	−2.014
10	2	1	1	1	1.258	−1.994	1.095	−0.788
11	2	1	2	2	1.805	−5.130	1.749	−4.856
12	2	1	3	3	2.088	−6.395	1.854	−5.362
13	2	2	1	2	1.258	−1.994	0.976	0.211
14	2	2	2	3	1.084	−0.701	0.992	0.070
15	2	2	3	1	1.96	−5.845	1.571	−3.924
16	2	3	1	3	0.943	0.510	0.695	3.160
17	2	3	2	1	1.329	−2.470	1.49	−3.464
18	2	3	3	2	2.099	−6.440	1.495	−3.493
19	3	1	1	1	1.446	−3.203	1.025	−0.214
20	3	1	2	2	1.694	−4.578	1.701	−4.614
21	3	1	3	3	1.635	−4.270	1.919	−5.661
22	3	2	1	2	1.357	−2.652	1.128	−1.046
23	3	2	2	3	1.234	−1.826	1.171	−1.371
24	3	2	3	1	1.863	−5.404	1.545	−3.779
25	3	3	1	3	0.751	2.487	0.135	17.393
26	3	3	2	1	1.352	−2.620	0.845	1.463
27	3	3	3	2	1.69	−4.558	1.188	−1.496

14.3.1 Optimization of Active Cutting Energy for Turning Operations

Figure 14.1 shows plots for a rough turning operation and Figure 14.2 for a finish turning operation for means and S/N ratios, respectively. The ACE upsurges with an enhancement in cutting speed and depth of cut, both in rough and finish turning operations; but for cutting speed, ACE decreases after level 2—a rise in cutting speed and depth of cut results in enhanced revolutions per minute of the motor. As a result, the motor consumes more current, and it increases active power consumption. The material removal rate increases with a rise in depth of cut. To remove more material from the workpiece, the load on the motor rises. So, maintaining the requisite speed, that is, revolutions per minute of the motor, consumes extra active power due to increased current requirements. With an increase in feed rate, time expended on the turning operation decreases; accordingly, the ACE also decreases because it is directly proportional to the turning cycle time.

Figures 14.1 and 14.2, for rough and finish turning operations, respectively, show that the first level of A and C and the third level of B and D results in the minimum value of ACE. The minimum value of ACE is desired as it is the "the smaller, the better" category of response. As a result, the optimal parameter/level blend A_1, B_3, C_1 and D_3 was given for ACE, both for rough as well as finish turning operations. Hence the optimum value of cutting parameters for ACE obtained are: for rough turning operation A_1 = 165.79 m/min., B_3 = 0.3 mm/rev., C_1 = 1 mm, D_3 = 1.2 mm, and the optimum value of ACE achieved is 0.722 Wh. The optimum value of cutting parameters for ACE obtained for the finish turning operation are: A_1 = 179.11-m/min., B_3 = 0.2-mm/rev., C_1 = 0.2-mm, D_3 = –1.2-mm,-and the optimum value of ACE achieved is 0.131 Wh.

The average, maximum, minimum and optimal active cutting energy for rough and finish turning operations are shown in Figure 14.3. In the rough turning operation, the ACE of 0.722 Wh is higher than the finishing turning operation at optimum turning parameters. This higher active energy consumption in the case of the rough turning operation is due to the higher feed rate values and depth of cut. The material removal requirement is more in the case of a rough turning operation. The average value of ACE observed for all sets of experiments for the finish turning operation is 1.108 Wh compared to the rough turning operation of 1.466 Wh. The 0.357 Wh is a more average value of ACE in the rough turning operation than the finish turning operation because of selecting a range of cutting parameters in both cases, as these parameters were chosen as per tool supplier suggestions and machine tool capacity.

Main effects plot for means of active cutting energy (Rough turning)
Data Means

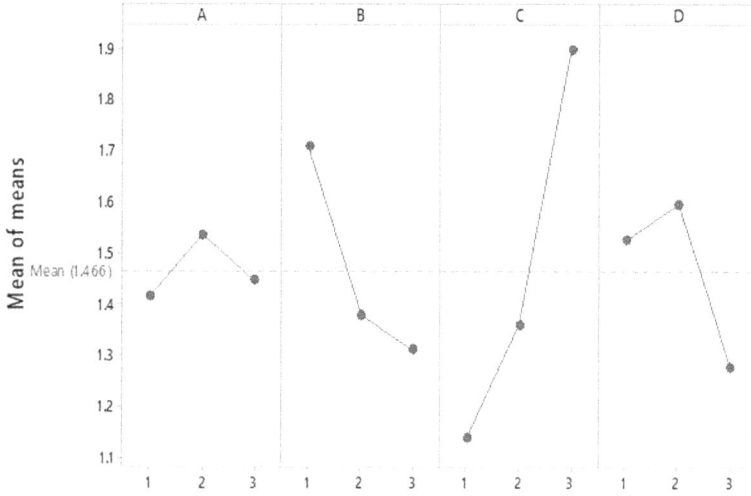

Main effects plot for SN ratios of active cutting energy (Rough turning)
Data Means

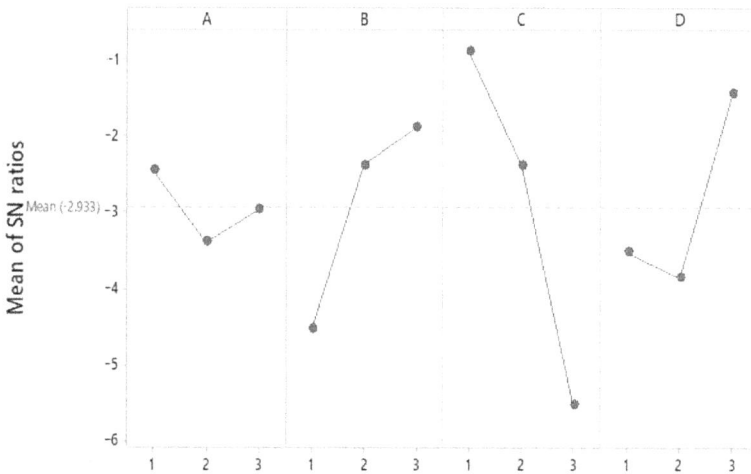

Signal-to-noise: Smaller is better

Figure 14.1 Main effects plot for means and S/N ratios of ACE for rough turning.

Main effects plot for means of active cutting energy (Finish turning)
Data Means

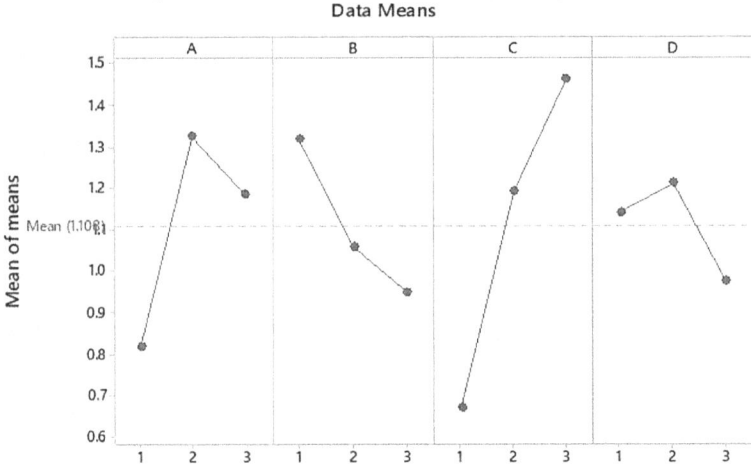

Main effects plot for SN ratios of active cutting energy (Finish turning)
Data Means

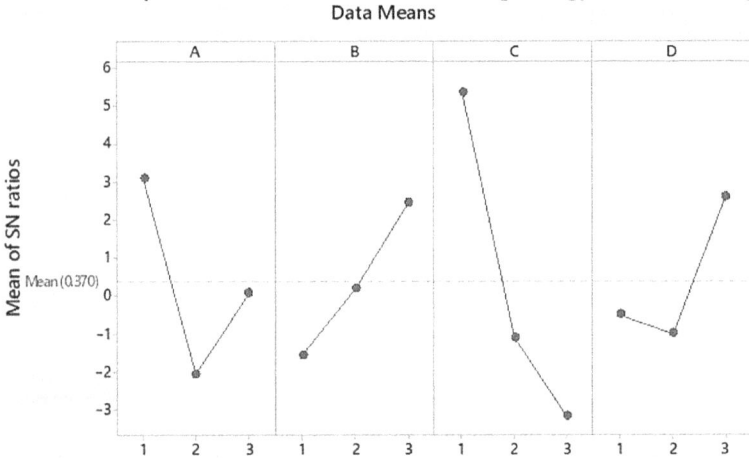

Signal-to-noise: Smaller is better

Figure 14.2 Main effects plot for means and S/N ratios of ACE for finish turning.

14.3.2 Analysis of Variance (ANOVA)

ANOVA is a multivariate statistical technique and estimating process for analyzing differences between means. ANOVA is governed by the law of amount of variance, which divides the perceived variance in a variable into components attributed to various causes of variation. In its most basic form,

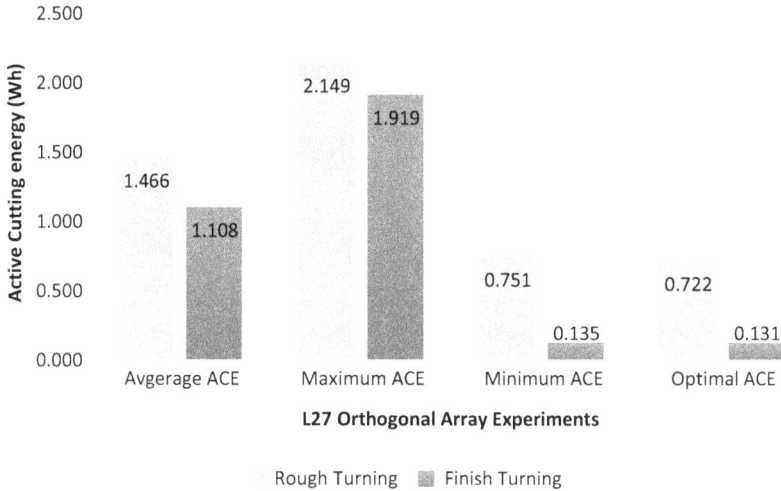

Figure 14.3 Active cutting energy for all sets of experiments on L_{27} OA for turning operations.

ANOVA is a statistical test that determines if two or more population means are equal and extends the t-test beyond two means. The p-value (p) is used to determine whether or not the input variables have a noticeable influence on the results. If the p-value is less than 0.05, then the variable has a noticeable effect on the result; otherwise, it does not (Bilga et al., 2016; Kumar et al., 2020). In quantitative computing, the degrees of freedom (Ď) indicates how often the simulation values have the liberty to vary. The mean square adjusted (м) is calculated using the Ď. It establishes how much independent data is available to compute each sequential sum of squares (S). Eq. (14.2) calculates the percentage contribution (PC) of the input parameters:

$$PC = \frac{\text{Sum of squared deviations}}{\text{Total sum of squared deviations}} \qquad (14.2)$$

The results of ANOVA for the means of ACE for a rough turning operation are shown in Table 14.5. The percentage participation of the depth of cut to ACE is highest (58.73%), followed by the feed rate (17.35%) and the nose radius (10.84%). The ANOVA outcome for the S/N ratios of ACE is shown in Table 14.6. The PC of the depth of cut to ACE is the largest (52.31%), followed by the feed rate (18.45%) and the nose radius (15.86%). The cutting speed has only a 1.52 and 2.01% contribution for the means and S/N ratios, respectively, and has no significant consequence on the ACE as the p-value is higher than 0.05. The residual error of 2.07 and 2.09% for the means and S/N ratios, respectively, is minimal. Subsequently, for the means and S/N ratios, the depth of cut is the most critical parameter for

Table 14.5 ANOVA for the Means of ACE for Rough Turning

Resource	$\check{D}$	S	M	F	p	PC
A	2	0.071	0.036	2.19	0.193	1.52
B	2	0.815	0.407	25.11	0.001	17.35
C	2	2.757	1.378	84.96	0.000	58.73
D	2	0.509	0.254	15.68	0.004	10.84
A*B	4	0.226	0.057	3.48	0.084	4.82
A*C	4	0.160	0.040	2.46	0.156	3.40
A*D	4	0.060	0.015	0.92	0.510	1.27
Residual Error	6	0.097	0.016			2.07
Total	26	4.694				100.00

Table 14.6 ANOVA for the S/N Ratios of ACE for Rough Turning

Resource	$\check{D}$	S	M	F	p	PC
A	2	3.882	1.9411	2.88	0.133	2.01
B	2	35.641	17.8204	26.48	0.001	18.45
C	2	101.046	50.5231	75.07	0.000	52.31
D	2	30.635	15.3175	22.76	0.002	15.86
A*B	4	10.499	2.6248	3.9	0.068	5.44
A*C	4	5.277	1.3193	1.96	0.220	2.73
A*D	4	2.136	0.5341	0.79	0.570	1.11
Residual Error	6	4.038	0.673			2.09
Total	26	193.155				100.00

ACE, followed by the feed rate for the rough turning operation. Two-way interactions A*B, A*C and A*D between parameters have minimal effect on ACE for the means and S/N ratios, and have no statistically significant impact. The p-value, which is lower than 0.05, shows that the depth of cut, nose radius and feed rate have a statistically significant effect on ACE for the means and S/N ratios for the rough turning operation.

The ANOVA outcome for the means of ACE for the finish turning operation is shown in Table 14.7. The percentage of cut to ACE's depth was the most extensive (48.09%), followed by the cutting speed (20.38%), feed rate (10.80%) and nose radius (4.37%). The ANOVA outcome for the S/N ratios of ACE is shown in Table 14.8. The percentage of the depth of cut to ACE (44.11%) is the largest, followed by the cutting speed (14.95%), feed rate (9.05%) and nose radius (8.52%). The residual error of 3.13 and 6.04% for the means and S/N ratios, respectively, is minimal. As a result, both for the means and S/N ratios, the depth of cut is the essential parameter for ACE

Table 14.7 ANOVA for the Means of ACE for Finish Turning

Resource	Ď	S	M	F	p	PC
A	2	1.233	0.617	19.55	0.002	20.38
B	2	0.654	0.327	10.36	0.011	10.80
C	2	2.910	1.455	46.11	0.000	48.09
D	2	0.264	0.132	4.19	0.073	4.37
A*B	4	0.732	0.183	5.8	0.029	12.10
A*C	4	0.024	0.006	0.19	0.936	0.39
A*D	4	0.045	0.011	0.35	0.834	0.74
Residual Error	6	0.189	0.032			3.13
Total	26	6.05088				100.00

Table 14.8 ANOVA for the S/N Ratios of ACE for Finish Turning

Resource	Ď	S	M	F	p	PC
A	2	119.74	59.871	7.42	0.024	14.95
B	2	72.5	36.249	4.5	0.064	9.05
C	2	353.29	176.646	21.91	0.002	44.11
D	2	68.25	34.123	4.23	0.071	8.52
A*B	4	92.45	23.112	2.87	0.12	11.54
A*C	4	34.06	8.515	1.06	0.453	4.25
A*D	4	12.32	3.079	0.38	0.815	1.54
Residual Error	6	48.38	8.064			6.04
Total	26	800.99				100.00

after cutting speed for a finish turning operation. Two-way interactions A*B, A*C and A*D between parameters have minimal effect on ACE for the means and S/N ratios and have no statistically significant impact for finish turning. The p-value, which is lower than 0.05, shows that the cutting speed, depth of cut, nose radius and feed rate have a statistically significant effect on ACE for the means and S/N ratios for a finish turning operation.

The effect of input parameters and the optimum parameter/level combination of ACE are similar, both for rough and finish turning operations. Nevertheless, the order of percentage contribution is different as the workpiece and machine tool is the same, though the cutting tool is of another material. In both cases, the depth of cut with a contribution of 58.73% in a rough turning operation and 48.09% in a finish turning operation is a vital parameter; refer to Figure 14.4. But this is followed by the feed rate of 17.35% in the case of a rough turning operation, and is followed by a cutting speed of 20.38% in the case of a finish turning operation. As in the case

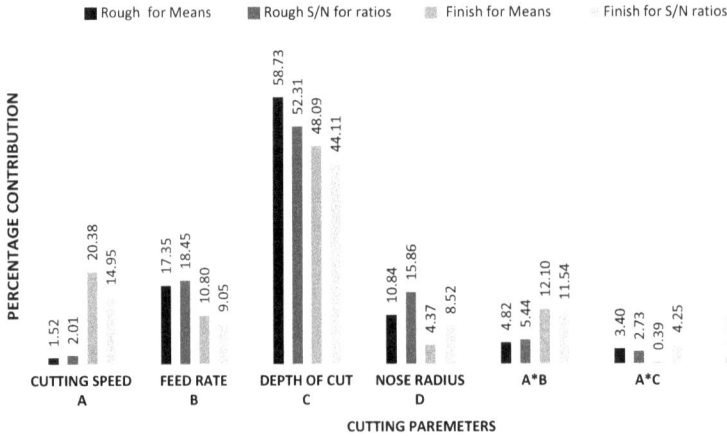

Figure 14.4 Percentage contribution of cutting parameters for rough and finish turning.

of the finish turning operation, the contribution of cutting speed is more than in the rough turning operation; this is because of the selection of a range of cutting parameters in both cases.

14.3.3 Response Surface Model of ACE for Turning Operations

The RSM model of ACE for turning operations was established based on the experimental information shown in Eq. (14.3) for a rough turning operation and in Eq. (14.4) for a finish turning operation. The model was developed based on the backward elimination method for quadratic, linear and square, linear and interaction, and linear terms. The coefficient of determination R-squared (R-Sq) for quadratic, linear and square, linear and interaction, and linear terms are shown in Table 14.9. The full quadratic model was analyzed for non-significant terms. The coefficient of determination values, that is R-squared, R-squared predicted and R-squared adjusted, were noted down while removing non-significant terms.

The coefficient of determination obtained as R-squared (94.79%), R-squared predicted (72.2%) and R-squared adjusted (80.31%) is a best fit for a model of ACE in terms of linear factors and the square of the depth of cut and nose radius for a rough turning operation:

$$ACE_{Rough} = 1.198 + 0.016A - 0.199B - 0.253C + 0.656D$$
$$+ 0.158C^2 - 0.195D^2 \tag{14.3}$$

The coefficient of determination obtained as R-squared (94.31%), R-squared predicted (89.12%) and R-squared adjusted (92.58%) is a best fit

Table 14.9 Coefficient of Determination of RSM Models for the Means of ACE for Turning Operations

Response	R-sq. (%)	Quadratic Full	Linear and Square	Linear and Interactions	Linear	Selected Model and Terms Included	
ACE	R-Sq	92.21	88.44	80.28	76.78	**84.85**	Linear, C^2,
Rough	R-Sq (pred)	60.05	73.98	49.33	66.16	**72.2**	D^2
	R-Sq (adj)	84.42	83.3	67.95	72.56	**80.31**	
ACE	R-Sq	96.45	83.64	83.09	98.81	**94.86**	Linear, A^2,
Finish	R-Sq (pred)	84.01	63.19	50.97	51.79	**89.12**	D^2, AB,
	R-Sq (adj)	92.9	76.37	72.51	63.14	**92.58**	BD

for a model of ACE in terms of linear factors and the square of cutting speed, nose radius and interaction between cutting speed and feed rate as well as feed rate and nose radius for a finish turning operation:

$$\text{ACE}_{\text{Finish}} = -13.39 + 3.649A + 1.395B + 0.307C + 1.836D \\ - 0.323A^2 - 0.152D^2 - 0.219A \times B - 0.175B \times D \tag{14.4}$$

The ANOVA technique was utilized to authenticate the response surface quadratic ACE model's adequacy for rough and finish turning operations. The ANOVA outcomes for RSM models are shown in Table 14.10 for a rough turning operation and Table 14.11 for a finish turning operation. Table 14.10 for a rough turning operation reveals that the regression of linear and square terms significantly affects ACE as the p-value is less than

Table 14.10 ANOVA for the RSM of ACE for Rough Turning

Resource	$\check{D}$	S	M	F	p
Model	6	3.983	0.664	18.67	0.000
Linear	4	3.604	0.901	25.34	0.000
A	1	0.005	0.005	0.13	0.720
B	1	0.713	0.713	20.05	0.000
C	1	2.606	2.606	73.31	0.000
D	1	0.280	0.280	7.88	0.011
Square	2	0.379	0.189	5.33	0.014
C*C	1	0.151	0.151	4.23	0.053
D*D	1	0.228	0.228	6.42	0.020
Error	20	0.711	0.036		
Total	26	4.694			

Table 14.11 ANOVA for the RSM of ACE for Finish Turning

Resource	$\breve{D}$	S	M	F	p
Model	8	5.740	0.717	41.53	0.000
Linear	4	3.631	0.908	52.54	0.000
A	1	1.202	1.202	69.57	0.000
B	1	0.397	0.397	22.96	0.000
C	1	1.065	1.065	61.62	0.000
D	1	0.236	0.236	13.65	0.002
Square	2	0.767	0.383	22.18	0.000
A*A	1	0.628	0.628	36.33	0.000
D*D	1	0.139	0.139	8.04	0.011
2-Way Interaction	2	0.810	0.405	23.43	0.000
A*B	1	0.579	0.579	33.49	0.000
B*D	1	0.231	0.231	13.38	0.002
Error	18	0.311	0.017		
Total	26	6.051			

0.05. The terms C^2 and D^2 show that there are significant quadratic effects. The interaction between terms has no significant impact, and these terms are not included in the model. Table 14.11 for the finish turning operation reveals that the regression of linear, square and two-way interaction terms A*B and B*D has a significant consequence on the response ACE as the p-value is smaller than 0.05. The terms A^2 and D^2 show that there are substantial quadratic effects.

The assumptions made for developing the model are checked by comparing the residual plots. The Minitab 16 software has a provision to draw the residual plot, which contains a normal-probability-plot, a residuals-versus-fitted-values-plot, a histogram plot and a residuals-versus-order-of-the-data-plot jointly in a single graph window. The residual is the difference between the observed value and the predicted value. The data are normally distributed or not; the normal probability plot confirms this. It also indicates whether the other parameters affect the response or not or whether there is any presence of outliers. Whether the variance is constant or not is evaluated by the residuals versus fitted values plot, and it also provides information about outliers or nonlinear association. The skewness of the data or the existence of an outlier is judged from the histogram plot. The scatter plot of residuals versus the order of the data plot provides information about the data collection order and time effects. The four-in-one residual plot for ACE means is displayed in Figure 14.5 for rough and finish turning operations. The normal probability plot for rough and finish turning operations of ACE demonstrates that the residuals seem to pursue a straight line. The majority of the points are grouped around the blue line. This implies that the errors

Residual plots for active cutting energy rough turning

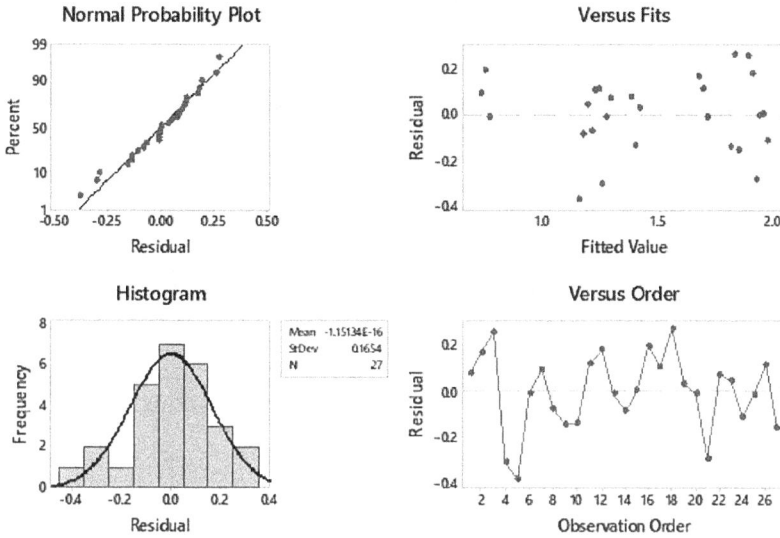

Residual plots for active cutting energy finish turning

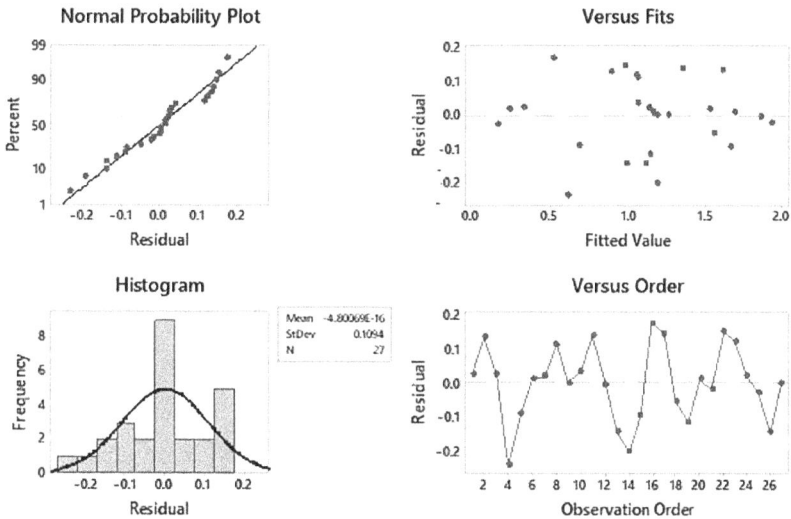

Figure 14.5 Four-in-one residual plot for rough and finish turning operations.

are distributed normally. So, there is no indication of non-normality, the existence of unknown parameters, skewness or outliers.

Thus, the assumption of normality is valid. The residuals versus fitted values for ACE show that the residuals are scattered randomly about the zero line. About half of the error terms are on top, and half are lower than the zero line. This indicates the validity of the error terms having a mean of zero. The histogram plot of the residuals of ACE for all observations shows that nearly all the residuals fall under a bell shape. Residuals versus the ACE plot's order show that the residual of ACE is scattered randomly near the zero line, and there is no proof that the error terms correlated refer to Figure 14.5. The contour and surface plots of active cutting energy for rough turning is shown in Figure 14.6 and finish turning in Figure 14.7.

14.4 COMPARATIVE ANALYSIS AND CONFIRMATION OF RESULTS

The confirmative experiment is the final step in validating the outcomes formed using Taguchi's approach. This is the most important phase, and the Taguchi method strongly advises the confirmation of the results' accuracy (Bilga et al., 2016; Chandel et al., 2021; Kumar et al., 2020). For rough and finish turning operations, the Taguchi approach estimated ACE by employing Minitab software. Three confirming experiments were carried out, with the average of the findings matched to the expected outcomes. Finally, the Taguchi method's accuracy was tested using the statistical approach of absolute error and RSM prediction, as shown in Equation (14.5) (Kumar, Bilga, and Singh, 2021c):

$$\text{Absolute error} = |E_i - P_i| \qquad (14.5)$$

where E_i is the experimental value and P_i is the predicted value. Table 14.12 shows the Taguchi technique anticipated the means at their optimal values. The comparative results show no significant difference between corresponding experimental data for rough and finish turning operations. Each response's absolute error was small and within the confidence interval. Figure 14.8 for rough turning and Figure 14.9 for finish turning illustrate the comparison results of the experiments and the RSM predictions graphically for all L-27 OA tests. The graphs indicate a strong correlation between the experimental readings and the RSM projected for both rough and final turning.

14.4.1 Optimal Rough and Finish Turning Parameters Used in Industries

The cutting parameters for machining operations in industries are chosen according to the operators' experience or suggested by the cutting insert

Contour plots of active cutting energy for rough turning

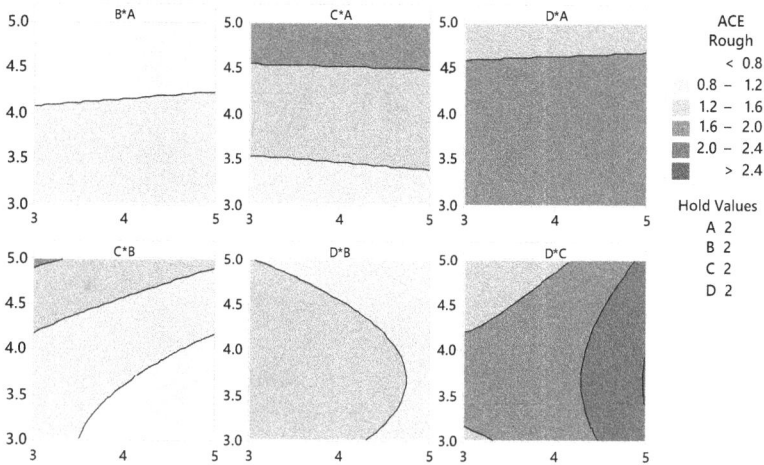

Surface plots of active cutting energy for rough turning

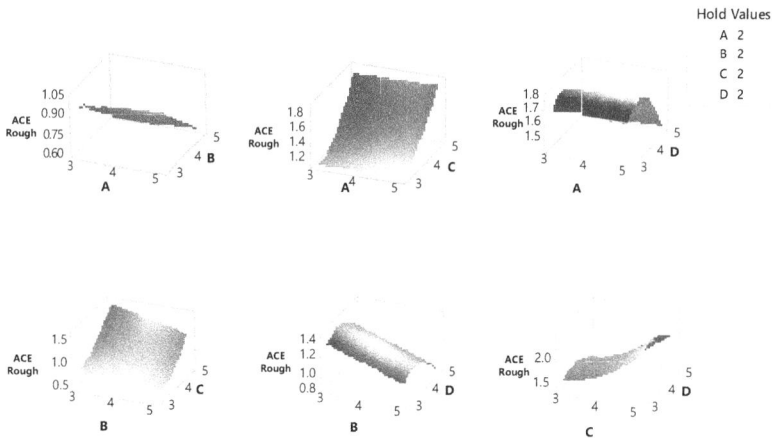

Figure 14.6 Contour and surface plots of active cutting energy for rough turning.

supplier or machining handbooks. Generally, selected cutting parameters are not set at optimum levels (Camposeco-Negrete, 2015). The turning parameters in everyday use for a rough turning operation are adapted (Bilga et al., 2016), namely a cutting speed of 165.79 m/min., feed rate 0.2 mm/rev., depth of cut 1.4 mm and nose radius 0.8 mm. The turning

Contour plots of active cutting energy for finish turning

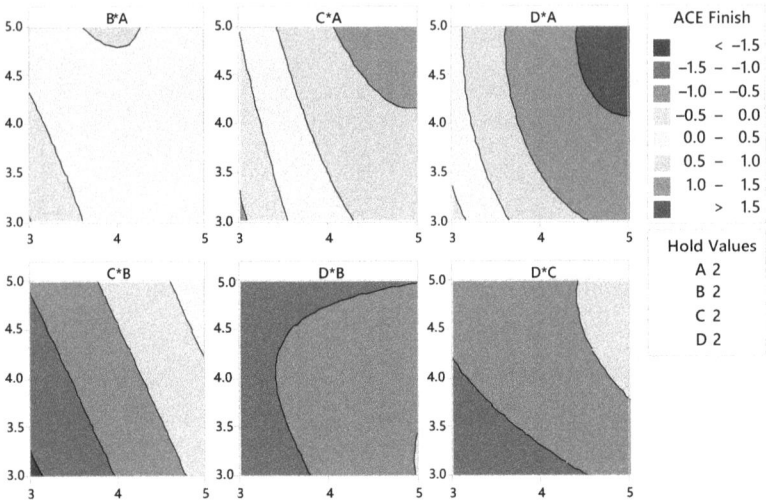

Surface plots of active cutting energy for finish turning

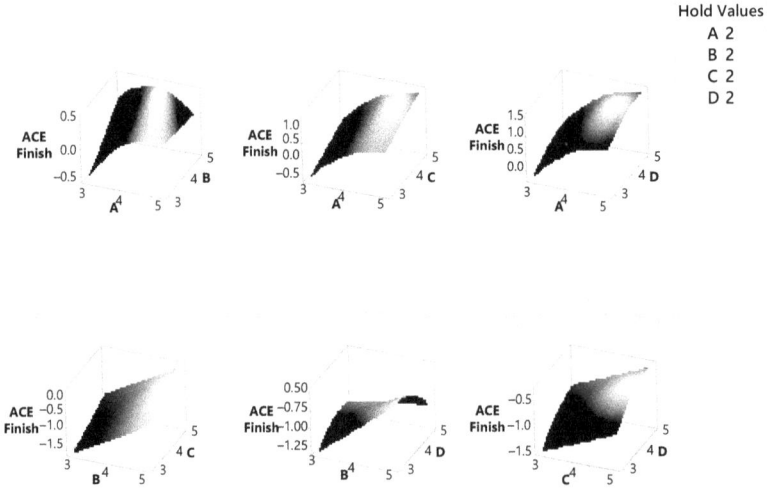

Figure 14.7 Contour and surface plots of active cutting energy for finish turning.

Table 14.12 Taguchi Method Projected Means and the S/N Ratios at Optimal Settings

Turning Operation	Response	Optimal Parameter/ Level	Taguchi Method Mean	Trial Mean	Absolute Error
Rough	Active Cutting Energy (Wh)	A_1, B_3, C_1, D_3	0.706	0.722	0.016
Finish	Active Cutting Energy (Wh)	A_1, B_3, C_1, D_3	0.257	0.131	0.126

R-SQ. = 94.86%

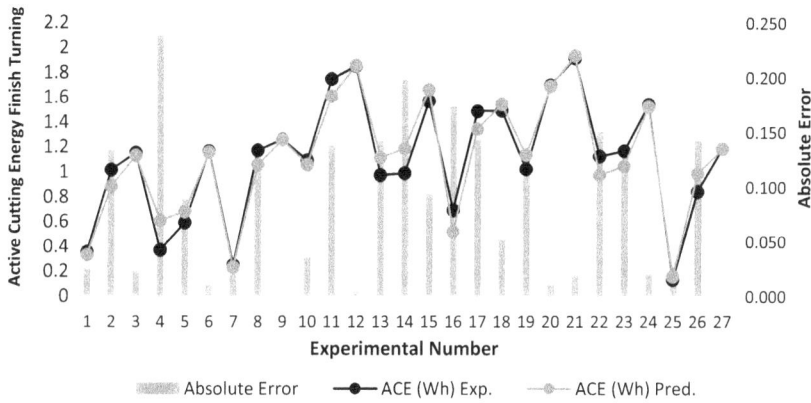

Figure 14.8 Experimental and RSM predicted readings of active cutting energy for rough turning.

R-SQ. = 94.86%

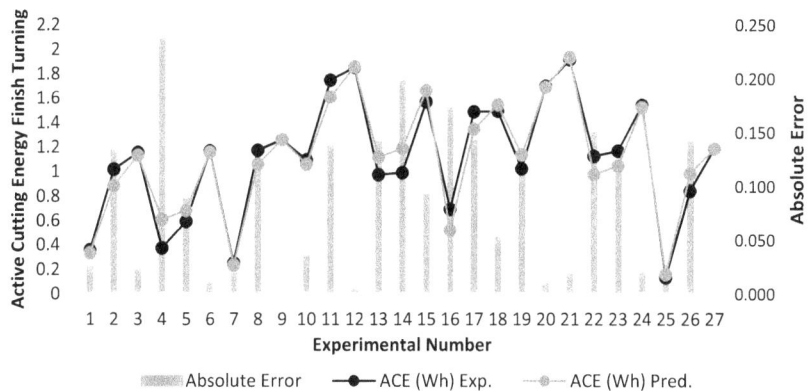

Figure 14.9 Experimental and RSM predicted readings of active cutting energy for finish turning.

Table 14.13 ACE in Industry Use and Optimum Turning Parameters

Response	Turning Parameters Used in Industry	Optimum Turning Parameters	Age Improvement (%)
ACE (Wh) Rough Turning	1.723	0.722	58.10
ACE (Wh) Finish Turning	0.925	0.131	85.84

parameters in common use for a finish turning operation suggested by several cutting insert suppliers are utilized in industries, namely a cutting speed of 236.882 m/min., feed rate 0.15 mm/rev., depth of cut 0.4 mm and nose radius 1.2 mm. The comparative outcomes of turning parameters are utilized in industry and an optimal setting for ACE has been achieved in the current research for a rough as well as a finish turning operation, as shown in Table 14.13. There is an improvement of 58.10 and 85.84% in ACE compared to turning parameters in industry use for rough and finish turning operations, respectively.

14.5 CONCLUSIONS

The experimental data of the rough and finish turning of EN 353 alloy steel were collected using Taguchi's L-27 OA. ANOVA was conducted to identify the significant parameters driving ACE. RSM was used to establish empirical models for ACE. The optimal parameter setting achieved with Taguchi's method was compared with the cutting parameters utilized in industries. Based upon the results, the following conclusions have been drawn.

1. ACE has the same optimum control parameter levels for rough and finish turning operations. The optimized control parameter setting of ACE for a rough turning operation are: A_1 cutting speed-165.79-m/min., B_3 feed rate-0.3-mm/rev., C_1 depth of cut-1-mm and D_3 nose radius-1.2-mm, with a maximum contribution of the depth of cut at 58.73%. The optimized control parameter setting of ACE for a finish turning operation are: A_1 cutting speed 179.11 m/min., B_3 feed rate 0.2 mm/rev., C_1 depth of cut 0.2 mm and D_3 nose radius 1.2 mm, with a maximum contribution of the depth of cut at 48.09%. Hence, the depth of cut was revealed to be the most critical parameter for ACE, both for rough and finish turning operations.
2. There is a 58.10% improvement in ACE at the optimum rough turning parameters compared to the rough turning parameters utilized in industries. In addition, there is an 85.84% improvement in ACE at the optimal finish turning parameters compared to the finish turning parameters used in industries.

3. Based on the results at the optimum turning parameters, the ACE consumption of 0.591 Wh is lower than the rough turning operation in the case of a finish turning operation. Also, there is a 27.74% greater ACE improvement in the case of a finish turning operation than in a rough turning operation, when compared with the turning parameters in industry use. So, the finish turning operation is more energy efficient. Therefore, the parametric optimization of ACE can reduce the operating cost of machine tools, as a reduction in ACE reduces electricity bills.

4. The nose radius contributes 10.84% for rough turning and 4.37% for finish turning operations. The nose radius with a variation from 0.4 to 1.2 mm does not contribute too much for ACE for rough and finish turning operations. The interactions between parameters have less contribution than the main parameters and have no significant effect on ACE. However, the main parameters statistically impact ACE, both for rough and finish turning operations.

5. An ANOVA result signifies that the residual error was minimal (less than 5%), all necessary parameters were included and there is no high estimation error. Furthermore, the confirmation experiments show no noteworthy difference between predicted and experimental data for ACE.

6. The coefficient of determination for the ACE regression model is relatively higher for rough turning (84.85%) and finish turning (94.86%), indicating the model's ability to make the right predictions. These models can be used to directly assess ACE under various combinations of cutting parameters within the limits of the parameters investigated and where other cutting conditions remain constant, e.g. cutting tool, workpiece material and machine tool. Optimizing different energy responses and other quality features may be carried out for more realistic results.

REFERENCES

Abhang, L., & Hameedullah, M. (2010). Power prediction model for turning EN-31 steel using response surface methodology. *Journal of Engineering Science and Technology Review*, 3(1), 116–122. https://doi.org/10.25103/jestr.031.20

Aggarwal, A., Singh, H., Kumar, P., & Singh, M. (2008). Optimizing power consumption for CNC turned parts using response surface methodology and Taguchi's technique—A comparative analysis. *Journal of Materials Processing Technology*, 200(1), 373–384. https://doi.org/10.1016/j.jmatprotec.2007.09.041

Balogun, V., Edem, I., & Mativenga, P. (2015b). The effect of auxiliary units on the power consumption of CNC machine tools at zero load cutting. *International Journal of Scientific and Engineering Research*, 6(2), 874–879.

Balogun, V. A., Edem, I. F., Bonney, J., Ezeugwu, E., & Mativenga, P. (2015a). Effect of cutting parameters on surface finish when turning nitronic 33 steel alloy. *International Journal of Scientific and Engineering Research*, 6(1), 560–568.

Bhushan, R. K. (2013). Optimization of cutting parameters for minimizing power consumption and maximizing tool life during machining of Al alloy SiC particle composites. *Journal of cleaner production*, *39*, 242–254. https://doi.org/10.1016/j.jclepro.2012.08.008

Bilga, P. S., Singh, S., & Kumar, R. (2016). Optimization of energy consumption response parameters for turning operation using Taguchi method. *Journal of Cleaner Production*, *137*, 1406–1417. https://doi.org/10.1016/j.jclepro.2016.07.220

Camposeco-Negrete, C. (2013). Optimization of cutting parameters for minimizing energy consumption in turning of AISI 6061 T6 using Taguchi methodology and ANOVA. *Journal of cleaner production*, *53*, 195–203. https://doi.org/10.1016/j.jclepro.2013.03.049

Camposeco-Negrete, C. (2015). Optimization of cutting parameters using response surface method for minimizing energy consumption and maximizing cutting quality in turning of AISI 6061 T6 aluminum. *Journal of Cleaner Production*, *91*, 109–117. https://doi.org/10.1016/j.jclepro.2014.12.017

Camposeco-Negrete, C., De Dios Calderón Nájera, J., & Miranda-Valenzuela, J. C. (2016). Optimization of cutting parameters to minimize energy consumption during turning of AISI 1018 steel at constant material removal rate using robust design. *The International Journal of Advanced Manufacturing Technology*, *83*(5), 1341–1347. https://doi.org/10.1007/s00170-015-7679-9

Cetin, M. H., Ozcelik, B., Kuram, E., & Demirbas, E. (2011). Evaluation of vegetable based cutting fluids with extreme pressure and cutting parameters in turning of AISI 304L by Taguchi method. *Journal of Cleaner Production*, *19*(17), 2049–2056. https://doi.org/10.1016/j.jclepro.2011.07.013

Chandel, R. S., Kumar, R., & Kapoor, J. (2021). Sustainability aspects of machining operations: A summary of concepts. *Materials Today: Proceedings*. https://doi.org/10.1016/j.matpr.2021.04.624

Diaz-Elsayed, N., Dornfeld, D., & Horvath, A. (2015). A comparative analysis of the environmental impacts of machine tool manufacturing facilities. *Journal of Cleaner Production*, *95*, 223–231. https://doi.org/10.1016/j.jclepro.2015.02.047

Draper, N. R. (1992). Introduction to Box and Wilson (1951) on the experimental attainment of optimum conditions. In *Breakthroughs in Statistics* (pp. 267–269). Springer, New York, NY.

Garg, G. K., Garg, S., & Sangwan, K. S. (2018). Development of an empirical model for optimization of machining parameters to minimize power consumption. *IOP Conference Series: Materials Science and Engineering*, *346*, 012078. https://doi.org/10.1088/1757-899x/346/1/012078

He, W., Goodkind, D., & Kowal, P. R. (2016). An aging world: 2015. *International Population Reports United States Census Bureau*, *16*(1), 95.

Karim, M. R., Tariq, J. B., Morshed, S. M., Shawon, S. H., Hasan, A., Prakash, C., … Pruncu, C. I. (2021). Environmental, economical and technological analysis of mql-assisted machining of Al-Mg-Zr alloy using PCD tool. *Sustainability (Switzerland)*, *13*(13). https://doi.org/10.3390/su13137321

Kumar, R., Bilga, P. S., & Singh, S. (2017). Multi objective optimization using different methods of assigning weights to energy consumption responses, surface roughness and material removal rate during rough turning operation. *Journal of cleaner production*, *164*, 45–57.

Kumar, R., Bilga, P. S., & Singh, S. (2018). *Optimization and Modeling of Active Power Consumption for Turning Operations*. Paper presented at *the ISME 19th*

Conference on advances in mechanical engineering (mechanical systems and sustainability), Dr. B. R. Ambedkar National Institute of Technology Jalandhar, Punjab, India.

Kumar, R., Bilga, P. S., & Singh, S. (2020). Optimization of active cutting power consumption by taguchi method for rough turning of alloy steel. *International Journal of Metallurgy Alloys,* 6(1), 37–45. http://materials.journalspub.info/index.php?journal=IJM&page=article&op=view&path%5B%5D=632

Kumar, R., Bilga, P. S., & Singh, S. (2021c). An investigation of energy efficiency in finish turning of EN 353 alloy steel. *Procedia CIRP,* 98, 654–659. https://doi.org/10.1016/j.procir.2021.01.170

Kumar, R., Singh, S., Bilga, P. S., Jatin, Singh, J., ...Pruncu, C. I. (2021b). Revealing the benefits of entropy weights method for multi-objective optimization in machining operations: A critical review. *Journal of Materials Research and Technology,* 10, 1471–1492. https://doi.org/10.1016/j.jmrt.2020.12.114

Kumar, R., Singh, S., Sidhu, A. S., & Pruncu, C. I. (2021a). Bibliometric analysis of specific energy consumption (SEC) in machining operations: A sustainable response. *Sustainability (Switzerland),* 13(10). https://doi.org/10.3390/su13105617

Montgomery, D. C. (2017). *Design and analysis of experiments.* John wiley & sons.

Mouleeswaran, S., Babu, D. M., & Jothi Prakash, V. (2012). Optimisation of cutting parameters for CNC turned parts using Taguchi's technique. *Annals Of Faculty Engineering Hunedoara International Journal of Engineering,* 10(3), 493–496. doi:http://annals.fih.upt.ro/pdf-full/2012/ANNALS-2012-3-86.pdf

Rosa, S. d. N., Diniz, A. E., Andrade, C. L. F., & Guesser, W. L. (2010). Analysis of tool wear, surface roughness and cutting power in the turning process of compact graphite irons with different titanium content. *Journal of the Brazilian Society of Mechanical Sciences and Engineering,* 32(3), 234–240. https://doi.org/10.1590/S1678-58782010000300006

Ross, P. J., & Ross, P. J. (1988). *Taguchi techniques for quality engineering: Loss function, orthogonal experiments, parameter and tolerance design.* McGraw-Hill New York.

Roy, R. K. (2001). *Design of experiments using the Taguchi approach: 16 steps to product and process improvement.* John Wiley & Sons.

Sidhu, A. S., Singh, S., & Kumar, R. (2021a). Bibliometric analysis of entropy weights method for multi-objective optimization in machining operations. *Materials Today: Proceedings.* https://doi.org/10.1016/j.matpr.2021.08.132

Sidhu, A. S., Singh, S., & Kumar, R. (2021b). Optimization and modelling of active power consumption of ST52.3 alloy steel during a drilling operation. *Materials Today: Proceedings.* https://doi.org/10.1016/j.matpr.2021.09.340

Sidhu, A. S., Singh, S., Kumar, R., Pimenov, D. Y., & Giasin, K. (2021). Prioritizing energy-intensive machining operations and gauging the influence of electric parameters: An industrial case study. *Energies,* 14(16), 4761.

Suresh, R., Basavarajappa, S., & Samuel, G. L. (2012). Some studies on hard turning of AISI 4340 steel using multi-layer coated carbide tool. *Measurement,* 45(7), 1872–1884. https://doi.org/10.1016/j.measurement.2012.03.024

Wakjira, M. W., Altenbach, H., & Ramulu, P. J. (2020). Cutting mechanics analysis in turning process to optimise product sustainability. *Advances in Materials and Processing Technologies,* 1–17. https://doi.org/10.1080/2374068X.2020.1785207

Index

For Product Safety Concerns and Information please contact our EU
representative GPSR@taylorandfrancis.com
Taylor & Francis Verlag GmbH, Kaufingerstraße 24, 80331 München, Germany

www.ingramcontent.com/pod-product-compliance
Lightning Source LLC
Chambersburg PA
CBHW060332220326
41598CB00023B/2684